The Origins of Human Nature

The Origins of Human Nature

EVOLUTIONARY DEVELOPMENTAL PSYCHOLOGY

DAVID F. BJORKLUND &
ANTHONY D. PELLEGRINI

American Psychological Association, Washington, DC

Published by
American Psychological Association
750 First Street, NE
Washington, DC 20002
www.apa.org

To order
APA Order Department
P.O. Box 92984
Washington, DC 20090-2984
Tel: (800) 374-2721;
Direct: (202) 336-5510
Fax: (202) 336-5502;
TDD/TTY: (202) 336-6123
Online: www.apa.org/books/
Email: order@apa.org

In the U.K., Europe, Africa, and the Middle East, copies may be ordered from
American Psychological Association
3 Henrietta Street
Covent Garden, London
WC2E 8LU England

Typeset in Goudy by World Composition Services, Inc., Sterling, VA

Printer: Sheridan Books, Ann Arbor, MI
Cover Designer: Naylor Design, Washington, DC
Technical/Production Editor: Casey Ann Reever

The opinions and statements published are the responsibility of the authors, and such opinions and statements do not necessarily represent the policies of the American Psychological Association.

Library of Congress Cataloging-in-Publication Data
Bjorklund, David F., 1949–
  The origins of human nature : evolutionary developmental psychology/
David F. Bjorklund & Anthony D. Pellegrini.
     p. cm.
Includes bibliographical references and index.
  ISBN 1-55798-878-1
  1. Genetic psychology.   2. Developmental psychology.   I. Pellegrini,
Anthony D.   II. Title.
  BF701.B57 2002
  155.7—dc21

                              2001053572

British Library Cataloguing-in-Publication Data
A CIP record is available from the British Library.

Printed in the United States of America
First Edition

This book is dedicated to the memory of Robert Cairns,
scholar, teacher, and friend

TABLE OF CONTENTS

PREFACE

Can evolutionary theory inform contemporary developmental psychology? We think so, but we may be considered to be unexpected proselytizers for such an argument. Our early careers were devoted to rather conventional topics in developmental and educational psychology, including children's memory, strategies, literacy, and play, and provided little hint of the somewhat unconventional perspective we are advocating here. But evolutionary thinking had been part of our early training, and although it may not have manifested itself in our research, it was never far from the surface in our thinking.

For example, during David Bjorklund's first semester in graduate school at the University of North Carolina at Chapel Hill, he was fortunate to partake in a Developmental Proseminar taught by Robert Cairns, Gilbert Gottlieb, and Harriet Rheingold. There he learned about developmental psychobiology, the importance of developmental timing, the bidirectionality of structure and function, the concept of probabilistic epigenesis, behavioral neophenotypes, and the possible role of neoteny in human evolution.

Tony Pellegrini's work in evolutionary psychology began as an extension of his research in children's play. Over the past 20 years, he spent 2 years in Sheffield, England, working with Peter Smith on ethological studies of play. During this time, he learned the value of examining human behaviors in terms of their immediate causes and consequences as well as their wider ontogenetic and phylogenetic contexts. As with Bjorklund, Pellegrini's attraction to evolutionary science stemmed, in part, from the way in which it integrated individuals interacting with their environments.

With the advent of evolutionary psychology, promulgated by people such as John Tooby, Leda Cosmides, David Buss, Martin Daly, and Margo Wilson, among others, we, and other developmental psychologists, were reminded that evolution, or phylogeny, was indeed a developmental phenomenon, different from ontogeny in its time scale but well-worth incorporating into our research and theorizing. But while our training and professional biases perhaps prepared us to look favorably on the emerging field of evolutionary psychology, upon closer examination we found that contemporary views of evolutionary and developmental psychology did not mesh perfectly, perhaps not even well. The primary reason for this misfit, we believed, was the lack of a specific model of how evolved and thus genetically transmitted psychological mechanisms get expressed. Although evolutionary psychologists contended that such mechanisms are sensitive to environmental contingencies and do not reflect a form of genetic determinism, this argument was lost on most developmental psychologists, who rightly pointed out evolutionary psychology's lack of a specific model for explaining gene–environment interactions over the course of development. Moreover, the pioneers of contemporary evolutionary psychology came from fields that focused on psychological functioning in adults, primarily from social psychology. While it is clear that it is the adults of a species that find mates, reproduce, and raise offspring—the paramount criteria of success in any Darwinian perspective—for a long-lived, slow-developing species such as humans, there is a long road to travel between birth and adulthood, and we believed that psychological characteristics of juveniles were as surely shaped by natural selection operating over geologic time as were those of adults. Although evolutionary psychologists had not totally ignored the preadult period of human life in their research and theorizing, it certainly took a back seat to the concerns of the adult.

In this book we develop a view that integrates many of the key aspects of contemporary evolutionary and developmental psychology. Although our original intent was to apply many of the new ideas in evolutionary psychology to the study of human development, we realized early on that this cannot be done. There are too many points of contention between mainstream views of evolutionary and developmental psychology. Instead, we present the field of evolutionary developmental psychology, which is neither the sum nor the average of its two parent disciplines, but a hybrid. Much research has already been done that fits within this perspective, some of which we review in this book. Our hope is that this perspective will serve as an overarching metatheory for the field of developmental psychology and as a guide for future research and theorizing.

ACKNOWLEDGMENTS

We would like to thank the following people for their constructive comments on drafts of our manuscript: Marc Bekoff, Jay Belsky, Jesse Bering, Barbara Bjorklund, Carlos Hernandez Blasi, Gordon Burghardt, Jeffrey Gredlein, Judith Rich Harris, Robert Lickliter, Angeline Lilliard, Susan Taylor Parker, Todd Shackelford, Peter Smith, and Jennifer Yunger.

The Origins of Human Nature

1

EVOLUTIONARY DEVELOPMENTAL PSYCHOLOGY

We have heard it said that evolutionary psychology is a "theory with an attitude," because it is based on the *fact* of evolution. We can be confident that humans are related to other primates and did evolve from earlier hominid species over the past 5–7 million years. But does this fact mean that psychological functioning must also be viewed through the lens of evolution? Others have made this argument (Barkow, Cosmides, & Tooby, 1992; Buss, 1995; Daly & Wilson, 1988b), proposing that any account of human psychological functioning that excludes evolution is ignoring an important piece of the picture about what it means to be human. Many psychologists have adopted modern evolutionary theory to explain human behavior—to serve as an overarching perspective, or metatheory—and we believe that this approach has been fruitful. But despite being enamored with the promise of evolutionary psychology, as developmentalists we believed that there were shortcomings. First, despite our confidence that humans are the product of a long evolutionary history, we are still unsure of the specific progression that took place from ape ancestor to modern *Homo sapiens*, as well as of the factors responsible for these changes. Second, although evolutionary psychology did not ignore development, it failed to treat it seriously in its theorizing. Third, although much had been written about psychological developmental issues from an evolutionary perspective (see Belsky, Steinberg, & Draper, 1991; Fernald, 1992), we found no overarching evolutionary perspective in developmental psychology.

This book is intended to fill this niche. It is, of course, not the first book to take a serious look at the relation between evolution and psychological development. For example, developmental psychologist Harold Fishbein's (1976) *Evolution, Development, and Children's Learning* had, in many ways, a similar intent as ours but was written before the "invention" of evolutionary psychology. Developmental psychologist Gilbert Gottlieb's (1992) important book *Individual Development and Evolution* focused on how evolutionary theorists have applied (or failed to apply) concepts of development (ontogeny) to explain evolution (phylogeny). (Some of Gottlieb's ideas are discussed in chapter 4.) Ours is the first book-length attempt to define the field of evolutionary developmental psychology and is intended to provide a developmental perspective for evolutionary psychologists and an evolutionary perspective for developmental psychologists.

EVOLUTIONARY PSYCHOLOGY VERSUS EVOLUTIONARY DEVELOPMENTAL PSYCHOLOGY

We believe that the basic tenet of evolutionary psychology is this: The human mind has been prepared by natural selection, operating over geological time, for life in a human group. Within this framework, we define *evolutionary developmental psychology* as the application of the basic principles of Darwinian evolution, particularly natural selection, to explain contemporary human development. It involves the study of the genetic and environmental mechanisms that underlie the universal development of social and cognitive competencies and the evolved epigenetic (gene–environment interactions) processes that adapt these competencies to local conditions; it assumes that behaviors and cognitions that characterize not only adults but also children are the product of selection pressures operating over the course of evolution. And although it holds that some characteristics of childhood serve to adapt children to that time in development only, other experiences, such as some aspects of play, may serve to add flesh to evolved skeletal structures that become functional only in adulthood (Geary, 1999).

But evolutionary developmental psychology is not simply evolutionary psychology applied to infants and children. Although we adopt many of the basic assumptions of contemporary evolutionary psychology (e.g., the importance of natural selection, a functional perspective), taking an evolutionary developmental perspective causes one to view aspects of psychological functioning differently than an evolutionary perspective that treats development only superficially. These aspects include, among others, how gene–environment interactions are interpreted, the role of domain-general mechanisms in explaining behavior, the significance of individual differences, an examination of the role of behavior and development in evolution (as opposed to only the inverse), and a belief that higher order cognitions should also be examined from an evolutionary perspective.

Model for Gene–Environment Interaction

Although evolutionary psychologists go to great pains to declare that an evolutionary perspective does not imply genetic determinism—that different environments will result in different patterns of adaptive adult behavior—with rare exceptions, they provide no specific mechanisms or models for how evolved genetic programs interact with local ecologies to produce adult behaviors. In fact, when we told one developmental colleague that we were writing this book, she commented that she begins to piloerect when she even sees the words *evolutionary psychology* and *development* together on the same page. The reason for her skepticism (revulsion?) was that evolutionary psychology is perceived (incorrectly, but understandably, we believe) as involving preformed genetic programs that become activated in adulthood given the proper environmental conditions. Such a perspective provides no room for development, and developmental scientists should view such accounts with suspicion. In this book we adopt a specific model, the developmental systems approach (Gottlieb, 2000; Oyama, 2000a), to show how such interaction does indeed take place, strengthening considerably, we believe, evolutionary psychologists' argument that genes are not necessarily destiny (see chapter 2).

Importance of Domain-Specific and Domain-General Mechanisms

Evolutionary developmental psychology also departs slightly from what is perhaps the central dogma of evolutionary psychology, that what has evolved are domain-specific mechanisms designed by natural selection to deal with specific recurrent problems faced by our ancestors in ancient environments (Tooby & Cosmides, 1992). Although we concur that domain-specific mechanisms underlie many (perhaps most) adaptive behaviors and cognitions, we also argue that natural selection has similarly shaped domain-general information-processing mechanisms (e.g., working memory, speed of processing) and that such processes must be considered seriously in any evolutionary psychological account of human behavior (see especially chapter 5).

Individual Differences

Whereas mainstream evolutionary psychologists emphasize universals—patterns that characterize all members of a species—evolutionary developmental psychologists are also concerned with how individuals adapt their behavior to their particular life circumstances and suggest that there are alternative strategies to recurrent problems that human children faced in our evolutionary past. Such a perspective suggests that individual differences in developmental patterns are not necessarily the result of idiosyncratic experiences but are predictable, adaptive responses to environmental pressures.

Take, for example, the well-known individual differences in attachment style. Research has shown that secure attachment in infancy is related to parental behavior, is often stable over time, and is generally associated with better subsequent adjustment than is insecure attachment (Ainsworth, Blehar, Waters, & Wall, 1978; Waters, Merrick, Treboux, Crowell, & Albersheim, 2000). However, insecure attachment and the later behaviors associated with it may not only reflect an ineffective style of interacting, but may also, in some environments, serve an adaptive function. Belsky and his colleagues (1991) proposed that an early family environment associated with insecure attachment, including high stress and possible father absence, may adapt children to life in a world where people and resources are unpredictable, leading them to different mating and parenting strategies than children reared in more supportive and predictable environments. Some of this research is discussed in chapter 9.

Role of Behavior and Development in Evolution

Mainstream evolutionary psychology is primarily concerned with how selective pressures in our species' past led to adaptive cognitions and behaviors and how such adaptations influence contemporary people. Evolutionary developmental psychologists share this emphasis but are also interested in how behavioral and especially developmental processes may have affected evolution (and, in turn, contemporary functioning). For example, how might factors that affect the timing of development serve to modify the organism, which in turn generate differences among individuals (variation) on which natural selection can work? (See chapters 3 and 4.) This concern with evolution qua evolution is likely related to evolutionary developmental psychology's greater reliance on comparative data, particularly evidence from the closest relatives of *Homo sapiens*, the great apes. Whereas we can only make educated guesses about the environment of evolutionary adaptedness in which humans as a species evolved (see chapter 2), we know with some certainty the environments and social and cognitive abilities of our contemporary simian relatives, with whom we last shared a common ancestor 5–7 million years ago. Investigations of the social, emotional, and cognitive abilities of these animals and their development can provide important insights into human behavior (see research reviewed in chapters 5, 6, and 7).

Higher Level Cognitions

A central theme in evolutionary psychology is that the evolved cognitive mechanisms that so influence our behavior are generally beyond awareness. That is, most adaptive cognitive processes and "strategies" are unavailable to consciousness, reflecting implicit, as opposed to explicit, cognition. We concur. This approach permits psychologists to explain the actions of humans in the same way that the actions of nonhuman animals are explained.

No special pleading for the uniqueness of humans is required to explain their behavior. However, as a result, higher level cognition is rarely considered. Evolutionary developmental psychology, in large part because of the influence of Piaget and Vygotsky in studying children's cognition, also views higher order cognitive mechanisms as being important to human functioning and being the product of natural selection. This is perhaps best exemplified by research into theory of mind and related forms of social cognition, which is examined in chapter 7.

It could be argued that many of the differences we have listed between mainstream evolutionary psychology and evolutionary developmental psychology are overstated. Some evolutionary psychologists, for example, understand the nature of gene–environment interactions, acknowledge the possible role of domain-general mechanisms in explaining behavior, see individual differences as more than just noise, are as concerned with the evolution of the species as they are with the effects of evolutionary processes on contemporary humans, and believe in the significance of higher order cognitive operations. But none of these perspectives, we argue, reflects the canonical position of evolutionary psychology, although they do reflect that of evolutionary developmental psychology.

ASKING "WHY" QUESTIONS: NECESSARY BUT NOT SUFFICIENT

The job of developmental psychologists is to describe and explain changes in structure and function over time. We have done a relatively good job of describing change over time but have far to go before we have an adequate explanation of development. Conventionally, psychologists look for proximal explanations. That is, they look for the immediate cause (e.g., reinforcement history, information-processing operations, patterns of neuronal firings) for a particular behavior or pattern of development. These explanations tell us something about how a behavior or developmental pattern comes about. There are other levels of explanation, however, one of which is the *distal. Why* does a behavior or pattern of development exist? In fact, we believe that the "whys" of development will help us acquire a better understanding of the "hows" and "whats" of development.

When we were being trained we were told not to ask why questions, only what and how questions. The latter were the purview of scientists, whereas the former was the business of philosophers. However, why a behavior exists turns out to be an important question, because it relates to the function of a behavior. In fact, one well-accepted theory in contemporary science specializes in such why questions: Darwinian evolutionary theory. Charles Darwin (1859) proposed that successful organisms, those that survive to reproduce, are generally well adapted to their local environments and that different species evolved different adaptive mechanisms to deal with the worlds in which they found themselves.

Although we believe that evolutionary psychology can provide an overarching perspective relevant to all students of development, we do not see evolutionary theory replacing other theoretical accounts of development. An evolutionary perspective is necessary, we believe, for understanding human psychological functioning and development, but it is not sufficient (cf. Wachs, 2000). Rather, we argue that an evolutionary perspective can inform other theories designed to explain the more proximal causes of development. There is no question in our minds, for instance, that cultural changes over the past 10,000 years have drastically affected the way humans behave and develop. Cultural influences such as nutrition and health care directly affect brain growth and development, as do aspects of material culture, including modern tools, methods of communication, and formal education. Humans did not evolve in a world anything like the one most people experience today, and to assume that our behavior and development is simply the product of our evolutionary past, with only passing reference to the impact of culture, is missing the big picture. Our claim, rather, is that modern humans still enter this world with a nervous system evolved over eons of time and possess psychological mechanisms that bias them toward certain behaviors and interpretations of their environment at different points in development. Whatever the effects of modern culture are on the human mind and behavior, they are substantial, and they are mediated through a brain evolved for life in very different times.

We believe that developmental psychology can benefit substantially by adopting an evolutionary perspective, in part because other, more domain-specific theories can be built on this edifice. Moreover, the concept of development is at its core a biological one. Indeed, aspects of evolutionary theory help us to understand the integrity of each developmental period. Development, however, is also historical science. Like history, developmental psychologists believe that knowing the past helps us to understand the present and to predict the future. This is no less true, we argue, for phylogeny. Knowing the evolutionary history of a species can help explain present and future behavior, including the amelioration of at least some "problem" behaviors (e.g., child abuse, male-on-male violence, and reading or math disabilities).

OVERVIEW OF THIS BOOK

It has been said that biology makes no sense unless viewed through the lens of evolution. Need we say the same about behavior and about development? As we have argued here, we think so, believing that an understanding of human evolution provides the framework for an understanding of human psychological functioning and development. We hope to convince readers of this in our book.

Chapter 2, "A New Science of the Developing Mind," presents the assumptions of mainstream evolutionary psychology, including the role of evolved psychological mechanisms in contemporary behavior, the concept of the environment of evolutionary adaptedness, and adaptationist thinking. We also discuss the developmental systems approach as a model for explaining gene–environment interactions. This model serves as the foundation in the remainder of the book for explaining how evolved dispositions become converted into behavior. The chapter also discusses the potential adaptive nature of developmental immaturity and the influence of natural selection at different times in ontogeny.

Chapter 3, "History and Controversy," puts the study of the relation between evolution and development in historical perspective. We show how ideas about the role of development in evolution (and the role of evolution in development) have changed over the past 200 years or so. We also investigate from a historical perspective the sometimes controversial concept of innateness and the interface among evolutionary psychology, behavioral genetics, and the study of individual differences.

In Chapter 4, "The Benefits of Youth," we examine how the timing of ontogeny may have influenced the course of human evolution. We also explore the interacting roles of an extended juvenile period, a big brain, and a complex social structure in human cognitive evolution.

Chapter 5, "Classifying Cognition," examines some of the ways cognition is conceptualized by evolutionary psychologists. Specifically, we look at the distinction between implicit versus explicit cognition, cognitive abilities that have been selected over the course of evolution versus those that are primarily shaped by culture, immediate versus deferred benefits of cognitive abilities in infants and children, and domain-general and domain-specific cognitive abilities from an evolutionary developmental perspective.

The first five chapters provide the theoretical framework for evolutionary developmental psychology and how it is similar to and different from mainstream evolutionary psychology. The remaining chapters focus on particular topics within evolutionary developmental psychology. Chapter 6, "Prepared to Learn," examines the learning abilities of infants and young children, specifically what contents are easily acquired (e.g., language) and what the constraints may be on human learning and cognition. Chapter 7, "Social Cognition," investigates social learning, including theory of mind and social reasoning, and examines the development of these abilities in our close primate relatives as well as in children. In Chapter 8, "All in the Family: Parents and Other Relations," factors that influence parenting and other aspects of family life (e.g., dealing with siblings) are examined, all under the umbrella of parental investment theory. Social relationships from infancy through adolescence are examined in Chapter 9, "Interactions, Relationships, and Groups," and the role of play is assessed from an evolutionary perspective in Chapter 10, "*Homo ludens*: The Importance of Play." In the final chapter, "Epilogue: Evolution and Development," we attempt to connect the broad range of topics covered in the previous chapters into a comprehensible framework and propose six principles of evolutionary developmental psychology.

We began this book with the intention of applying the new ideas found in evolutionary psychology to human development. On our way to completing the book, we recognized that many of the underlying assumptions held by evolutionary and developmental psychologists were at odds, making any simple application of the ideas from one field to the other impossible. Yet we continued to believe that each field had much to offer the other and that evolutionary developmental psychology would not be the average or the sum of the two disciplines, but instead a hybrid. We present this hybrid approach to evolution and development in the remainder of this volume and hope that the offspring of this pairing will be a fertile one.

2

A NEW SCIENCE OF THE DEVELOPING MIND

In this chapter we present the basic tenets of the new science of evolutionary psychology as they have been developed since the mid-1980s. Our emphasis, however, is on evolutionary psychology as it relates to development and vice versa. Although leaders of this burgeoning field have certainly not ignored development or factors from humankind's evolutionary history that affected the ontogenetic process most directly, development has taken a peripheral role in research and theory among the new cadre of evolutionary psychologists. Here we place development on center stage and argue that our understanding of human development and behavior is greatly benefited by looking at the relation between *phylogeny* (development of the species) and *ontogeny* (development of the individual; see Bjorklund & Pellegrini, 2000; Fishbein, 1976; Geary & Bjorklund, 2000; Keller, 2000).

We first introduce the similarities and differences between the two types of development that are the focus of this book: phylogeny and ontogeny. We next examine some of the basic tenets of evolutionary psychology and then introduce the developmental systems approach as a model for explaining gene–environment interaction. We then discuss some basic ideas of evolutionary developmental psychology, such as the concept of ontogenetic adaptations and the influence of natural selection at different times in ontogeny.

PHYLOGENY AND ONTOGENY

Phylogeny and ontogeny are both forms of development—changes in structure and function over time. *Phylogeny* refers to changes in a species (or of a particular genetic line that might include a series of related species) over geological time. Phylogeny is used synonymously with *evolution* (of a species). *Ontogeny* refers to changes over the lifetime of an individual and is what most people mean most of the time when they use the term *development*. Phylogeny and ontogeny are both historical concepts, in the sense that events early in the life of an individual or of a species can have consequences for what happens later.

Many differences exist between phylogeny and ontogeny, the most obvious being the time scales involved. Evolution (phylogeny) typically occurs over hundreds of thousands or millions of years (although it can occur over shorter time periods; Grant & Grant, 1989); it can only be inferred from the fossil records and from examining behavioral, morphological, and genetic similarities and differences among extant species. Ontogeny occurs on a much shorter time scale and can be witnessed directly and repeatedly, in different members of a species, across different cultures, and at different historical times. Also, species evolve, whereas individuals develop.

Other than being types of development, do phylogeny and ontogeny share anything? Might knowing something about one form of development help us better understand the other? For example, are the underlying mechanisms that drive changes in these two types of development similar? Are the processes affecting phylogeny and those affecting ontogeny related in any meaningful way? For instance, because each member of a species shares the same phylogenetic history, might knowing something about the evolution of a species help us explain contemporary development (ontogeny)? Or might we be able to apply our knowledge about the ontogenetic process to the phylogenetic one? After all, our ancestors, whose ancient genes we carry, also developed from conception to adulthood, and perhaps insights from ontogeny can aid us in getting a better grasp of phylogeny.

These are not new questions; developmental psychologists and biologists have asked them before. For example, the mechanism believed to drive evolution is natural selection, Charles Darwin's seminal contribution to the life sciences. Basically, Darwin (1859) proposed that variability in characteristics exists among individuals of a species and that some of those characteristics fit better than others with current environmental conditions (are "selected" by the environment) and increase the chance of survival of those individuals possessing those features. (More is discussed about natural selection later in this chapter.) Might similar mechanisms of variation and selection also be applied to individual development (see Cziko, 1995; Hull, Langman, & Glenn, 2001)?

Other aspects of evolutionary theory applied to human development have not weathered the test of time as well as has "selectionist thinking." For example, one idea shared by turn-of-the-20th-century scientists interested in evolution and development was *progress*. Ontogeny can be seen as a progression from immature to mature forms, and it occurs in essentially the same pattern for all biologically normal members of the species. It is teleological, in that the end product (the adult) is known. Once a child is born (or even conceived), the general course of development is predictable. Evolution was similarly seen as being progressive, leading ever onward and upward to increasing perfection and to *Homo sapiens*. Such views were an outgrowth of the Enlightenment and the Industrial Revolution in the 18th and 19th centuries; for several hundred years there had been continuous technological progress, and the dominant religions of Western Europe held that humans were instructed by God to exert control over the Earth (Charlesworth, 1992; Morss, 1990). This is not a view that is shared by biologists today, although it remains implicit in much of the discussion of human evolution.

Despite the shortcomings of previous attempts to "biologize development" (see Morss, 1990), we believe that modern evolutionary theory can be applied to the study of human development and will provide a better understanding of both phylogeny and ontogeny. Our primary focus is on how taking an evolutionary (phylogenetic) perspective can inform us about development (ontogeny), specifically human development. We do not ignore the possible influence of ontogeny on phylogeny, however, because our ancestors certainly developed over their life spans, and many species-level changes surely occurred by varying important, if often subtle, aspects of the ontogenetic process (see chapters 3 and 4).

To understand the value of behavior in ontogeny, we must consider the place of that behavior in the development of the species (phylogeny; Tinbergen, 1963). The value of a behavior can be understood in terms of "ultimate" function (i.e., "fitness," or producing offspring who survive to reproduce) or in terms of beneficial consequences of that behavior to the organism during its life span (Hinde, 1980). The Nobel-prize-winning ethologist Niko Tinbergen (1963) stated that we must ask four questions to understand the value of behavior: What are the immediate benefits (internal and external) to the organism? What are the immediate causes? How does it develop within the species (ontogeny)? How did it evolve across species (phylogeny)? Thus, to understand the value of a behavior during childhood, we must understand the place of that behavior in the life span of an individual and in the evolution of the species. As part of this logic, we must examine the values of behavior during different periods of ontogeny and recognize that different features may be selected at different times.

EVOLUTIONARY PSYCHOLOGY

Although many psychologists still regard it with disdain, evolutionary theory has captured the attention and imagination of psychologists of many schools of thought, who have recognized the explanatory and predictive power of Darwin's great idea (Buss, 1995; Daly & Wilson, 1988b; Geary, 1998; Tooby & Cosmides, 1992). Darwin's (1859) theory of evolution, as presented in *The Origin of Species*, is probably the best and most enduring general explanation of the human condition and our adaptation to the world.

The basic principles behind Darwin's theory are relatively simple. First, many more members of a species are born in each generation than will survive, a situation termed *superfecundity*. Second, all members (at least in sexually reproducing species) have different combinations of inherited traits; that is, *variation* in physical and behavioral characteristics exists among individuals within a species. Third, this variation is heritable. Fourth, characteristics that result in an individual surviving and reproducing tend to be selected as a result of an interaction between individuals and their environment and thus are passed down (via genes) to future generations, whereas the traits of the nonsurvivors are not. That is, genetically based variations in physical or psychological features of an individual interact with the environment and, over many generations, these features tend to change in frequency, resulting eventually in specieswide traits.[1] Thus, through the process of natural selection, adaptive changes in individuals, and eventually in species, are brought about.

Although scientists have never witnessed the emergence of a new species of animal via natural selection, there have been many examples of natural selection producing changes in the phenotype of a species (referred to as *microevolution*). For instance, British scientists have studied the pepper moth (*Biston betularia*) for nearly 150 years. This moth is found in two basic colorations, light and dark. In 1848, prior to heavy industrialization in one area of Great Britain, approximately 99% of these moths were light in color and rested on lichen-covered trees, which camouflaged them from birds. There were occasional dark moths, but they were rare because birds could easily spot and eat them. When industry increased in these areas, sooty smoke darkened the trees, and then the dark moths were better camouflaged from bird attacks; their numbers correspondingly increased as the light moths became bird food. When antipollution laws were passed in the 1950s and the environment (and trees) became cleaner, natural selection again favored the light moths (Kettlewell, 1959). Thus, over very brief periods of time, changes in the environment can result in some individuals in a species having a better fit with their environment than others and, over several generations, the heritable characteristics of these individuals will increase in frequency in the population.

It is important to emphasize that although species evolve, natural selection works on individuals and not on the species as a whole. That is, natural selection does not necessarily yield what is "best for the group" but rather works on the level of the individual. Individuals who have

---

[1] Although modern evolutionary theory holds that heritable genetic changes are the basis of evolution, we argue later in this and following chapters that organisms also inherit environments as part of developmental systems and that phylogeny can be viewed as changes in developmental systems over generations.

characteristics that are well adapted to the current environment will produce more offspring than less well-adapted individuals. In many cases, however, there is overlap between benefits for individuals and for the group (Axelrod & Hamilton, 1981). After enough generations, these traits may be spread across the species, yielding a species-typical trait. But it was not for the benefit of the group (or the species) that these traits were selected, but rather for the benefit of individuals.[2] However, although natural selection may work for the benefit of individuals and not groups, evolution occurs within populations of individuals. Thus, a mutation or a novel variation resulting from recombining genes through sexual reproduction that results in some benefit to an individual must spread within a local population of similar individuals if a change at the species level is to occur (Tattersall, 1998).

Darwin referred to the reproductive success of individuals as reflecting their *reproductive fitness*, which basically refers to the likelihood that an individual will become a parent and a grandparent. Contemporary evolutionary theorists, taking advantage of the scientific advances that have occurred since Darwin's time (particularly in genetics), use the concept of *inclusive fitness* (W. D. Hamilton, 1964). Inclusive fitness includes Darwin's concept of reproductive fitness (in this case, having many offspring), but also considers the influence that an individual may have in getting other copies of his or her genes into subsequent generations. For example, by having one child, a woman passes 50% of her genes to the next generation. That is, half of the genes of a son or daughter are identical to those possessed by their mother. But by helping to rear her four nieces and nephews, each of whom shares, on average, 25% of her genes, a woman can further increase the copies of her genes in the next generation, thereby increasing her inclusive fitness.

The idea of inclusive fitness was developed to account for the phenomenon of *altruism*, individuals helping others, even at some expense to themselves. Evolutionary biologist William Hamilton (1964) developed a formula to describe conditions when natural selection would favor altruism. This formula, known as Hamilton's rule, is expressed as $c < rb$, where c is the cost to the actor, r is the degree of genetic relatedness between the actor and the recipient, and b is the benefit to the recipient. Both costs and benefits are conceptualized in terms of reproduction. Of course, no animal consciously makes these computations, and because no one can read the future, there is no way of knowing for certain whether altruism toward a daughter will result in more eventual offspring than altruism toward a niece or nephew. But assisting a daughter will result on average in greater reproductive

---

[2] Ideas of group selection have returned to biology in more sophisticated form and may account for some aspects of behavior in social species (D. S. Wilson, 1997).

benefit for one's genes than assisting a niece or nephew, and selection pressures must have been strong enough for animals of many species to unconsciously make these computations, making altruism both possible and predictable (see Dawkins, 1976, for numerous examples).

Evolutionary psychology takes these basic tenets of Darwin's theory and advancements made to it over the past 140 years (usually termed the *modern synthesis*, or *neo-Darwinism*; see chapter 3) and applies them specifically to human psychological functioning. Although, as in any fertile area of intellectual inquiry, some healthy disagreements exist about specifics of evolutionary theory applied to humans, there are certain aspects of this new paradigm to which most practitioners of the field adhere in one form or another. Foremost is the idea that psychological mechanisms underlie important social and intellectual behaviors and that these mechanisms have evolved (Buss, 1995; Tooby & Cosmides, 1992). These mechanisms evolved in the environment of evolutionary adaptedness (Bowlby, 1969), when our ancestors survived as hunters and gatherers (and before) and may not be associated with greater reproductive fitness today. Evolutionary psychological models look at the function of a behavior (why this mechanism evolved) as well as how the mechanism works. And, counter to some common misconceptions, evolved psychological mechanisms are in transactional relations with environmental factors. Believing that certain behaviors are under the influence of evolved psychological mechanisms does not imply that aspects of the physical and social environment do not play a critical role. In fact, quite the opposite is true; most evolved mechanisms are quite sensitive to variations in environments and are expressed differently depending on one's surroundings (see Gottlieb, 1992, 1998, 2000; Ho, 1998). We believe that factors associated with development are particularly important in the expression of evolved psychological mechanisms. As such, explanations of complex cognitive–behavioral abilities based on evolved mechanisms are no different in kind than explanations that do not invoke distal (evolutionary) causation.

Evolved Psychological Mechanisms

How is it that human beings around the globe deal with the complexities of the physical and social world? One explanation is that humans have an advanced general problem-solving ability and use this monolithic skill to deal with the people, objects, and situations they encounter. At the extreme, this is the "blank-slate" model, which proposes that experiences are imprinted on the developing human mind and that people learn to identify important contexts and modify their behavior appropriately as a result of experience. In other words, few or no cognitive biases are built into the human brain that would cause people to more readily make sense of some pieces of information relative to others, and few or no constraints exist on the type of information that humans can process (given our sensory systems, of course). Few, if any, serious scientists advocate this extreme version of the human mind today. But the idea that the human mind uses a *domain-general mechanism* continues to have advocates. From this perspective, "the mind is one": A single, general-purpose processing mechanism governs most aspects of cognition. This perspective can be seen in Piaget's theory of cognitive development (cognition is relatively homogeneous within a stage), as well as in more contemporary neo-Piagetian approaches (Case, 1992); in many information-processing models of cognition and cognitive development that posit a domain-general set of mental resources used for executing cognitive tasks (Kail & Salthouse, 1994); and in theories that propose a single mental factor (g) underlying all "intelligent" behavior (A. R. Jensen, 1998). In fact, we believe that ample evidence exists that humans do possess some domain-general cognitive skills. (See further discussion of domain-general skills in chapter 5.) However, human cognition consists of more than a domain-general mechanism, and the hallmark of evolutionary psychology is the existence of *domain-specific* cognitive skills.

Modules of the Mind

The idea that the human mind is composed of a set of relatively independent cognitive abilities goes back at least to the phrenologist Franz Joseph Gall, who mistakenly believed that he could detect individual differences in abilities by evaluating the bumps on peoples' heads. Given this disreputable beginning of domain-specific theories, it is no surprise that domain-general theories were taken more seriously by psychologists during the 20th century. However, over the past two decades or so, domain-specific theories have gained new respect and in fact now dominate the field of cognitive science. Much of the emphasis on the domain-specific nature of human cognition can be traced to the philosopher Jerry Fodor (1983), who proposed the concept of *modularity* in brain functions. By this, Fodor meant that certain areas of the brain are dedicated to performing specific cognitive tasks, such as language. Fodor referred to these areas of the mind/brain as modules that are relatively independent of one another. Each module is a special-purpose system that is *informationally encapsulated* (or "cognitively impenetrable," in the words of Pylyshyn, 1980), meaning that other parts of the mind/brain can neither influence nor access the workings of the module. Fodor did not reject the existence of a domain-general mechanism; in fact, whereas the operation of these modules is unavailable to conscious awareness, their output eventually becomes available to a central information processor, although the activity of this domain-general mechanism does not affect the domain-specific modules. Although Fodor's idea of domain-specificity in human cognition has been challenged (Elman et al., 1996; and see Fodor, 2000, for a revision of his earlier ideas), it is the dominant perspective of cognitive psychology today, and we are confident that many, if not most, important aspects of human cognition are domain-specific.

Information-Processing Mechanisms as the "Missing Link" in Evolutionary Explication

How can we explain evolved behavior patterns? What level of analysis is most useful for understanding how the human mind and behavior evolved over the past 5 million years? The answer provided by evolutionary psychologists is that cognitive-level information-processing mechanisms are the missing link in the evolution of human behavior. This is a position presented by evolutionary psychologists Leda Cosmides and John Tooby (1987), who proposed that cognitive processes "in interaction with environmental input, generate manifest behavior. The causal link between evolution and behavior is made through psychological mechanisms" (p. 277). According to Cosmides and Tooby, at least in humans, adaptive behavior is predicated on adaptive thought. Natural selection operates on the cognitive level—information-processing programs evolved to solve real-world problems.

But not all types of information-processing mechanisms are equally reasonable from an evolutionary perspective. Rather, mechanisms should have evolved that solve specific problems. These are domain-specific mechanisms, what Cosmides and Tooby (1987) referred to as *Darwinian algorithms*.[3] That is, rather than influencing general intelligence, for instance, Darwinian algorithms affect very specific cognitive operations, such as face recognition, language acquisition, or the processing of certain types of social relationships. Consistent with Fodor's (1983) view of human cognitive architecture, infants, children, and adults evolved specific cognitive abilities to deal with reasonably well-defined problems, and these abilities are modular in nature, relatively independent of a person's other cognitive skills. Cognitive scientist and evolutionary psychologist Steven Pinker (1997) captured this perspective succinctly:

> The mind is organized into modules or mental organs, each with a specialized design that makes it an expert in one area of interaction with the world. The modules' basic logic is specified by our genetic

---

[3] Although we use the phrase *Darwinian algorithms* here, we do not mean to imply that such information-processing mechanisms are literally algorithmic, in that they are always applicable, successful, and completable in the proper context (Davies, Fetzer, & Foster, 1995). Rather, as will be clear in our discussion of the developmental systems approach below, we believe that evolved psychological mechanisms can be better thought of as epigenetic programs, or, following E. O. Wilson (1998), as "rules of thumb."

program. Their operation was shaped by natural selection to solve problems of the hunting and gathering life led by our ancestors in most of our evolutionary history. (p. 21)

Although some theorists with an evolutionary psychological perspective have proposed that domain-general mechanisms can coexist with sets of domain-specific mechanisms (e.g., Bjorklund & Kipp, 1996; MacDonald & Geary, 2000; Mithen, 1996), most evolutionary psychological accounts dismiss, inappropriately we believe (see chapter 5), the existence of any domain-general mechanisms.[4]

If humans possess domain-specific mechanisms for solving specific problems, the implication is that our mind is not a general-purpose problem solver and that some things will be very difficult or impossible to learn. Stated differently, this perspective proposes that there are constraints on learning (Gelman & Williams, 1998). Constraints imply restrictions, and restrictions are usually thought of negatively. The human mind is notable for its flexibility. We, more than any other species, live by our wits, and we have been able to adapt to a more varied range of environments than any other large animal. But constraints, from this perspective, enable learning rather than hamper it.

Children enter a world of sights, sounds, objects, language, and other people. If all types of learning were truly equiprobable, they would be overwhelmed by stimulation that bombards them from every direction. Instead, infants and young children are constrained to process certain information in "core domains" (such as the nature of objects, language) in certain ways. They come into the world with some notion of how the world is structured, and this leads to faster and more efficient processing of information within specific domains. According to Gelman and Williams (1998),

> From an evolutionary perspective, learning cannot be a process of arbitrary and completely flexible knowledge acquisition. In core domains, learning processes are the means to functionally defined ends: acquiring and storing the particular sorts of relevant information, which are necessary for solving particular problems. (p. 600)

In chapter 6 we examine in greater detail the "constraints" that infants and young children come into the world with and how such constraints influence their thinking.

Evolutionary psychologists have found domain-specific mechanisms nearly everywhere they have searched for them. For example, children

---

[4]When aspects of a species' habitat are unstable over generations, the selection of domain-specific abilities would not be promoted. Rather, when climatic changes are unpredictable over generations, as they apparently were for hominids (Potts, 1998), the selection of domain-general cognitive mechanisms, adaptive for dealing with novel environments, may be promoted (MacDonald & Geary, 2000). This issue is discussed in greater detail in chapter 5.

acquire language over 4 or 5 years. For adults, learning a second language is often very difficult, and the ease with which children learn a first language seems at odds with (i.e., independent of) their other, more general, cognitive abilities. Specific evolved psychological mechanisms have been proposed to explain children's acquisition of language (see Pinker, 1994; see also chapter 6), although other evolutionary-friendly proposals that posit a domain-general mechanism have also been suggested (see Elman et al., 1996). Similarly, aspects of children's understanding of social functioning, including theory of mind and social reasoning, have been hypothesized to be modular in nature (see Baron-Cohen, 1995; Cummins, 1998a; see also chapter 7).

Within the realm of social development, Bugental (2000) proposed five domains of social life and social development: attachment, reciprocity, hierarchic power, coalitional group, and mating. Each general area follows its own developmental course, is governed by neurohormonal regulators (opiods and oxytocin for attachment; testosterone for hierarchic power), and has *social algorithms* associated with it to achieve specified goals (e.g., proximity maintenance and protection for attachment). These algorithms have evolved to deal with recurrent social problems infants and children have faced over the millennia. Although the behaviors Bugental described are not traditionally considered to be "cognitive" in nature, what underlies them are hypothesized to be domain-specific information-processing programs, designed by natural selection to solve relatively specific problems our ancestors encountered.

The evolved psychological mechanisms proposed by evolutionary psychologists have some things in common with the innate-releasing mechanisms, or fixed-action patterns identified by ethologists to explain the often-complex behaviors of animals in response to specific environmental conditions, but are also different. For example, Tinbergen (1951) described the aggressive behavior of male stickleback fish in response to the red belly of another male stickleback (the red belly being an indication of readiness to mate). Stereotypic aggressive behavior was displayed to any red stimulus that closely resembled the underbelly of another male fish. Such behavior is adaptive, in that it limits access of other males to a prospective mate. Such responses could be thought of as evolved psychological mechanisms, but most (if not all) such mechanisms possessed by humans are more flexible, reflecting general propensities to respond in certain ways depending on environmental conditions and developmental history. It is not the case, for example, that human males act aggressively toward any male stranger who enters their territory (red belly or not). This pattern may be found in some cultures, however, and depending on the social organization of the group and the developmental history of the individual, how a person (male or female) responds to a stranger will vary. Nonetheless, according to evolutionary psychological theory, what underlie such responses are evolved psychological mechanisms, which find their expression as a result of interaction with the environment over the course of development.

In sum, evolved psychological mechanisms were adaptive to humans' ancestors; they helped solve specific problems. It would be extreme to claim, of course, that all adaptive behaviors or thought processes have been explicitly selected for their fitness value; some may have been associated with another adaptive trait and not selected for themselves, and others may simply have not been sufficiently maladaptive to result in extinction. But a core assumption of evolutionary psychology is that these mechanisms evolved to solve specific problems and that they are modular in nature. In addition, these mechanisms did not evolve in response to the problems of contemporary humans, who have only recently abandoned a nomadic lifestyle for one in villages, towns, and cities. Rather, they evolved over the past several million years to handle the problems faced by our hominid ancestors.

Our proposal that animals are "prepared" for certain types of experiences and are "constrained" in their ability to process information should not be read to mean that these abilities are "prepacked," "in the genes," or "innate." Our contention is that evolution, through the process of natural selection, has resulted in organisms who, when provided with a species-typical environment, are biased to process some pieces of information more easily than others. What have evolved are developmental systems, which include genes and the internal and external environments of an organism (Lickliter & Berry, 1990; Oyama, 2000a, 2000b). The products of such evolution may appear to have predetermined abilities, with development playing no significant role, but this is not the case. Some outcomes are highly likely in certain environments and unlikely in others. Whether a mechanism is seen as domain specific or domain general, by itself, does not imply anything about the role of development in its expression.

Environment of Evolutionary Adaptedness

A basic tenet of evolutionary psychology is that psychological mechanisms evolved to solve adaptive problems our hominid ancestors faced. (*Hominids* refers to the class of bipedal or upright-walking animals that includes humans and our ancient ancestors.) This means that, although many evolved mechanisms may still serve humans' survival and reproductive interests well today, others may not. Cities and nation-states, birth control, handguns, and well-stocked grocery stores were foreign to our ancient ancestors. For example, our penchant for sweet and fatty foods, which signal a high-calorie meal, was surely adaptive for our nomadic ancestors. Such a proclivity is less adaptive today and in fact is often maladaptive in societies in which high-caloric food is readily available (Nesse & Williams, 1994).

How did humans' prehistoric ancestors live? We can never know their lifestyles with 100% certainty, of course, but analysis of the fossil and archeological records and examination of modern hunter–gatherer societies can give some idea of how our ancestors made a living and the conditions under which the human mind evolved (see Johanson & Edgar, 1996; Kaplan, Hill, Lancaster, & Hurtado, 2000; Mithen, 1996; Tattersall, 1998).

The Fossil Record

Homo sapiens is a relatively new species, in geological time. The genetic line that eventually led to our species diverged from the ancestor we share with common chimpanzees (*Pan troglodytes*) and pigmy chimpanzees (or bonobos, *Pan paniscus*) between 5 and 7 million years ago. The genetic similarity between humans and chimpanzees is striking: We share approximately 99% of our DNA with chimps, making humans and chimpanzees closer genetically than horses and zebras.

Although the fossil record of human evolution extends back nearly 5 million years, it is incomplete. There has been and continues to be much debate among paleoanthropologists concerning which species evolved from what other species and even how many different hominid species there were at any one time. However, the basic sequence of events over the past 5 or 6 million years is generally agreed on, even if it is currently impossible to be precise about the human family tree. One plausible phylogenetic tree is presented in Figure 2.1.

The earliest identified hominid that was distinct from the ancestor we share with chimpanzees has been called *Ardipithecus ramidus*, but little is known about these animals, which were first seen in the fossil record about 4.5 million years ago. (Claims of yet another hominid species, *Orrorin tugenesis*, dating back 6 million years have recently been reported; see Balter 2001, but the place of this animal in the human lineage and whether or not it is a hominid are much debated.) Fragmentary evidence exists dating to about 4 million years ago for a species called *Australopithecus anamensis*, which was presumably the direct ancestor for the earliest species for which good fossil evidence exists, *Australopithecus afarensis* (of which the famous Lucy was a member, Johanson & Edgar, 1996). *A. afarensis* were small animals, the females being about 3.5 feet tall, who walked upright much as modern humans do, but who had small skulls that were more apelike than human. The fossil skulls show that *A. afarensis* had a cranium of about 400 cc, and it follows that they also had small brains. Several distinct australopithecine species evolved over the next several million years, and one of these lines led to the *Homo* (true human) genus beginning about 2.5 million years ago. Some evidence exists that one recently identified species of australopithecine may have used tools (*Australopithecus garhi*,

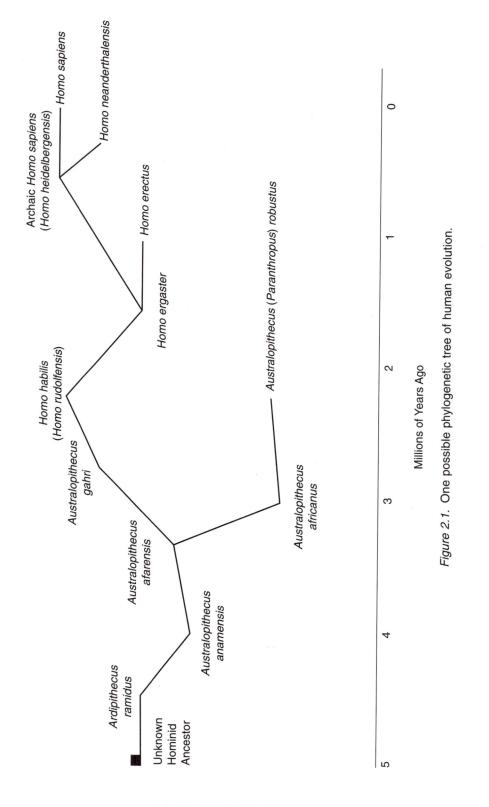

Figure 2.1. One possible phylogenetic tree of human evolution.

Asfaw et al., 1999). (There is also recent evidence of a hominid species from a different genus, *Kenyanthropus platyops* [Leakey et al., 2001], that existed 3.5 million years ago, but its relation with australopithecines, and thus with the *Homo* genus, is currently unknown.)

The earliest identified species of the *Homo* line has been called *Homo habilis*. These animals had larger brains than their australopithecine ancestors (about 650 cc), used primitive tools, and, like the australopithecines, were confined to Africa. *Homo ergaster* apparently evolved from *Homo habilis* (or from *Homo rudolfensis*, a related *Homo* species living at about the same time as *Homo habilis*) about 1.6 million years ago, about the beginning of the Pleistocene geological age. One scenario of human evolution suggests that *Homo ergaster* evolved into *Homo erectus*, who had larger brains (about 900 cc), may have used fire, and created more complex tools than their ancestors. About 1.7–2 million years ago, *Homo erectus* (or perhaps *Homo ergaster*) left Africa and made their way to Europe and Asia (Gabunia et al., 2000). By most accounts, *Homo erectus* became extinct about 250,000 years ago (although there is evidence that they may have existed in some locations as recently as 26,000 years ago [Swisher et al., 1996], which would make them contemporaries of fully modern humans). The ancestral *Homo ergaster* species that stayed in Africa was replaced about 300,000 years ago by *Archaic Homo sapiens* (classified by some as *Homo heidelbergensis*), who showed substantial increases in brain size (relative to body size; Ruff, Trinkaus, & Holliday, 1997) and possessed characteristics of both *Homo erectus* and modern *Homo sapiens*. Modern humans with large brains (about 1,300 cc) apparently evolved in Africa within the past 50,000–100,000 years (Ingman, Kaessman, Pääbo, & Gyllensten, 2000). They migrated out of Africa and, according to most paleoanthropologists, replaced the aboriginal *Homo* species they encountered (e.g., Neanderthals in Europe), either by killing or by out-competing them (Eccles, 1989; Johanson & Edgar, 1996; Wood, 1994).

The unity of modern *Homo sapiens* is reflected in the pattern of genetic diversity found among groups of humans around the world. For example, greater than 80% of all genetic differences are among individuals within the same population; in contrast, variation among populations from different continents accounts for only about 10% of human genetic diversity. This pattern of data suggests that most human genetic variation occurred before the migration of modern humans out of Africa (Owens & King, 1999). These data also suggest that "race" represents superficial rather than biologically important differences among humans.

As indicated by the fossil evidence, humans have not changed that much as a species over the past 100,000 years, at least physically, and certainly little at all over the past 35,000 years or so. But this conservatism of morphology has been accompanied by radical changes in how we live as a species. The advent of agriculture and life in stationary communities beginning about 12,000 years ago changed drastically how most human beings lived.

Lifestyles in the Environment of Evolutionary Adaptedness

How did humans' ancestors live, and what were the conditions in which the human mind evolved? Based on how our close genetic relatives, the chimpanzees, live today, it is a safe bet to say that hominids, back to *Ardipithecus ramidus*, have always been social species. However, based on fossil evidence and how people in modern hunter–gatherer groups live, an intact group was likely small (probably 30–60 individuals). One picture that emerges consistently when examining human evolution is the increasing social complexity of groups. Modern humans, in both information-age and hunter–gatherer societies, cooperate and compete with members in their own group and with people from outside groups. In fact, warfare of one type or another is a characteristic of all human groups. The only other mammal that displays behavior at all similar to that of human war parties—attacking and killing members of another group of their own species—is the chimpanzee (Goodall, 1986), suggesting that the roots of warfare run deep in our species' genetic history (Wrangham & Peterson, 1996).

Our ancestors made their living on the savannas of Africa gathering fruits, nuts, vegetables, and tubers; scavenging food left from the kills of large predators; and hunting. (Given patterns in modern hunter–gatherer societies, gathering was most likely the work of women, and hunting most likely the work of men, although men likely gathered as well.) Since the migration of humans from the African savannas 100,000 years or so ago, they surely experienced a wide variety of environments.

For the first 4 or 5 years of life, children of early modern humans were most likely cared for almost exclusively by their mothers (as is the case in most mammals), although fathers likely provided protection and support in the form of food and other tangible resources for their children and their children's mothers (Kaplan et al., 2000). It is likely that some males had multiple "wives," whereas others had little or no access to females. In other words, as modern humans are, ancient humans were likely a marginally monogamous/marginally polygamous species, in which males competed with one another for access to females, and females selected males who could provide resources for themselves and their offspring.

That modern humans' marginally monogamous/marginally polygamous mating style also typified our ancestors is supported by recent comparative immunological research. Nunn, Gittleman, and Antonovics (2000) examined the number of white blood cells for 41 species of nonhuman primates and found that sexually promiscuous species (those in which females have multiple mating partners) have significantly more white blood cells of various types than monogamous species. Nunn and his colleagues proposed that promiscuous species are more susceptible to sexually transmitted diseases, requiring a more complex immune system than monogamous species. *Homo sapiens'* immune system is more like that of the polygamous, harem-based gorilla (*Gorilla gorilla*) and the monogamous gibbon (*Hylobates lar*) than that of the promiscuous chimpanzee (*Pan troglodytes*). This suggests that the marginally monogamous relationships that characterize modern and historic humans also characterized our species' prehistoric forebears.

Females probably reached puberty relatively late (late teens) and gave birth every 3–5 years, with pregnancy often following the cessation of nursing a previous child. Infant and childhood mortality were surely high, and, even for those who did make it to adulthood, life was relatively brief by contemporary standards, with few people living past 40 years (Austad, 1997), although it is likely that there were always some "old" people in a group.

Females likely relocated to the villages of their "husbands." This is reflected not only by lifestyles of contemporary hunter–gatherers, but also by patterns of genetic diversity. Worldwide variation in mitochondrial DNA, which is passed to offspring only from their mothers, is similar to the variation found for genes on autosomal (nonsex) chromosomes. An average population anywhere in the world possesses about 80%–85% of all variation in mitochondrial DNA. (This figure is comparable for genes on autosomal chromosomes.) In contrast, most populations contain only 36% of the possible genetic variation found on the Y (male) chromosome, with 53% of this variation being attributed to the continent on which men reside (Owens & King, 1999; Pennisi, 2001; Seielstad, Minch, & Cavalli-Sforza, 1998). This pattern suggests that women migrated more than men, leaving their birthplaces to live with their mate. Most female migration would not have involved long distances, but over hundreds of generations, the genetic effects would accumulate.

We should note that the common depiction of the Pleistocene environment as consistent and stable is likely inaccurate. Analyses of ice-core records show rather drastic climatic changes over the past 2 million years, with large-scale instability rather than progressive change being the primary environmental challenge hominid populations faced (Potts, 1998, 2000). Those individuals who could deal with unpredictable changes in climate and habitat were the ones who reproduced and became our ancestors. Variable environments suggest variable behavioral strategies, making any claim that ancient humans followed any signal lifestyle invalid. Yet, despite the environmental and presumably the behavioral instability, some characteristics of hominid lifestyle, such as its social complexity, division of labor, and tool use, have likely always characterized our ancestors, and these may have served as the foundation for domain-specific and species-universal psychological mechanisms.

Life has changed much over the past 100,000 years or so. Humans are still a highly social species, mothers and fathers still cooperate to rear their children, and we continue to use tools to solve problems both great and small. However, the social groups of most contemporary humans are far different from those of our ancestors, the tools we use are far more sophisticated, infant mortality is low, and longevity is greatly extended. Nevertheless, our brains possess psychological mechanisms that evolved to solve problems our ancestors faced. We have lived as a species under the influence of civilization only 10,000–12,000 years, not long enough for natural selection to have substantially affected our minds. Our current genetic makeup, then, most likely reflects adaptations to our hunter–gatherer past, which may or may not be adaptive for us today.

This insight is relevant to formal educational practices. Given that the modern human mind evolved to solve problems faced by small groups of nomadic hunters and gathers, it is no wonder that many children balk at attending school. From the perspective of evolutionary psychology, much of what we teach children in school is "unnatural" in that teaching involves tasks never encountered by our ancestors (Bjorklund & Pellegrini, 2000; R. D. Brown & Bjorklund, 1998; Pellegrini & Bjorklund, 1997). For example, although the human species has apparently been using language for tens of thousands of years, reading is a skill that goes back only a few thousand years, and only in this century have a majority of people on the planet become literate. Evolutionary developmental psychologist David Geary (1995) has referred to cognitive abilities that were selected in evolution, such as language, as "biologically primary abilities," and skills that build on these primary abilities but are principally cultural inventions, such as reading, as "biologically secondary abilities." Biologically primary abilities are acquired universally, and children typically have high motivation to perform tasks involving them. Biologically secondary abilities, on the other hand, are culturally determined and often tedious repetition and external motivation are necessary for their mastery. It is little wonder that reading, a supposed "language art," and higher mathematics give many children substantial difficulty (see further discussion in chapter 5).

Reading and higher mathematics may be the best examples of skills that humans' ancestors never acquired, but the "unnaturalness" of school also extends to the social and behavioral realms. For example, most modern schools' emphasis on seat work and extended periods of focused attention may conflict with children's natural tendencies toward high activity and exploration, particularly in boys. This can be seen in the high incidence of attention deficit hyperactivity disorder (ADHD) among school children in some countries today (P. S. Jensen et al., 1997). Several researchers have suggested that most children diagnosed with ADHD may simply be highly active and playful youngsters who have a difficult time adjusting to the demands of school (Panksepp, 1998). Although we agree with Goldstein and Barkley (1998) that individuals who truly have ADHD would be at a disadvantage in any environments, high levels of activity and frequent switching of attentional focus may have been adaptive in some contexts for our ancestors (P. S. Jensen et al., 1997) and may still be in some environments today. Unfortunately, such behavior, even when not extreme, conflicts with "proper" school conduct and is often treated with psychostimulants that reduce hyperactivity but may also reduce the desire and opportunity to play, which may, in turn, reduce neural and behavioral plasticity.

A Functional Analysis

Pregnancy Sickness as an Adaptation

Evolutionary psychological explanations focus on adaptationist thinking, which stresses the function of a behavior or trait. As an example of adaptationist thinking, let us examine the case of pregnancy sickness. Pregnancy sickness occurs during the early months of pregnancy in the majority of women around the world (Tierson, Olsen, & Hook, 1986). Symptoms include nausea, vomiting, and food aversions. Given these symptoms, pregnancy sickness is understandably considered an illness. Is it possible, however, that such an "illness," timed as it is during the first trimester of pregnancy, might have a function? This was investigated by Profet (1992) and has been expanded on by Flaxman and Sherman (2000; see also Hook, 1978).

The first trimester of human pregnancy is a time of rapid development for the embryo or fetus. All of the major organ systems are developed during this time, although they are not perfected. Limbs, hands, and feet are formed, and although the proportions of the fetus will change over the remainder of prenatal development (with the size of the head diminishing relative to the rest of the body), the fetus is easily recognized as human by the end of the third month of gestation. Particularly during the first two months or so, the unborn child is most susceptible to the effects of *teratogens*, agents that may interfere with development. This is seen in the well-documented cases of deformed or impaired infants of women who take certain drugs (e.g., thalidomide) or contract certain diseases (e.g., rubella) early in pregnancy. When these drugs are taken or diseases contracted later in pregnancy, deleterious effects are usually absent. Pregnant women are also more susceptible to food-borne parasites and microorganisms during early pregnancy, because their immune system is naturally suppressed to prevent them from rejecting the fetus as a foreign tissue (Flaxman & Sherman, 2000).

Teratogens that adversely affect prenatal development are not a curse cast only on modern women as a result of manufactured pharmaceuticals or civilization-born diseases; they can be produced naturally in plants and animal flesh. In fact, all plants seemingly produce toxins to ward off predators (see Profet, 1992). Women in ancient times who avoided foods containing elements that might prove injurious to their unborn infants would have been rewarded by an increased chance of having healthy babies. Profet (1992) hypothesized that pregnancy sickness is an adaptation, with women who have an aversion to toxin-containing foods being more likely to have healthy babies.

This is an interesting hypothesis, but what makes it more than speculation, or a "just so" story, concocted to fit an outcome (pregnancy sickness) we already know? Several lines of evidence converge to make the hypothesis a sound one. First, as we have noted, pregnancy sickness corresponds to the period when an unborn child is most susceptible to the effects of teratogens. It typically commences between 2 and 4 weeks after conception, when organ development is most rapid, and usually wanes by 12 weeks, when the fetus is less susceptible to the adverse effects of outside agents. Second, modern women acquire aversions to food that are highest in toxins and tend not to develop aversions to foods that are more apt to be toxin-free. Meat often contains fungus or bacteria, particularly as it decomposes, and many vegetables, such as cabbage, cauliflower, and Brussels sprouts, contain toxins that can prove harmful to an unborn fetus. Evidence for this was provided by Tierson, Olsen, and Hook (1985), who interviewed 400 women in their 12th week of pregnancy about food aversions, most of which had passed by the time of the interview. Eighty-five percent of the women reported experiencing some food aversion. Although there was substantial variability in the types of foods women developed aversions to, the list of "aversive foods" generally consisted of items high in toxins. For example, 32% of women reported an aversion to coffee, 31% to meat and poultry, 20% to alcoholic beverages, and 11% to vegetables. In contrast, very few women reported an aversion to bread (1%) and none to cereals (see also Dickens & Trethowan, 1971). Third, despite variability among cultures, pregnancy sickness appears to be universal. Although little research has examined pregnancy sickness in nonindustrial cultures, it has been found in hunter–gatherer cultures in Africa and South America. The few cultures in which it has not been reported rarely include animal products as dietary staples and subsist on a bland maize diet (see Flaxman & Sherman, 2000; Profet, 1992).

Perhaps the most compelling evidence in support of the adaptive function of pregnancy sickness derives from research demonstrating a reduction in spontaneous abortions for women who experience pregnancy sickness versus those who do not (Profet, 1992). In a meta-analytic review that looked at the relationship between pregnancy sickness and miscarriage in more than 20,000 pregnancies, Flaxman and Sherman (2000) confirmed the findings of an earlier review (Weigel & Weigel, 1989) that reported that the spontaneous abortion and fetal death rates for women who do not experience pregnancy sickness are significantly greater than for women who do.

The functional interpretation of pregnancy sickness also provides an interpretation for what has sometimes been viewed as pathological behavior among some groups of people. For example, *pica*—the eating of nonnutritive substances such as clay—is a common practice for pregnant women in some parts of the world and in parts of the United States. Research has shown that when clay is eaten along with certain toxic foods, it prevents the toxins from being absorbed into the blood system and thus reduces pregnancy sickness and the effects of teratogens on the fetus.

A functional analysis of pregnancy sickness demonstrates the benefit that an evolutionary perspective can have. What has typically been viewed as a dysfunctional state, for which medication is sometimes prescribed, is actually a well-adapted mechanism that fosters the development of the unborn child. Although the discomfort associated with pregnancy sickness is real, its consequence is an embryo or fetus protected from environmental toxins that would impair its development. It is worth noting that thalidomide, the drug that led to serious deformations of children's limbs when taken early in pregnancy, was sometimes prescribed to alleviate pregnancy sickness.

Adaptations, By-products, and Noise

Not all current aspects of cognition, behavior, or morphology are the result of adaptation. Evolution yields at least three products (Buss, Haselton, Shalelford, Bleske, & Wakefield, 1998): adaptations, by-products, and noise. *Adaptations* refer to reliably developing, inherited characteristics that came about as a result of natural selection and helped to solve some problems of reproduction or survival in the environment of evolutionary adaptedness. The umbilical cord is an example of an adaptation: It solved the problem of how to get nutrients from a mammal mother to her fetus. *By-products* are characteristics that did not solve some recurring problem and have not been shaped by natural selection but are a consequence of being associated with some adaptation. The belly button would be an example of a by-product. Finally, *noise* refers to random effects that may be attributed to mutations, changes in the environment, or aberrations of development, such as the shape of one's belly button. As this tripartite classification indicates, a characteristic may have evolved in a species but not have been designed by the forces of natural selection. The evolutionary psychologist's task is to identify and describe psychological mechanisms that may have solved survival or reproductive problems in humans' evolutionary past and to differentiate those mechanisms from characteristics that may be better classified as by-products or noise.

Furthermore, some adaptations may have negative effects (by-products) associated with them. For example, the enlarged skull of a human fetus is surely an adaptation (to house a large brain, associated with greater learning ability and behavioral flexibility); however, the size of the baby's head makes birth difficult (due to limits on the width of a woman's hips because of bipedality), and many women and infants have died in childbirth. However, the benefits of an enlarged brain outweighed the detriments of neonatal and maternal death.

Exaptations

A relatively new concept related to that of by-products is *exaptations* (S. J. Gould, 1991, 1997; S. J. Gould & Lewontin, 1979; S. J. Gould & Vrba, 1982), defined by evolutionary biologist Stephen Jay Gould (1991) as "a feature, now useful to an organism, that did not arise as an adaptation for its present role, but was subsequently co-opted for its current function" (p. 43). Further, exaptations are "features that now enhance fitness, but were not built by natural selection for their current role" (p. 47). For example, evidence exists that feathers in birds, necessary for flight, initially evolved for thermoregulation. As other morphological changes occurred in the ancestors of modern birds, some of the features of their heat-retaining feathers made flight possible. Thus, although someone trying to discern the function of feathers in contemporary birds could easily assume that they evolved to permit flight (an adaptation), the assumption would not be completely correct. Rather, according to Gould, feathers are more properly classified as an exaptation. An issue for evolutionary psychology is the extent to which psychological mechanisms (or more properly, brain structures underlying such mechanisms) are the result of specific adaptations or are, rather, exaptations, the by-product of earlier and seemingly unrelated adaptations. Gould (1991), for instance, has argued that many aspects of modern human intelligence do not represent domain-specific evolved psychological mechanisms but rather are the by-products of an enlarged brain (see also Finlay, Darlington, & Nicastro, 2001).

The concept of exaptations is attractive and has been adopted by many evolutionists (Tattersall, 1998), although it has not been generally accepted by evolutionary psychologists (Buss et al., 1998). We believe that the basic idea behind exaptations is solid—that many of the products of evolution arose based on by-products of other adaptations or features that, initially, had no inherent function for an organism. We firmly believe that evolution works by co-opting brain structures, for example, for purposes for which they had not initially been selected. However, we also believe that, once co-opted, any new function must pass through the sieve of natural selection. That is, if a new function that arises based on previous structures is to spread through the population and become part of the genome of the species, it must past the muster of natural selection. For example, although feathers may not have originally evolved for flight, they became important (likely necessary) for bird flight, and thus it is reasonable to examine the adaptive value of feathers for this co-opted purpose. Similarly, even if, for instance, many human intellectual abilities are seen as the by-product of a large brain, one is still left with the question, For what purpose did the human lineage evolve this inflated neural cortex? The onus, thus, is still on the scientist to discern the likely function of characteristics that affect functioning now and in our evolutionary past, regardless of whether the characteristic initially evolved to serve the function that it now serves.

Although our own tendency is to view exaptations as a special case of adaptations, the concept of exaptations does cause one to look at the evolution of structure and behavior a bit differently and to ask slightly different questions. Keeping in mind that evolution is blind—that it does not have a final product, such as *Homo sapiens* in mind—and that it works only with what features currently exist within a species, we realize that many contemporary and ancient adaptations may have been co-opted from other seemingly unrelated functions. This may be particularly true for psychological functions associated with the expanding neocortex over the past 5 million years. However, believing this does not preclude discovering the adaptive problem that any function may have solved for humans' ancestors and does not dictate the nature of adaptations (e.g., whether adaptive psychological mechanisms are domain specific or domain general in nature). It does, however, cause us to probe a bit more deeply than we might have otherwise.

ROLE OF THE ENVIRONMENT IN EVOLUTIONARY PSYCHOLOGICAL PERSPECTIVE: THE DEVELOPMENTAL SYSTEMS APPROACH

If evolved psychological mechanisms underlie contemporary behaviors and thought patterns, what role can culture, or experience in general, play? A common misperception about evolutionary explication is that if an ability is said to have "evolved" or to have an innate component, the result is one of biological, or genetic, determinism. In other words, if it is in the genes (which it seemingly must be if it evolved), it cannot be changed. This is not the case, and evolutionary psychologists are explicit in the role that the environment plays (and played in our evolutionary past) in the expression of evolved psychological mechanisms. However, despite the insistence that explanations based on evolutionary theory are not ones of genetic determinism, few evolutionary psychologists have proposed a detailed model of how evolved mechanisms interact with the environment to produce behavior.

Providing such a model is one significant contribution of evolutionary developmental psychology.

As noted above, evolutionary psychologists assume that organisms adapt and evolve, through natural selection, through their transactions with the environment. Organisms affect their environment (by choosing and then "furnishing their niches"), and environments, in turn, affect the organism (by changing behaviors to meet the particular demands of a setting). Because of this transactive relation between organism and environment, we must study organisms interacting with their environments if we want to understand adaptation and development. This position rejects any simplistic biological determinism such as the theory that genetic endowment has a main effect on cognitive functioning (see Pellegrini & Horvat, 1995, for a discussion) or on social development (Pellegrini & Smith, 1998). More specifically, we believe that the developmental systems approach provides a proper appreciation of how biology and environment, at a variety of levels, interact to produce behavior and development, and that such a model can be used to explain how evolved psychological mechanisms are translated into behavior (Gottlieb, 1991a; Kuo, 1967; Lickliter & Berry, 1990; Oyama, 2000a).

The core concept of the developmental systems approach is that of *epigenesis*, which developmental psychologist Gilbert Gottlieb (1991a) defined as "the emergence of new structures and functions during the course of development" (p. 7). Gottlieb (1991a, 1996, 1998, 2000; Gottlieb, Wahlsten, & Lickliter, 1998) stated that epigenesis reflects a bidirectional relationship among all levels of biological and experiential factors, such that genetic activity both influences and is influenced by structural maturation, which is bidirectionally related to function and activity. This relationship can be expressed as

$$\text{genetic activity (DNA} \leftrightarrow \text{RNA} \leftrightarrow \text{proteins)}$$
$$\leftrightarrow \text{structural maturation} \leftrightarrow \text{function, activity.}$$

According to Gottlieb (1991b),

> Individual development is characterized by an increase of complexity of organization (i.e., the emergence of new structural and functional properties and competencies) at all levels of analysis (molecular, subcellular, cellular, organismic) as a consequence of horizontal and vertical coactions among the organism's parts, including organism–environment coactions. (p. 7)

From this perspective, functioning at one level influences functioning at adjacent levels. For example, genes code for the production of protein molecules, which in turn determine the formation of structures such as muscle or nerve cells. But activity of these and surrounding cells can turn on or off a particular gene, causing the commencement or cessation of genetic activity. Also, self-produced activity or stimulation from external sources can alter the development of sets of cells. From this viewpoint, there are no simple genetic or experiential causes of behavior; all development is the product of epigenesis, with complex interactions occurring among multiple levels. This bidirectional approach to development is expressed in Figure 2.2. This point was made clear by developmental neuropsychologist Mark Johnson (1998) in his review of the neural basis of cognitive development:

> Since it has become evident that genes interact with their environment at all levels, including the molecular, there is no aspect of development that can be said to be strictly "genetic," that is, exclusively a product of information contained in the genes. (p. 4)

Evolved psychological mechanisms can be thought of as genetically coded "messages" that, following epigenetic rules, interact with the environment to produce behavior. Areas of the newborn brain are biased to process certain types of information more effectively than others (e.g., language). Such information-processing biases derive from genetically specified features of neurons as well as endogenous factors associated with neurons' own functioning and growth (Changeux, 1985; Edelman, 1987). With species-typical experience, these sets of neurons become increasingly specialized, enhancing the efficiency of these brain areas for processing specific information, while decreasing their ability to process other information (M. H. Johnson, 2000).

The experiences of each individual are unique, beginning before birth, and if the relations expressed in Figure 2.2 closely mirror reality, there should be substantial plasticity in development. This probabilistic and unpredictable nature of development is captured by Oyama's (2000a) statement that "Fate is constructed, amended, and reconstructed, partly by the emerging organism itself. It is known to no one, not even the genes" (p. 137). This perspective

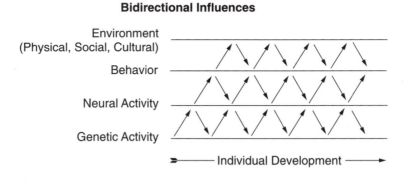

Bidirectional Influences

Environment (Physical, Social, Cultural)

Behavior

Neural Activity

Genetic Activity

Individual Development →

Figure 2.2. A simplified schematic of the developmental systems approach, showing a hierarchy of four mutually interacting components.
Note. From *Individual development and evolution: The genesis of novel behavior* by G. Gottlieb, *Oxford University Press,* p. 186. Copyright 1992.

seems to be at odds with evolutionary psychology's contention for universal, "innate" features (as in the claims of neonativists such as Pinker, 1997). In fact, despite the fact that genes will be expressed differently in different environments, almost all members of a species (human or otherwise) develop in a species-typical pattern. How can this be so, and the developmental systems perspective, as presented in Figure 2.2, still be valid?

The answer lies in the fact that humans (or chimpanzees or ducks) inherit not only a species-typical genome but also a species-typical environment. According to Robert Lickliter (1996),

> The organism–environment relationship is one that is structured on both sides. That is, it is a relation between a structured organism *and* a structured environment. The organism inherits not only its genetic complement, but also the structured organization of the environment into which it is born. (pp. 90–91)

To the extent that an organism grows up under conditions similar to that in which its species evolved, development will follow a species-typical pattern. Tooby and Cosmides (1992) have argued that complex, psychological mechanisms evolve only when the environments are relatively stable over many generations. Thus, over long periods of time, members of a species could "expect" certain types of environments and evolved species-typical solutions to deal with such stable environments.

If one takes seriously this perspective of development, it requires a modification in the canonical definition of "evolution" presented earlier in this chapter. According to conventional wisdom, evolution is defined as changes in gene frequencies over generations. It is genes that evolve, and anyone who disputes this position must have had his or her head in the sand over the past half century. However, organisms also inherit environments, both internal and external, and it is within these environments that genes are expressed and phenotypes derived. From this perspective, what evolve are developmental systems, which include genes but also the varied and interacting environments in which genes exist. According to Oyama (2000b),

> What is transmitted between generations is not traits, or blueprints, or symbolic representations of traits, but developmental *means* (or *resources*, or *interactants*). These means include genes, the cellular machinery necessary for their functioning, and the larger developmental context, which may include a maternal reproductive system, parental care, or other interaction with conspecifics, as well as relations with other aspects of the animate and inanimate worlds. This context, which is actually a system of partially nested contexts, changes with time, partly as a result of the developmental processes themselves. (p. 29)

Evidence for the influence of species-typical environments on species-typical and presumably "innate" behavior comes from research with precocial birds. For example, in the wild, a mother duck will lay several eggs in a nest and stay close by until they hatch. While in the egg, the ducklings begin to vocalize, and so hear themselves, the vocalizations of their brood mates, and those of their mother. How might these "experiences" influence later species-typical behavior?

The answer is "substantially." In a procedure developed by Gottlieb (1976, 1991b, 1996), ducklings, while still in the egg, were isolated from other eggs and their mother so that they could not hear the vocalizations of other animals. Their vocal chords were also treated so that they could not produce any sound (a condition that wears off several days after hatching). Following hatching, these animals were placed in a large container and heard the maternal call of two species—their own and another—played through speakers on opposite sides of the container. Most untreated birds in this situation will approach the call of their own species, seeming to know "instinctively" which call is that of their species and which is not. However, ducklings who were prevented from hearing any duck vocalizations—their mothers, those of other ducklings still in their eggs, or their own—failed to make this discrimination and were just as likely to approach the call of an alien species as that of their own. Thus, prehatching experience, particularly the duckling hearing its own vocalizations, plays a critical role in posthatching species-typical behavior. The reason that nearly all ducks approach the species-typical call after hatching is that nearly all ducks inherit not only the genetic disposition to make such a selection but also the species-typical environment that provides the necessary experiences for such a pattern to develop.

Results such as these demonstrate that behaviors (here related to infant–mother attachment) that are found in almost all normal members of a species are influenced by often-subtle characteristics of the environment. Evolved psychological mechanisms at the human level can be viewed similarly. Strong specieswide biases may exist for certain behaviors, but how any particular evolved mechanism is expressed will vary with environmental conditions experienced at certain times in development.

How early experience can reorganize the way the brain responds to certain stimuli is illustrated by a study of the effects of painful stimulation in neonatal rats (Ruda, Ling, Hohmann, Bo Peng, & Tachibana, 2000). Neurons in the spinal cord of rats develop during embryonic and early postnatal times, typically in the absence of noxious stimuli. When newborn rats were caused to experience pain, however, the nerve circuits that respond to pain were permanently rewired, altering their response to sensory stimulation and making the animals more sensitive to pain as adults. Only within the past 20 years have physicians bothered to give young human infants anesthesia for some medical procedures, believing that their perception of painful stimuli was minimal. These findings suggest that such unexpected stimulation, even if it is not immediately perceived as keenly as by an older individual, can adversely alter the species-typical course of development. Central to the perspective of evolutionary developmental psychology is the realization that the environment and the organism interact in different ways at different periods of ontogeny. An implication of this position is that different features may be selected at different points of development (W. D. Hamilton, 1966). We now examine aspects of evolutionary psychology that are particularly pertinent to development.

EVOLUTIONARY PSYCHOLOGY AND DEVELOPMENT

Earlier in this chapter we introduced the four questions proposed by Tinbergen (1963) that must be asked to understand the value of a behavior: What is the immediate benefit to the organism? What is the immediate consequence? How does it develop within the species (ontogeny)? How did it evolve across species (phylogeny)? To answer these questions, we must take a developmental perspective; we must appreciate the adaptive value of a particular behavior at a specific time in development. This implies that different behaviors or characteristics of an animal may be selected at different times in ontogeny. In other words, over the course of evolution, natural selection has adapted organisms to their current environments, and the environments and selective pressures humans' ancestors experienced early in their ontogeny were different from the environments and selective pressures they experienced later in their life spans.

Ontogenetic Adaptations and Adaptive Immaturity

An evolutionary account of ontogeny cannot ignore situations in which characteristics of children's behavior or cognitions serve as preparations for later life. Surely, natural selection would favor conditions in which children are able to practice in a safe environment behaviors that will be important to them in adulthood. For example, easily identifiable and culturally universal differences exist in the play styles of boys and girls, and these styles are related to sex differences in adult behavior (Geary, 1998, 1999; Pellegrini & Smith, 1998; and see also chapter 10). Boys usually demonstrate more rough-and-tumble (R&T) play than girls, which has been seen as preparation for male–male competition in adulthood (at least in the environment of evolutionary adaptedness and arguably in many contemporary societies). Girls, in contrast, usually engage in more play parenting, such as doll play, which has been seen as preparation for real parenting.

Although we believe that many features of childhood serve as preparations for adulthood, it is inappropriate to believe that all adaptive aspects of childhood need to be associated with adult functioning. Rather, we propose, some aspects of infancy and childhood are not preparations for later adulthood but evolved to serve an adaptive value for that specific time in development (Bjorklund, 1997a; Oppenheim, 1981). Developmental psychobiologists, whose typical subjects are birds or infrahuman mammals, have long held the view that immature aspects of an animal often have adaptive value and were selected in evolution to help keep the young animal alive at that time in development (Gottlieb et al., 1998; Spear, 1984; Turkewitz & Kenny, 1982). This perspective has been less popular with developmental psychologists who study human ontogeny, whose focus has often been to find behaviors or traits early in development that are predictive of later development (see Kagan, 1996).

Many adaptations are limited to a particular time in development, facilitating the young organism's chances of surviving to adulthood and eventually reproducing. This is reflected by the concept of *ontogenetic adaptations*—neurobehavioral characteristics that serve specific adaptive functions for the developing animal (see Oppenheim, 1981). These are not simply incomplete versions of adult characteristics but have specific roles in survival during infancy or youth and disappear when they are no longer necessary. For example, embryos of most species have specializations, such as the yolk sac, embryonic excretory mechanisms, and hatching behaviors in birds, that keep them alive but disappear or are discarded once their purpose is served (Oppenheim, 1981).

Such adaptations are not limited to prenatal behaviors. Infant reflexes, such as the sucking reflex in mammals, are obvious postnatal behaviors that serve a specific function and then disappear. Some aspects of human infants' cognition have also been interpreted as serving a specific function, only to disappear or to become reorganized later in life. For example, newborns' imitation of facial gestures (Meltzoff & Moore, 1977, 1985) has been characterized by some as an ontogenetic adaptation (Bjorklund, 1987). Under the appropriate conditions, newborn infants will imitate a range of facial gestures, although imitation of facial expressions decreases to chance levels by about 2 months (Abravanel & Sigafoos, 1984; Fontaine, 1984; Jacobsen, 1979). Several researchers have speculated that neonatal imitation, rather than serving to acquire new behaviors, which seems to be the primary function of imitation in later infancy and childhood, has a very different and specific function for the neonate. For example, Jacobsen (1979) suggested that imitation of facial gestures is functional in nursing, Legerstee (1991) proposed that it is a form of prelinguistic communication, and Bjorklund (1987) suggested that it facilitates mother–infant social interaction at a time when infants cannot control their gaze and head movements in response to social stimulation. Heimann (1989) provided support for these latter interpretations, reporting significant correlations between degree of neonatal imitation and quality of mother–infant interaction 3 months later. Thus, early imitation appears to have a specific adaptive function for the infant (to facilitate communication and social interaction) that is presumably different from the function that imitation will serve in the older infant and child (but see Meltzoff & Moore, 1992, for a different interpretation). Presumably, these different functions for similar behavior at different times in ontogeny were selected over evolutionary time.

Similar examples exist from social development. Take the case of boys' R&T play during childhood. One interpretation that we discussed earlier, probably the dominant one (Fagen, 1981; Groos, 1901), is that R&T play is an incomplete form of adult fighting. The value of R&T play, consequently, has been seen in terms of benefits deferred until after childhood (Smith, 1982). More recently, however, it has been suggested that play generally, and R&T play specifically (Martin & Caro, 1985), has benefits during childhood. In the case of R&T play, boys may use it as a way in which to learn and practice social signaling; for example, exaggerated movements and a play face communicate playful intent. Further, it is used as a way in which boys establish leadership in their peer group and assess others' strength (Pellegrini & Smith, 1998). R&T play also has immediate nonsocial benefits; it provides opportunities for the vigorous physical exercise that is important for skeletal and muscle development. (The role of play in development from an evolutionary perspective is the topic of chapter 10.)

From this perspective, cognitive or social immaturity is viewed in a different light (Bjorklund, 1997a; Bjorklund & Green, 1992; Wellman, in press). For instance, seemingly "immature" behavior, such as play, may have been selected as a way in which young organisms can negotiate the niche of childhood. After all, organisms must survive childhood if they are to mature and then reproduce (Pellegrini & Smith, 1998). As an example from the cognitive domain, young children's poor metacognition, particularly their ability to judge the competency of their own performance, may be more of a blessing than a curse. Children who overestimate their own abilities may attempt a wider range of activities and not perceive their less-than-perfect performance as failure (Bjorklund, Gaultney, & Green, 1993; see chapter 7). Certain aspects of immaturity may thus be adaptive, and attempts to accelerate intellectual development, frequently advocated in the United States, may be counterproductive (see Bjorklund & Schwartz, 1996; Goodman, 1992).

These and other findings (see Bjorklund, 1997a; chapter 6) indicate that infants and young children respond to experiences differently than older children and adults. This interpretation, we argue, is consistent with an evolutionary developmental psychological perspective and is apt to be missed or interpreted otherwise without such a perspective.

Influence of Natural Selection at Different Times in Ontogeny

Related to the concepts of ontogenetic adaptations and the adaptive value of immaturity is the idea that natural selection has had greater impact on some phases of ontogeny than on others. Natural selection operates so that individuals with characteristics that "fit" well with the current environment reach adulthood and leave more offspring than less-well-fit individuals. The criterion for success (for reproductive fitness) is leaving more, as opposed to fewer, copies of your genes (usually in the form of offspring) in subsequent generations. For humans, this means reaching puberty, having children, and seeing that those children survive to become parents themselves. (It can also mean that one's close relatives, such as brothers and sisters, have offspring, who will share a large proportion of your genes.) Thus, characteristics that increase the chances of an individual reaching adulthood, procreating, and rearing children to independence should all have positive selective value. In contrast, characteristics that promote good health beyond one's ability to reproduce likely have little selective value. The genes that promote long life, for example, as well as those that promote midlife (post-reproduction) death, will not be selected for or against because they are expressed only after a person has had children. Although aspects of longevity may be inherited, from this perspective it is unlikely that they have been selected in evolution.

Perhaps the best documented example of this is Huntington's disease, a fatal, genetic disease caused by brain degeneration that, unlike most other fatal genetic diseases, is passed on by a single dominant gene. That is, only one gene rather than two recessive genes is necessary to display the symptoms of Huntington's disease. This means that the likelihood of an individual with the Huntington's gene passing it to his or her offspring is 50%. Huntington's disease occurs in about 1 in every 15,000 people of European descent. Contrast Huntington's disease with another disease also caused by a single dominant gene, progeria, which results in premature aging and early death. Progeria occurs in about 1 in every 8 million births, or about 500 times less frequently than Huntington's disease (Austad, 1997).

Why the difference in frequency between the two diseases? The primary reason is that most people with progeria die before reaching their teenage years and thus rarely reproduce. The defective, dominant gene dies with them. New cases of the disease are the result of a random mutation and are thus quite rare. The gene for Huntington's disease, in contrast, is expressed only later in middle adulthood. By the time the disease strikes a person, he or she has often had children and thus has a 50% chance of passing the dominant gene to offspring. Natural selection works against genes that prevent individuals from reaching reproductive age (negative selection). The possessors of genes such as the one responsible for progeria die before reproducing. In contrast, natural selection has no impact on genes that are expressed late in life, such as the one responsible for Huntington's disease, after individuals have already reproduced.

One way of expressing this effect is to say that the power of natural selection to influence genes wanes with age, or as life span psychologist Paul Baltes (1997) put it, "the benefits resulting from evolutionary selection evince a negative age correlation" (p. 367). As a result, as humans age, there should be more deleterious genes that are expressed. This is seen in contemporary times with the rapid increase in occurrences of Alzheimer's dementia late in life. During earlier times in human history, people possessing these genes died of other causes long before these late-life genes could be expressed. As a result, there were no selection pressures against such genes. Their presence today, and that of other late-life genes that contribute to illness and death, is evidence of the failure of natural selection to influence characteristics in the latter part of the life span over human history.

Another possible reason for the phenomenon of senescence and the mitigation against longevity is that genes that promote good health and contribute to one's selective advantage early in life may have deleterious effects later in life (G. C. Williams, 1957). Genes that have multiple functions are referred to as *pleiotropic*, and it seems likely that not only are multiple genes implicated in affecting any single characteristic of an animal, but also many genes likely influence multiple characteristics. If a gene expression affords some benefit early in life, that gene will be selected for, even if the expression of that same gene later in life (or the long-term effects of a gene's earlier functioning) has negative consequences for survival. For example, there is some evidence that testosterone, so critical for the development of male secondary sexual characteristics and associated with aspects of social status and dominance among males, impairs the immune system (see Geary, 1998) and increases the risk of certain cancers. Yet, the genes that promote high levels of testosterone may be selected because of the early benefit they afford a young male animal at the expense of increased vulnerability to disease later in life.

It is possible that characteristics of old age may have some positive selective value, if, by living longer, more copies of one's genes, usually in the form of grandchildren, survive. That is, perhaps long-lived people, although no longer reproducing themselves, may foster their grandchildren's survival. This possibility has been proposed (e.g., Euler & Weitzel, 1996) and is discussed in chapter 8.

DEVELOPMENTAL ORIGINS OF HUMAN NATURE

Evolutionary psychologist David Buss (1999) referred to evolutionary psychology as "the new science of the mind." An evolutionary perspective provides a common ground for interpreting all aspects of human behavior—social, emotional, and cognitive—and may integrate the often-disparate subfields of psychology. As developmental psychologists, we have long believed that the best way to understand any aspect of human functioning is to look at its ontogeny. But developmental psychology has been as fractionated as its parent discipline. Many developmental scientists, even within what some people would consider to be a relatively "narrow" area (e.g., cognitive development), often talk past one another because they fail to share a common view of what is important about development (see Bjorklund, 1997b). Evolutionary psychology, we believe, provides a metatheory for developmentalists assessing a wide range of topics and ages (Baltes, 1997; Bjorklund, 1997b). As evolutionary theory is the foundation for modern biology, we believe that it needs to be the foundation for modern psychology (Daly & Wilson, 1988a; Tooby & Cosmides, 1992; E. O. Wilson, 1998), without necessarily replacing more traditional theories that account for the proximal causes of behavior and development.

We also believe that an explicitly developmental perspective can positively influence the field of evolutionary psychology. Much of evolutionary psychology has been concerned with the natural selection of "mature" behaviors. Given this perspective, it is easy to see why some evolutionary psychologists have not looked at development for interesting phenomena. Evolution proceeds when successful individuals reproduce. These are the most advanced members of the species, and factors that promote their reproduction obviously characterize adulthood and not infancy and childhood.

We must admit that, on the surface, this seems to be a reasonable argument and, obviously, much of what contributes to individual success at reproduction, both today and in our evolutionary past, is found in the adult. But our ancestors also developed, and before organisms can reproduce to get their genes into the next generation, they must first reach adulthood, and for a slow-developing species such as humans, that can be a long and treacherous path. How people develop is important to eventual reproductive success and, as we have noted previously in this chapter, we have every reason to believe that evolution has worked to select characteristics of infancy and childhood that are adaptive to surviving to adulthood, just as it has worked to make adults responsive to the appropriate social and sexual cues that are so important to getting one's genes into the next generation. Moreover, important characteristics of adulthood, such as different "sexual strategies" of men and women, should not be seen as preformed, springing into existence with the first blast of pubertal hormones. Rather, even these characteristics have a developmental history that can alter the expected course of adult behavior.

Contemporary evolutionary psychologists often see themselves as providing answers to the ancient question, What is human nature? One important contribution of evolutionary developmental psychology, we believe, is to propose that human nature is not something that is simply "innate," "instinctive," or "within the organism"; rather, human nature develops. Human beings (and all animals) are born with genetically influenced dispositions that have been shaped by millions of years of natural selection. These dispositions interact with all levels of the environment, producing, in most cases, species-typical patterns of development. What we call human nature has not only a phylogenetic history, but also an ontogenetic one. The origins of human nature lie in the immortal genes and in the developmental processes that are required for their expression.

In the remainder of this book we examine issues and topics that we believe are particularly germane to an evolutionary developmental perspective of human behavior. Much of what we have to say, we believe, is compatible with mainstream evolutionary psychology. However, a developmental approach has its own intellectual history and its own unique concerns, which sometimes are at odds with the canonical view of evolutionary psychology. In chapter 3 we examine some of the history and controversy in the study of the relationship between phylogeny and ontogeny, as well as several other controversies that will need to be resolved before we have a single evolutionary science of human behavior.

3

HISTORY AND CONTROVERSY

Evolutionary thinking is not exactly new to developmental psychology and is certainly not new to biology. But it would be a mistake to think that contemporary evolutionists are simply modern-day Darwinians with 140 years of additional data. Although Charles Darwin's basic idea of natural selection, discussed in chapter 2, remains the cornerstone of modern evolutionary theory, there have been many changes to how we view evolution, and particularly, for our purposes, the connection between evolution and development. Many of the debates that are central to evolutionary developmental psychology have histories dating back to the time of Darwin. In this chapter, we address some of those controversies in a historical perspective. Readers should not think of this chapter as history alone, however. Although many of the controversies have been resolved (or more appropriately, reformulated), earlier perspectives continue to persist in the minds of many, despite new findings and theories to the contrary.

This chapter is divided into five major sections. We first provide a brief description of evolutionary thinking from pre-Darwinian theory through the modern synthesis. We next explore some of the ideas pertaining to the influence of development on evolution; in the following section we look at ideas pertaining to the influence of evolution on development. In the fourth section we discuss what it means for something to be innate, and in the final section we explore how evolutionary psychology and behavioral genetics deal (or fail to deal) with individual differences. These are by no means the only controversies pertinent to evolutionary developmental psychology, but they are controversies that have long histories and important consequences for contemporary theory.

BRIEF HISTORY OF EVOLUTIONARY THINKING

Charles Darwin (1809–1882) did not invent the concept of evolution (or "descent with modification," to use his term). To explain the similarity of extant species and the fossils of extinct species, many pre-Darwinian scientists had argued that species had evolved over time, in opposition to the Genesis-based theory that all species were created by God "in the beginning." Darwin's major contribution was to provide a plausible mechanism for evolution, natural selection, as well as abundant documentation. (Evolution by natural selection was simultaneously discovered by Alfred Russel Wallace [1823–1913], although it was Darwin who provided most of the ammunition for the revolution in biological thinking that the discovery would create.)

The most influential of the pre-Darwinian evolutionists was French naturalist Jean Baptiste Lamarck (1744–1829). The essential features of Lamarck's theory were as follows: (a) different environments required different types of adaptations; (b) adaptations are related to the behavior of animals; (c) as a result of use or disuse of an organ, bodily changes are brought about; and (d) these bodily changes are inherited. It is for the latter idea of the inheritance of acquired characteristics that Lamarck is best remembered (and derided) today. Lamarck's emphasis on behavioral adaptability recognized the importance of the modifiability of behavior in development for evolution, and Darwin and later evolutionists adopted his notion of the inheritance of acquired characteristics (see Gottlieb, 1992; S. J. Gould, 1977; Schwartz, 1999).

A major sticking point for Darwin (1859) and all other 19th-century evolutionists was that he did not have a solid theory of inheritance. Natural selection worked on heritable variation in physical or behavioral characteristics, but how those characteristics were inherited was not known. In addition to advocating the inheritance of acquired characteristics, Darwin believed, as did most of his contemporaries, that traits of the parents blended to produce an "average" offspring. Thus, a male bird with a short beak and a female bird with a long beak would produce offspring with a medium-sized beak. Of course, Darwin also recognized that two parents produced offspring who varied both from the parents and from their siblings (variation was the necessary condition for natural selection); but exactly how this occurred escaped him and others of his age. It was not until the rediscovery in 1900 of Austrian botanist Gregor Mendel's (1822–1884) research with pea plants that a plausible mechanism for passing on traits from one generation to the next was available.

Mendel postulated that traits do not blend but rather that heredity is *particulate*, that is, a particular trait would be represented by discrete alternatives, what today we call different *alleles* of a gene. For example, the shape of a seed may be round or wrinkled. One of these characteristics may be dominant relative to the other, so that if an allele for a round seed is paired with an allele for a wrinkled seed, the phenotype would not be an average, or blending, of the two characteristics, but would be determined by the dominance relation of the alleles (in this case, round is dominant). Moreover, when forming gametes, these alleles segregate so that half of the sex cells of an individual would possess one allele and half the other. The seven traits Mendel identified in his pea plants all had only two levels (e.g., round vs. wrinkled seed; yellow vs. green unripe pod), and all showed a simple dominance relationship. Most characteristics are influenced by many different genes, and traits are not necessarily dichotomous in their distribution. Nonetheless, the genetic age was born, and evolutionary theory has not been the same since.

Another important figure at this time was German biologist August Weissman (1834–1914), an ardent opponent of the inheritance of acquired characteristics who may have been the first to understand the significance of Mendel's discoveries for evolutionary theory. Weissman articulated the distinction between somatic (body) cells and reproductive cells (gametes, or the germ plasma). Whatever happens to an individual during development will affect only its somatic cells. The germ plasma, being independent of somatic cells, is not affected. The same genes that individuals received from their parents are passed along to their offspring (in different combinations, of course), unaffected by life experiences.

Exceptions to this were found in the form of *mutations*, variations in genes in the germ line that could occur spontaneously or be induced by abnormal environmental conditions (e.g., heat shock to fruit fly larvae). Research on fruit flies (*Drosophila*) led by American geneticist Thomas Hunt Morgan's (1866–1945) lab at Columbia University illustrated dramatically the phenomenon of mutation and genetic inheritance more generally.

Initially, the new field of genetics was seen as providing an alternative to Darwin's ideas of evolution by natural selection: Species change as a result of mutations, with natural selection playing only a minor role (see Schwartz, 1999). But researchers from a variety of disciplines within biology began to merge on a common view of evolution, one that incorporated the new science of genetics with Darwin's idea of natural selection. This new perspective is termed the *modern synthesis*, or neo-Darwinism, and it became the dogma of evolutionary theory (Dobzhansky, 1937; Mayr, 1942; Simpson, 1944; see also Gottlieb, 1992; Gould, 1977; Schwartz, 1999). The modern synthesis adopted Weissman's idea of the separation of the gametes and somatic cells. Inheritance, and thus genetic variation, is found only within the germ line and is not influenced by experience. Evolution takes place gradually over eons by the accumulation of random mutations, each of which is maintained or excluded in the genome by the process of natural selection. An important new insight of the modern synthesis concerned the focus on populations. Evolution must be considered in terms of changes in the frequencies of individual genes in populations of organisms. This should not be confused with *group selection*, in which natural selection operates for the "good of the group." Individuals possess genes, but it is populations of individuals that reproduce and thus evolve. Mutations existing in individuals would not have any consequences to evolution unless the resulting trait is expressed in populations of similar organisms (G. C. Williams, 1966). Evolution, then, becomes the change in genetic composition within populations.

The modern synthesis's emphasis on the separation of the somatic and germ line essentially afforded no role for development in evolution. Development may result from the differential expression of genes in interaction with the environment, but such effects are governed directly by the genome, and variations in development, brought about by variations in the environment, cannot affect the germ line and thus cannot exert any influence on evolution (see Gottlieb, 1992).

The modern synthesis remains intact today, although evolutionary theory has not stood still. For example, the theorizing of evolutionary biologist William Hamilton (1964) and the concept of inclusive fitness (see chapter 2), and the theories of evolutionary biologist Robert Trivers (1971, 1972, 1974; see chapters 8 and 9), among others, have changed the focus of evolutionary theory. The advent of sociobiology (Dawkins, 1976; E. O. Wilson, 1975), with its emphasis on explaining complex social behavior, such as altruism, in terms of evolutionary principles, focused attention on the evolution of behavior, something that many developmental psychologists found appealing (MacDonald, 1988). But the basic tenets of the modern synthesis (other than the possibility that evolution is not as gradual as Darwin originally postulated, see discussion of punctuated equilibrium below) have not been seriously questioned, and development has not been given a prominent role in evolutionary explication. Evolutionary psychology has been no exception (see Barkow, et al., 1992; Buss, 1999).

ROLE OF DEVELOPMENT IN EVOLUTION

Opinions about the role development might have in influencing evolution have shifted considerably ever since the publication of Darwin's magnum opus. In the pregenetic era, before the rediscovery of Mendel's work and the reformulation of evolutionary thinking into the modern synthesis, development figured prominently in the theories of several major evolutionary scientists.

Before the Modern Synthesis

One popular concept about evolution among scientists in the 19th century and among many laymen today was that of progress. As we mentioned in the previous chapter, evolution was viewed as being progressive, with humans at the pinnacle, just below angels in a divine plan (Charlesworth, 1992; Morss, 1990). This was captured by the principle of *orthogenesis*, the belief that there is an inherent perfecting force in all of organic life that makes evolution always moving "forward." More "advanced" species, such as humans, relative to less advanced species, such as chimpanzees (or more properly, our ape-like ancestors), evolved by adding something to the adult stages of ancestors. So, for example, humans evolved "more" brain.

Recapitulation Theory and the Biogenetic Law

This progressive spirit was captured by German biologist and philosopher Ernst Haeckel (1835–1919), who applied findings in embryology to evolutionary thinking to postulate the biogenetic law (see Gottlieb, 1992; S. J. Gould, 1977; Mayr, 1982; Schwartz, 1999, for historical reviews). The biogenetic law was the basis for *recapitulation theory*, which can be captured by the phrase "ontogeny recapitulates phylogeny." This means that the development of the individual (ontogenetic development) goes through, or repeats, the same sequences as the evolutionary development of the species (phylogenetic development), with evidence of these ancestral stages most clearly seen during embryological development. From this perspective, the entire phylogenetic past of a species can be discerned by looking at (primarily) embryological development, which is essentially a much speeded-up version of evolutionary history. What is new in evolution is what is added to the end states of ontogeny.

This was an attractive theory, in part because it simplified the study of evolution. Studying old bones or the behavior of extant animals to get clues of ancient ancestors would eventually become unnecessary. All that was seemingly needed to understand evolution was a detailed knowledge of embryological development. The data, however, were not always consistent with the theory. For example, the order in which a feature appeared in phylogeny did not necessarily follow the same path in ontogeny. The development of teeth and tongues provides a good example. Teeth are an earlier evolutionary invention than tongues, but they appear later than tongues in the embryological development of present-day mammals (de Beer, 1958). By the mid-1920s, it was becoming clear that there were just too many exceptions. Although no one doubted that many evolutionary innovations were added to the end states of an ancestor, there were many other paths that were also taken, some involving the retardation of certain aspects of development (Davidson, 1914; Thorndike, 1913; see also Hinde, 1983).

Further, the notion that behavior at a given point in time is a result of past experiences, both "internal" and "external," violates a widely held assumption that behavior is a result of a complex contemporaneous transaction between individuals and their surroundings (Gottlieb et al., 1998;

Lewontin, 1982). From this view, individual organisms actively "cause" behavior; consequently, one's phenotype is in a constant state of flux. Ontogeny, then, can be thought of not as a simple product of genotype and the current environment, but as a "first-order Markov process" in which the subsequent behavior depends on the immediately preceding phenotype and genotype (Lewontin, 1982, p. 279).

Neoteny: Evolving by Starting Over

In large part as a reaction to the excessive and often erroneous claims of recapitulation theory, several theorists in the early part of the 20th century proposed that evolution often proceeds by taking a step backward, so to speak. Haeckel's proposal that evolution occurred by making additions to the end point of ancestral forms represents one way in which differences in developmental timing can influence the course of phylogeny. Genetic-based differences in developmental timing have been referred to as *heterochrony* (de Beer, 1958; Gould, 1977; McKinney, 1998, 2000; Shea, 1989, 2000). Different systems or parts of an organism can develop at different rates, and these rates may be accelerated (as in the case of recapitulation) or retarded relative to the developmental rates experienced by one's ancestors. Changes in ontogenetic timing produce different phenotypes, and if these phenotypes are adaptive (or at least not maladaptive), they eventually produce phylogenetic changes (i.e., new species; de Beer, 1958). One type of developmental retardation is *neoteny* (literally "holding youth"), which refers to the retention into adulthood of ancestral embryonic or youthful stages.

The idea of evolutionary change brought about by modifications in ontogenetic rates is attractive to evolutionary biologists. Genetic mutations that produced seemingly small changes in the onset or offset of a developmental process (mutations in regulatory as opposed to structural genes) can have cascading effects, yielding substantial changes in a phenotype. "Evolution by heterochrony" presents a more parsimonious approach to phylogenetic change than the multiple-point mutations that are presumably required to produce the same degree of phenotypic change in a species. Modifications of developmental rates can also result in relatively rapid changes (termed *saltations*) in a species. Although conventional neo-Darwinian perspectives (and evolutionary psychological perspectives) argue that phylogenetic changes must be gradual, evidence exists in the fossil record for long periods of stability (*stasis*) in a species followed by rapid and substantial change (termed *punctuated equilibrium*; Eldredge & Gould, 1972). Heterochronic changes provide a mechanism for such saltations. Natural selection will, of course, act on the heterochronically modified organisms, so the inclusion of such changes to explain patterns of evolutionary change does not conflict with the principal tenet of evolutionary biology (i.e., natural selection).

In the early part of the 20th century, evolutionary biologists such as Great Britain's Gavin de Beer (1958) and Walter Garstang (1922) and the Netherlands' Louis Bolk (1926) proposed that the driving force of evolution is the change in the timing of ontogeny, with neoteny playing an important role. Of the early neoteny theorists, Bolk was most concerned with human evolution. Bolk saw as the primary difference between humans and apes the fetal character of the human body, believing that neoteny (or, using his term, *fetalization*), was the essence of humankind. People are apes who, bodily, have never grown up. Bolk's claims for human evolution were extreme. He proposed that all "essential" features of modern humans were retarded, or neotenous, ignoring the fact that different systems or structures show different rates of development. According to Bolk,

> There is no mammal that grows as slowly as man, and not one in which the full development is attained at such a long interval after birth . . . What is the essential in Man as an organism? The obvious answer is: The slow progress of his life's course. (p. 470)

Bolk went so far as to suggest that "man, in his bodily development, is a primate fetus that has become sexually mature" (p. 470).

Evolutionary theory after the modern synthesis tended to ignore issues of developmental timing as a force in evolution. However, the idea was kept alive by a hearty few scientists (de Beer, 1958; M. F. A. Montagu, 1962), and in 1977 was fully revived in a book titled *Ontogeny and Phylogeny* by evolutionary biologist Stephen Jay Gould. Since that time, many researchers have viewed heterochrony, and specifically neoteny, as playing an important role in evolution, particularly human evolution (Hattori, 1998; A. Montagu, 1989; Schwartz, 1999; Thomson, 1988; Wesson, 1991). For example, Wesson (1991) suggested that neoteny seems to be a good strategy for evolutionary innovation, permitting "a new beginning and relatively rapid change as the organism backs up evolutionarily to get a better start" (p. 205). On a similar note, S. J. Gould (1977) stated that

> The early stages of ontogeny are a storehouse of potential adaptations, for they contain countless shapes and structures that are lost through later allometries. When development is retarded, a mechanism is provided (via retention of fetal growth rates and proportions) for bringing these features forward to later ontogenetic stages. (p. 375)

One well-known example of the role neoteny plays in development is that of the salamander species *Axolotl* (see S. J. Gould, 1977). As with salamanders in general, they start life in the water as tadpoles and then metamorphose into air-breathing, land-dwelling newts. But under certain conditions, when life in the water is good and looks to stay that way for a while, the tadpoles will mature sexually and reproduce, still in the larval state. That is, the developmental timing of their reproductive system and their gill-to-lung/water-to-land "systems" are independent, with the reproductive system maturing while the organism is still morphologically in the juvenile (larval) state. Some of the offspring may then go through the "normal" developmental sequence, from tadpoles to salamanders, while their parents remain larva (albeit sexually active ones).

Several contemporary scientists have seriously questioned the still-popular view that *Homo sapiens* is a neotenous species (Langer, 1998; McKinney, 1998, 2000; Parker & McKinney, 1999; Shea, 1989, 2000). We discuss in greater detail issues related to heterochrony, and specifically neoteny, with respect to human evolution in chapter 4 and with regard to cognitive evolution in chapter 5. However, despite the controversy about the role that developmental retardation may have played in human evolution, heterochrony, including both acceleration and retardation, is viewed as an important mechanism by which differences in rates of ontogeny can affect patterns of phylogeny.

Heterochrony should not be thought of as the cause of human (or any species') evolution, but rather as a description of how changes in ontogeny can contribute to changes in phylogeny. When growth is retarded, for example, other avenues, lost to faster developing organisms, can be explored. A similar argument can be made when development is accelerated. Modified developmental rates thus permit evolutionary innovations rather than cause them. In many cases, they may be necessary for evolutionary changes to have occurred, but they are not, per se, sufficient to bring about phylogenetic modifications (cf. Wachs, 2000). The pressures for changing the pace of development must be found in the environment, with heterochrony being a response to some of those pressures. And once changes are brought about by changes in developmental rates, they are subject to the pressures of natural selection.

Epigenetic Theories of Evolution

Epigenetic principles, such as those advocated by the developmental systems perspective favored here, emphasize the continuous interaction of the environment, broadly defined, with the biology of the individual. Genes are never activated in isolation, and how (or whether) a gene is expressed is dependent on a host of interacting factors that vary over time (development). Epigenetic theories of evolution (e.g., Ho, 1998) view a developing organism's response to environmental changes as a mechanism for phylogenetic change. Natural selection still plays an important role in evolution, but it is the developmental plasticity of an organism that provides the creative force for evolution.

The Baldwin Effect

One major contribution of developmental theory to evolutionary formulation during the pre-synthesis days concerned the possible transmission of acquired responses to stress via non-Lamarckian routes. This process, referred to as *organic selection*, was derived independently around the turn of the 20th century by comparative psychologist Conway Lloyd Morgan (1852–1936), biologist Henry Fairfield Osborn (1857–1935), and comparative developmental psychologist James Mark Baldwin (1861–1934). Baldwin (1902), however, seems to have had the better press secretary, and the phenomenon became known as the *Baldwin effect* (for discussion, see Gottlieb, 1992; Waddington, 1975). Baldwin proposed that when a population of individuals experiences some environmental stress, many in the population will die. Others, however, will be able to cope with the stress. These latter individuals will reproduce and pass on these tendencies to their offspring. What is novel about Baldwin's proposal is that the environmental stressors produce new phenotypes, physical or (more likely) behavioral changes in an organism that are transmitted to the next generation. Although these modifications could not be transmitted directly to progeny via genetics, they could be transmitted socially. These mainly behavioral modifications kept members of the species alive until a genetic variation came about. *Organic selection* refers to internal forces that stabilize the change in subsequent generations. Eventually, the surviving animals will be subject to an appropriate *congenital variation*, a term Baldwin (seemingly) used to refer to mutations. From this perspective, evolutionary novelty can first arise as developmental modifications in response to a changing environment that somehow becomes inherited.

Baldwin's account of evolution emphasized individual differences in the adaptability of organisms. That is, some individuals are more susceptible to modification given a novel environment.

Waddington's Genetic Assimilation

Baldwin's idea has always been out of the mainstream of the modern synthesis (as have all developmental explanations) and has often been seen as being Lamarckian or, at best, neo-Lamarckian. There were several difficulties with the theory, the most prominent being the lack of evidence that any such phenomenon exists. Experimental evidence consistent with the Baldwin effect was demonstrated in the 1950s by the British biologist Conrad H. Waddington (all references to Waddington's work are from his 1975 collection of essays, *The Evolution of an Evolutionist*). In a classic experiment, Waddington exposed pupal fruit flies (*Drosophila melanogaster*) to heat shock. Some of the surviving flies responded to this treatment by developing wings with few or no cross veins. Waddington then selectively bred flies without cross veins and exposed the pupa of the next generation to heat shock, which yielded a second generation of flies with few or no cross veins. He continued this procedure for 14 generations, at which time some of the flies developed the no-cross-veined-wing phenotype without being exposed to the heat shock. That is, a new phenotype, which was initially elicited only by exposure to an extreme environment (that killed many of those exposed), was eventually displayed spontaneously by offspring of no-cross-veined-winged parents, in the absence of the initiating environmental event. Waddington referred to this phenomenon as *genetic assimilation*, which he defined as "the conversion of an acquired character into an inherited one; or better, as a shift (towards a greater importance of heredity) in the degree to which the character is acquired or inherited" (p. 61).

Having no cross veins in their wings has no apparent adaptive value for a fruit fly but reflected merely a convenient trait Waddington could manipulate. However, in a later study, Waddington demonstrated genetic assimilation for a characteristic that could, conceivably, have an adaptive function. Flies were fed food with added salt. This produced a high mortality rate, and surviving flies were bred with one another. Flies exposed to high-salt diets responded by developing larger anal papillae, which facilitate the excretion of salt from the body. After 21 generations, eggs were placed on media containing various degrees of salt concentration. Compared with control flies, flies whose parents (and 19 generations of grandparents) had been fed high concentrations of salt developed larger anal papillae, even when grown on the low-salt concentration media. Again, a characteristic that had been acquired in response to an environmental stressor had come to be expressed in the absence of the initial event, this one being adaptive for survival.

Other researchers, both before and after Waddington, have demonstrated the phenomenon of genetic assimilation on a wide range of organisms (see Jablonka & Lamb, 1995; Waddington, 1975). In fact, an early demonstration of the effect was reported for the pond snail *Limnaea stagalis* by Jean Piaget (1896–1980; 1976). Piaget observed that the form of the snail's shell varied with the wave action it experienced during development, which varied in different parts of the lake. When bred in the laboratory, the snails' shells retained their unique shape over many generations. Piaget (1929/1976) himself advocated a form of genetic assimilation, taking an explicitly epigenetic perspective of the relation between development and evolution. As a more recent example, researchers have noted that an asexually reproducing species of water flea (*Daphina cuncullata*) grows a large protective helmet when raised in the presence of the larvae of a potential predator. This enhanced defense is then passed on to daughters and, to a lesser extent, granddaughters raised in safe environments (Agrawal, Laforsch, & Tollrian, 1999).

With respect to evolution, genetic assimilation and the Baldwin effect[1] operate according to a three-step process. First, members of a species experience a modified environment, and as a result, some survive by developing novel responses. This is presumably accomplished by expressing genes that are not normally expressed. This individual difference among members of the population can be seen in terms of adaptability: Those that can more readily adapt to environmental change are more likely to survive. Second, offspring of the survivors selectively breed among themselves, continuing to show adaptive responses to the now stable environment. Third, the response becomes genetically assimilated in that it is now expressed even in the absence of the environmental events that originally precipitated the change.

Mechanisms of Change

Waddington was never clear on the specific mechanism behind genetic assimilation. He was adamant that all aspects of an organism were the joint product of action of the phenotype and the environment, but obviously genetic assimilation must eventually produce some "genetic" change if a characteristic is to be expressed in the absence of the environmental event that initially evoked it. One alternative is the geneticists' favorite, mutation, similar to that proposed by Baldwin for organic selection. A second possibility is that the novel environment promotes the activity of only a select set of alleles for active genes that all members of the species possess. After many generations of selective breeding, only those alleles associated with an extreme value of a trait remain in the genotype. Thus, changes occur not in the production of new genes (via mutations), but in terms of the frequencies of different alleles for a particular trait, precisely as any trait can change in frequency in a population following conventional Mendelian analyses. (Of course, any single gene can have multiple effects—*pleiotropy*—and many different genes are likely associated with a single characteristic, making this scenario more complicated than it appears on the surface.) Third, the environmental event could have activated heretofore-dormant genes or, relatedly, served to deactivate certain genes, the end result being similar (a different combination of genes are involved in a particular response, relative to individuals in the pre-stressed environment). Fourth, mutations can occur in cells in the immune system, which in turn influence the germ cells, resulting in acquired immunity. Contemporary research in genetics suggests that each of these alternatives is

---

[1] Waddington (1975, pp. 88–91) dissociated his theory from Baldwin's, believing that Baldwin, unlike himself, provided no role for natural selection. Although it can be difficult to interpret the precise meaning of Baldwin (1902), particularly considering his use of pre-Mendelian language, we see great similarities between genetic assimilation and the Baldwin effect and, for our purposes here, make no meaningful distinction between them.

possible, at least under laboratory conditions (see Gottlieb, 1998; Ho, 1998; Jablonka & Lamb, 1995; Steele, Lindley, & Blanden, 1998).

Another possibility concerns *cytoplasmic inheritance*, modifications that can be passed along from mothers to their offspring not through nuclear genes but by way of changes found in the cytoplasm of the mother's gametes (Ho, 1998; Jablonka & Lamb, 1995). Mae-Wan Ho and her colleagues (1983) demonstrated this with fruit fly embryos that were exposed to ether, which produced a double set of wings. After several generations, this condition was expressed even in unaffected flies, an example of genetic assimilation. To illustrate that the effect was transmitted via the cytoplasm, Ho and her colleagues bred groups of treated males with untreated females, and treated females with untreated males. Only the offspring of the treated females displayed the double-wing condition. Again, this is attributed to the fact that, although both males and females pass on nuclear genetic material to their progeny, only females pass on cytoplasm (i.e., the cell body) to their offspring. Environmental changes that induce chemical changes in the cytoplasm can thus be inherited through the mother but not through the father.

Cytoplasmic inheritance should not be thought of as nongenetic. Although Ho and colleagues were not able to specify exactly what was being inherited in the cytoplasm, it necessarily expressed its effect on the genes. As we noted in chapter 2, events in the cell body (which communicates with other cells and eventually the external world) influence gene expression, turning genes on or off. Thus, heredity resides not only in the genes but, according to Ho (1998),

> It resides as well in an epigenetic cellular state—a dynamic equilibrium between interlinked genic and cellular processes. But . . . it cannot be assumed that heredity is exhausted at the boundary of cells or organisms. For as organisms engage their environments in a web of mutual feedback interrelations, they transform and maintain their environments, which are also passed on to subsequent generations as home ranges and other cultural artifacts. (p. 114)

Role of Behavioral Plasticity in Evolution

Despite the broad recognition that DNA is affected by events that occur outside of the nucleus and that there is no simple connection between genes and a trait, epigenetic perspectives on evolution have not yet entered mainstream biology or evolutionary psychology. The possible influence of behavioral novelty on evolution is not per se a new idea (Baldwin, 1902; de Beer, 1958; Garstang, 1922). But new findings in genetics, demonstrating that genes are part of a broader developmental system (Gottlieb, 1998, 2000; Ho, 1998), have prompted several contemporary psychologists to

Three possible stages in evolutionary pathway initiated by behavioral neophenotypes.

Stage I: Change in Behavior
First stage in evolutionary pathway. Change in ontogenetic development results in novel behavioral shift (behavioral neophenotype), which encourages new environmental relationships.

Stage II: Change in Morphology
Second stage of evolutionary change. New environmental relationships bring out latent possibilities for morphological–physiological change. Somatic mutations, cytoplasmic alteration, or change in genetic regulation may also occur, but a change in structural genes need not occur at this stage.

Stage III: Changes in Genes
Third stage of evolutionary change resulting from long-term geographic or behavioral isolation (separate breeding populations). It is significant to observe that evolution has already occurred phenotypically before Stage III is reached. Modern neo-Darwinism, however, does not consider evolution to have occurred unless there is a change in genes or gene frequencies.

Note. From "The Developmental Basis of Evolutionary Change," by G. Gottlieb, 1987, *Journal of Comparative Psychology, 101,* p. 269. Copyright 1987 by American Psychological Association. Reprinted/Adapted with permission.

reemphasize the role that developmental plasticity can have on evolution (Bateson, 1988; Gottlieb, 1987, 1992). However, whereas earlier theorists saw no alternative to mutations being the eventual cause of evolutionary change, modern theorists point to the vast untapped DNA in the genome (sometimes referred to as *junk DNA* or *intergenic DNA*) and suggest that the activation of such dormant DNA prompts change. In fact, the recent sequencing of the human genome indicated that 75% of the genome was comprised of intergenic DNA (Venter et al., 2001), suggesting that there is substantial underused DNA. Mutations may also occur, of course, but given the storehouse of presumably unused (or underused) DNA in the genome of many species, environmental changes that begin early in ontogeny may activate such genes, which lead to behavioral or morphological change, can then be genetically assimilated and become part of the broader inheritance of a population of individuals.

Developmental psychologist Gilbert Gottlieb (1987, 1992) has proposed a three-step process whereby novel behavioral phenotypes (behavioral neophenotypes; Kuo, 1967) can lead the way to evolutionary change (see Exhibit 3.1). The first step involves a modification of development, and thus behavior (behavioral neophenotype), brought about by environmental changes. The second step involves a change in morphology. As new organism–environment relations develop, latent genetic possibilities for morphological or physiological change may be expressed. Finally, genetic change may occur. According to Gottlieb (1992),

Genes are part of a very flexible and highly adaptable developmental system, but . . . genes do not determine the features of the mature organism. Consequently, from this point of view, evolution involves changes in the developmental systems (of which the genes are an essential part), but not necessarily changes in the genes themselves. (p. 177)

A possible epigenetic account of human evolution is discussed in chapter 4.

Epigenetic evolution should be viewed as an extension of the modern synthesis. Like the modern synthesis, evolution is still conceived of as changes in gene frequencies in populations of individuals, with populations being the primary target for evolution and natural selection being the driving force of evolutionary change. However, it is not just "genes" that change in frequency, but *developmental systems*, which include genes but also the internal and external environment of the organism (Lickliter & Berry, 1990; Oyama, 2000a, 2000b). And, because of its emphasis on natural selection and its proposed mechanisms for extragenetic evolutionary change, epigenetic evolution should not be confused with Lamarckian mechanisms of evolution. What epigenetic evolution adds to the modern synthesis is the idea that modifications in development play a significant role in evolution and that there is no simple relation between genes and traits. The implications of such an addition to the modern synthesis are potentially enormous.

ROLE OF EVOLUTION IN DEVELOPMENT

The principle focus of this book is how an evolutionary perspective can provide a better understanding of ontogeny, particularly human ontogeny. Although we argue that the perspective provided by evolutionary developmental psychology is a relatively new one, the founders of developmental psychology were also greatly influenced by evolutionary theory. Unlike other subdisciplines of academic psychology that trace their roots back to 19th-century physics, developmental psychology had its origins in 19th-century biology, particularly in the evolutionary thinking of that time (Cairns, 1998). Contemporary developmental psychologists are often unaware of this connection, or if they are aware of it, see it only as a piece of history and not as a factor influencing modern thought. That may be true to some extent, for explicitly evolutionary thought was excised from mainstream developmental psychology, as were all forms of biological explication, during the middle portion of the 20th century when behaviorism held sway. Nonetheless, Darwin had an enduring, if often unrecognized, influence on developmental psychology, and the current rediscovery of his ideas reflects only the latest reincarnation of evolutionary thinking in the field.

Charlesworth (1992) listed four general influences that Darwin had on psychology. The first is the recognition that there is substantial continuity in mental functioning between animals and humans. The use of animal models for human development, whether they involve the behavior of quail chicks in perceptual development (Lickliter, 1990) or theory of mind in chimpanzees (Call & Tomasello, 1999), relies on the implicit assumption of continuity in intellectual functioning across species at particular times in development. The second influence is an emphasis on the importance of individual differences (see the section on individual differences later in this chapter). The third influence is a focus on adaptive behavioral function, and the fourth, expanding methodologies beyond those of psychophysics and introspection popular in early psychology. Yet Darwin's influence on child developmental theory seems mostly indirect and, according to Charlesworth, weak. Few of the major developmental theorists gave more than lip service to Darwin's ideas (Baldwin being an exception). But "Darwinian" should not be viewed as equivalent to "evolutionary," particularly at the turn of the 20th century. There were many ideas about evolution that, even if they were associated with Darwin 100 years ago, were not associated with the modern synthesis that would come to define evolutionary thinking in the last two-thirds of the 20th century, and early developmental psychology was influenced by some of these subsequently discredited ideas (see Charlesworth, 1992; Morss, 1990; Surbey, 1998a).

Recapitulation in Psychological Development

Perhaps the evolutionary concept that most permeated the thinking of the earlier child developmentalists was that of recapitulation (Morss, 1990). Recall from our earlier discussion Haeckel's notion that "ontogeny recapitulates phylogeny"; that is, evolution of the species can be seen in the ontogeny of individuals, with evolutionary changes being expressed exclusively in terms of additions to the adult form. Whereas Haeckel was concerned with morphological development, psychologists proposed recapitulation theories of behavior and mind. In a history of the influence of evolutionary thinking on developmental psychology, Morss (1990) listed most of the great developmental theorists in the first half of the 20th century, including Wilhelm Preyer (1841–1897), Baldwin (1890–1968), G. Stanley Hall (1844–1924), Sigmund Freud (1856–1939), Arnold Gesell (1880–1961), Lev Vygotsky (1896–1934), Piaget (1896–1980), and Heinz Werner (1890–1964), as being unduly influenced by recapitulation thinking.

The most explicit of the recapitulation theorists was Hall (1904), who believed that humans were still actively evolving and that the period of adolescence, not adulthood, must provide the jumping-off point for further evolutionary advance. At first, this seems contradictory to Haeckel's claim that evolution proceeds by making additions to the adult stage; but Hall viewed adulthood as a period of decline. It was adolescence that was the true zenith of human development (a claim that may seem particularly odd for anyone rearing an adolescent). Hall acknowledged the identification of earlier stages of phylogeny in the human embryo and young child but extended this argument by proposing that human ontogeny parallels a more recent evolutionary past, specifically in terms of the behavior of humans' ancestors. Accordingly, Hall believed that the behavior and minds of modern children ages 8–12 years old corresponded to adult behavior and thought of prehistoric times. By understanding what children are like, we are able to understand what our ancestors were like. Consistent with this theme, Hall advocated hunting and fishing for boys ages 8–12, activities in which their prehistoric male predecessors would likely have engaged.

Although Hall's theory was the most extreme in terms of recapitulation thinking, Morss (1990) identified important elements of such thinking in other early theorists. For example, the ordering of Freud's five psychosexual stages—oral, anal, phallic, latency, and genital—was based on explicit recapitulation theorizing. According to Freud (cited in Morss, 1990), "Ontogenesis may be regarded as a recapitulation of phylogenesis, in so far as the latter has not been modified by more recent experience. The phylogenetic disposition can be seen at work behind the ontogenetic process" (p. 46). Piaget believed in the "cultural epoch" theory of education, whereby children are brought through progressively more recent stages of civilization. In general, Morss interpreted much of the biological thinking by early developmental psychologists as pre-Darwinian, relying on discredited ideas of Haeckel's recapitulation theory and Lamarckianism. He also viewed contemporary developmental theory (through the 1980s) as similarly flawed, stating that "developmental psychology is built upon foundations that are rotten" and based on "outdated notions of a biological–philosophical nature" (p. 227).

As we noted earlier, recapitulation theory has been making a bit of a comeback in cognitive evolutionary theory of late (Langer, 1998, 2000; McKinney, 1998; Parker & McKinney, 1999), based primarily on the observations that many aspects of primate cognition can be viewed in terms of Piaget's stages of cognitive development, with apes progressing further than monkeys and humans further than apes. (These ideas are discussed in greater detail in chapters 4 and 5.) Nonetheless, few contemporary recapitulationists would agree with the formulations of Hall, Freud, or likely even Piaget, regarding the extent to which ontogeny is proposed to recapitulate phylogeny.

Sociobiology, Neonativists, and Evolutionary Psychology

As we noted earlier, evolutionary thinking fell out of favor in psychology during the heyday of behaviorism and remained out of fashion through the early years of the cognitive revolution. There were exceptions, of course.

For example, in 1974 Freedman published *Human Infancy: An Evolutionary Perspective*, and two years later Fishbein published *Evolution, Development, and Children's Learning*. These books reviewed the principles of evolution, human phylogeny, and certain aspects of ontogeny from an evolutionary perspective, but they were ahead of their time, had little impact on the field, and did not find a supportive audience until years after they went out of print. Another important class of exceptions concerned psychologists who viewed development from an ethological perspective (Bowlby, 1969; McGrew, 1972; see also chapter 9 and discussion below). Generally, most psychologists essentially ignored evolution and biology when trying to understand child development. The advent of sociobiology (Dawkins, 1976; E. O. Wilson, 1975), based on the theories of Hamilton (1964, 1966) and originally developed to explain altruism, put an emphasis on behavior, and many developmental psychologists took notice and began research programs with an explicitly evolutionary perspective (see papers in MacDonald, 1988). However, for the most part, this research and theorizing stayed outside of the mainstream of developmental psychology.

During the 1980s, a movement apparently independent of ethology gained steam in cognitive development: *neonativism*. Extending the pioneering work of Chomsky (1965) on the nativism and the domain specificity of language, neonativists held, unlike Piaget or the behaviorists, that children come into the world with certain domain-specific knowledge. They know something already, for example, about language (Karmiloff-Smith, 1991; Pinker, 1994), objects (Gelman & Williams, 1998; Spelke & Newport, 1998), and social relations (Bowlby, 1969; Hinde, 1974). Children still require a supportive environment for these abilities to develop, but the seeds of knowledge, if not the knowledge itself, are within the child and present at birth. (More is said about innateness in the following section and about neonativism in chapter 6.)

Concurrent with the rise of neonativism was the field of evolutionary psychology (Barkow et al., 1992; Daly & Wilson, 1988a). As we have seen, development was not a major focus of this new discipline, although it was not explicitly excluded, either. Child developmentalists increasingly began to take notice of evolutionary psychology and advocated, as we are here, that development can be best understood when looked at from an evolutionary perspective (Bjorklund, 1997b; Bjorklund & Pellegrini, 2000; Geary, 1995; Geary & Bjorklund, 2000; Keller, 2000; Surbey, 1998a).

And this is the topic of the remainder of the book. Although we realize that there is more than one evolutionary perspective and that there remain many controversies about the relations between ontogeny and phylogeny (among other issues), we believe that biology has returned to the mainstream of developmental explication and with it a tolerance, at least, and an ardor, at best, for evolutionary thinking.

WHAT DOES IT MEAN WHEN WE SAY
SOMETHING IS "INNATE"?

As we noted earlier, it was not until the rediscovery of Mendel's work that a plausible mechanism for inheritance was available. Eventually, Darwin's ideas were integrated with those of the emerging field of genetics, producing the modern synthesis. Since that time, despite modifications to the modern synthesis (Hamilton, 1964), evolutionary biology has been based on genetics.

When discussing inheritance of physical characteristics such as fingers and toes, or blue or brown eyes, it is relatively easy to conceive of what is inherited (even if the actual biochemical mechanisms of inheritance may be beyond one's ken). It is more problematic, however, when conceiving of inherited behavior. Surely, if a duck's webbed feet are inherited, so are its quack and its tendency to attach to its mother hours after hatching. But behavior, particularly among vertebrates and especially among primates and humans, is flexible. It cannot be genetically programmed as rigidly as the color of one's eyes; there must be some room for learning. Yet, evolutionary psychology assumes that important aspects of behavior and cognition have a genetic component to them, although such components cannot be divorced from environmental effects (broadly defined) and complex gene–environment interactions. When we speak of inherited psychological characteristics we typically use the term *innate* or sometimes refer to *instincts*.

Opinions about how to conceive of innateness and what, if anything, is innate in humans, have varied considerably over the 20th century. In this section, we present a brief history of these debates. Although modern views differ considerably from those expressed by leaders of the field early in the last century, subtle differences in what "innateness" connotes continue to exist. To presage a bit, we take a position consistent with the developmental systems approach discussed in chapter 2; however, although we argue against a view of genetic determinism, we propose that species-wide genetic (inherited) biases exist that are expressed when individuals experience species-typical environments. Thus, the expression of a genotype must be considered in the historical contexts of ever-changing environments.

Ethologists and the Concept of Instincts

One of the first scientists to seriously apply Darwin's theory to behavior and mental abilities was George Romanes (1848–1894; see Gottlieb, 1992; Morss, 1990). Romanes emphasized the continuity of "mind," believing that the intellectual capacity of different animals, including humans, is a matter of degree. Romanes is credited with being the father of comparative psychology; he established a framework for comparing the behavior and cognitive abilities of different species. However, as was the case with Darwin, Romanes's work was limited by a lack of a theory of inheritance. But as the mechanisms of genetics became more widely accepted and as the modern synthesis took shape in the 1930s and 1940s, a new group of biologists, the ethologists, began to scrutinize animal behavior from an explicitly evolutionary perspective. *Ethologists* studied the behavior of animals, mostly in their natural habitat. Like contemporary evolutionary psychologists, they looked for species universals—behavior that characterizes all normal members of a species at some time in their development, or perhaps all members of one sex of a species—and interpreted those behaviors in an evolutionary framework, specifically, in terms of what adaptive function those behaviors may have. The careful observations, field experiments, and innovative theorizing gained the three pioneers of this field, Konrad Lorenz (1903–1989), Niko Tinbergen (1907–1988), and Otto von Frisch (1886–1982), the Nobel Prize in Physiology or Medicine in 1973, the only behaviorists ever to receive such a distinction.

We introduced some of the ideas of Tinbergen (1963) in chapter 2, specifically his four questions that we must ask to fully understand any behavior: What is the immediate benefit (adaptive function) of a behavior? What are its causes? How does it develop? How did it evolve? But ethologists' most important contribution with respect to the "innateness" issue was the concept of instincts. The classic demonstration of instincts comes from the work of Lorenz (1937, 1965) with geese. Lorenz observed that, hours after hatching, goslings would follow the first moving thing that they saw. (Later research indicated that auditory cues were of even greater importance; see Hess, 1973.) Usually the first moving and noise-making object in a young gosling's life would be that of its mother. In the wild, the goslings would stay close to their mother, which has obvious survival value. But if during this critical period the goslings see not their mother but a human being, they will follow that person, much as they would their mother. Lorenz called this phenomenon *imprinting*. No prior experience is necessary for this phenomenon, and once the image of the mother goose (or Lorenz himself) is "imprinted" in the gosling's brain, it cannot be altered. From this research, the idea that a complex and adaptive behavior could be "built into the genes" and elicited by a particular experience at a particular time, became the canonical definition of instinct.

Tinbergen (1951) described a particular class of instincts as *fixed-action patterns*, stereotypic behavior patterns that are elicited by a specific environmental event. This pattern is like a complicated reflex that, once evoked, runs its course. The difference between a fixed-action pattern and a reflex is mainly in the complexity of the behavior. A classic example of a fixed-action pattern was provided by Tinbergen (1951) for the male stickleback fish, described briefly in chapter 2. Tinbergen noticed that, at times, male stickleback fish behaved aggressively toward other males in their territory. Specifically, they acted aggressively when the invading male had a red belly, which signals readiness to mate. But he also noticed that males would make aggressive displays when other red objects entered their line of vision (including a red truck passing the window where the fish tank sat). Through a series of experiments, Tinbergen demonstrated that it was the red spot that elicited the stereotypic aggressive behavior of the stickleback fish. This is normally highly adaptive in the wild, in that the presence of a red spot is typically a sign of a sexual competitor. But it is a thoughtless "instinct," programmed into the brain of male stickleback fish that can be aroused under experimental circumstances that are totally unrelated to the purpose for which the response had evolved.

Instincts for the early ethologists were examples of genetic determinism, or genes → behavior. Complicated behaviors are preprogrammed and, by definition, no experience is necessary for these behaviors to be realized. Experience did play some role, of course. For example, goslings were not genetically programmed to imprint to their mother, per se, but rather were programmed to imprint to the first moving thing they saw at a particular time in development which, in the natural world, would inevitably be their mother. The genetic determinism of the early ethologists relegated relatively little role to development. Although certain experiences may be necessary at specified times in ontogeny (i.e., critical periods), behavior was "innate" and would be expressed when the animal was exposed to the proper environmental cues.

We saw in our discussion of the developmental systems approach in chapter 2 that even such "instinctive" behavior as imprinting in precocial birds is not uninfluenced by experience. Rather, as Gottlieb (1976, 1997) and many others (see Gottlieb et al., 1998; Kuo, 1967; Lehrman, 1970; Schneirla, 1960) have demonstrated, prenatal experience, specifically hearing a species-typical call, plays a critical role in whom a baby bird will approach and follow. Moreover, other research has demonstrated that imprinting is not as permanent as Lorenz had believed (see Cairns, 1979; Hess, 1973). Rather, later experiences (i.e., after the "critical period") can modify a previously "imprinted" response, illustrating the flexibility of such "instincts" (Bateson, 1981b).

Although the early ethologists' observations of complex behaviors elicited by specific environmental events advanced the understanding of animal behavior immeasurably, their interpretation of "no experience necessary" and "no flexibility" for such behaviors was incorrect, something that later ethologists recognized (Hinde, 1980). However, the ethologists' original definition of instinct is still with us today. Although we believe that most ethologists and comparative and developmental psychologists recognize that the "no experience necessary" definition of instincts is untenable, many other psychologists, and nearly all laypeople, implicitly continue to accept this view of instincts and thus of innateness.

However, for many people concerned exclusively with human behavior, this is not a substantial problem because many students of human behavior continue to hold the position, perhaps advocated by a majority of behavioral scientists only several decades ago, that humans have no "instincts"—that *Homo sapiens* possess no specific innate mechanisms other than a large brain and a general ability to learn.

Standard Social Science Model

Although ethological theory did influence the thinking of some scientists concerned with human behavior and development (Bowlby, 1969; Eibl-Eibesfeldt 1989; Hinde, 1974; McGrew, 1972), most research in ethology was devoted to nonhuman animals. Moreover, for much of the 20th century, any form of biological causation with respect to human behavior, including evolutionary accounts, was anathema to the academic mainstream. Evolutionary psychologists John Tooby and Leda Cosmides (1992) labeled the dominant antibiological perspective of human behavior the Standard Social Science Model (SSSM). According to Tooby and Cosmides, the SSSM basically denied the existence of a human nature. Human beings come into the world with the ability to gather and interpret perceptual information and a general capacity for learning. Biological evolution as a sculptor of human behavior and thought is replaced by cultural evolution.

Although we think that Tooby and Cosmides's (1992) depiction of the ubiquity of the SSSM is a bit overstated, we fully agree with their observation that, until relatively recently, majority opinion in Western academe has been one of cultural determinism. Our own undergraduate educations were steeped in this perspective. We were taught that humans have no instincts, or innate predispositions. Infants were essentially blank slates; even their smiles were just arbitrary behaviors that had become conditioned to positive events. All sex differences other than those directly related to reproduction were culturally arbitrary. People in different cultures around the world were far more different than they were similar, and the pressure to conform to local social norms and mores determined such behavior.

The hegemony of the cultural-relativism perspective cut across the social sciences, including developmental psychology. One reason for the dominance of this position in developmental psychology was the emphasis that the field put on early experience. Learning theories such as those of John Watson (1878–1958) and B. F. Skinner (1904–1990) equated development with learning and held that that which is learned early sets the course for later development. Even theories that assumed that children possessed a biological nature, such as Freud's theory of psychosexual development, stressed the role that early experience plays in forming personality. And when Piaget's cognitive theory began to replace behaviorism as a model for intellectual development, the emphasis, in the United States at least, was on early development, particularly how parents or educators could best establish intellectual competencies in children, which would presumably be maintained thereafter (Hunt, 1961; White, 1978). There was little room for any hint of biological influence in such perspectives.

Related to this viewpoint of developmental psychologists is what Charlesworth (1992) referred to as *meliorism*. Child developmentalists have long been concerned with fostering development in children, beginning with the child study movement established by Hall at the beginning of the 20th century (Cairns, 1983). The goal of improving children's lives fits well with a strong behaviorist perspective. The "child welfare" agencies that were established at Iowa and Minnesota were an outgrowth of this orientation. If early experience is responsible for establishing social, emotional, and intellectual competence, science can discern what those experiences are and develop programs to provide all children with them. If, in contrast, important aspects of behavior are determined by genes, behavioral intervention can have little influence. If it is in the genes, it is presumably not subject to modification. Moreover, any suggestion that socially consequential behaviors such as aggression, personality, and intelligence are innate implies the existence of biologically defined categories of people. Smart people and aggressive people are that way "by nature," raising the possibility of a genetically based caste system, the antithesis of modern liberal democracy. This is also related to the *naturalistic fallacy*—the false belief that if something has evolved and is "natural" it must be morally "right." Such a view understandably makes the acceptance of evolutionary interpretations for certain aspects of human behavior (e.g., aggression) undesirable. However, evolutionists do not equate what is "natural" with what is "right," and neither should anyone else, thus making untenable the argument that evolutionary explanations for human behavior or development are morally reprehensible (see Thornhill & Palmer, 2000; also chapter 6).

Even strong proponents of a "gene's eye view" of behavior (i.e., proponents of gene → behavior viewpoints) do not support the idea of genetic determinism. For example, Richard Dawkins (1982), one of the architects of sociobiology and a believer that our social behavior (such as altruism), is greatly influenced by genetics, discounts such a claim:

> People seem to have little difficulty in appreciating the modifiability of "environmental" effects on human development. If a child has had bad teaching in mathematics, it is accepted that the resulting deficiency can be remedied by extra good teaching the following year. But any suggestion that the child's mathematical deficiency might have a genetic

origin is likely to be greeted with something approaching despair: If it is in the genes, "it is written," it is "determined" and nothing can be done about it. This is pernicious rubbish on an almost astrological scale. Genetic causes and environmental causes are in principle no different from each other. Some influences of both types may be hard to reverse; others may be easy to reverse. Some may be usually hard to reverse but easy if the right agent is applied. The important point is that there is no general reason for expecting genetic influences to be any more irreversible than environmental ones. (p. 13)

Cultural relativism continues to hold a strong position within the social sciences but, over the past several decades, emphasis on biologically based accounts of behavior and development has increased. This can be seen in research in language development that has demonstrated that children do not learn language via the rules of operant and classical conditioning, but rather are "prepared" to acquire the language spoken around them (Pinker, 1994; see also chapter 6); in work in behavioral genetics that illustrates that many aspects of cognition and personality are highly heritable (McGuffin, Riley, & Plomin, 2001); and in evidence showing that infants are not born as blank slates, but come into the world with certain "knowledge" and expectations about what type of environment they can expect (Spelke & Newport, 1998; see also chapter 6).

This increased acceptance of biological causation requires that researchers develop more precise theories about what it is that is inherited. That is, when child developmentalists essentially rejected any idea of innate abilities in their subjects, it mattered little how they defined innateness. Now that biologically based accounts of development, including evolutionary accounts, are being taken more seriously, it is necessary to examine more closely what we mean when we say that something is innate.

Contemporary Conceptions of Innateness

For contemporary developmental psychologists, the term *innate* can be contentious, and many would just as soon not see it used at all. This is because "innate" is often seen as being equivalent to "instinct," which is interpreted by many not much differently from the way the early ethologists interpreted it, that is, as complex behavior requiring no prior experience for its expression. As should be clear by now, this is not what developmental psychologists today mean when they speak of "innateness," when they speak of it at all. For evolutionary developmental psychologists, the term can have a range of meanings, depending on the characteristic under study. As with physical characteristics, some innately influenced behaviors can be highly canalized, that is, they are likely to be expressed in all but the most extreme environments, whereas others can be less specifically directed by the genome.

Our own perspective is consistent with the developmental systems approach described in chapter 2. This perspective views development as being the result of the bidirectional interaction at all levels of organization, from DNA $\leftrightarrow$ RNA, through the individual and his or her culture (see Gottlieb, 1998, 2000; Gottlieb et al., 1998; Ho, 1998; M. H. Johnson, 1998). The functioning of the organism itself, as in the firing of neurons, is a form of experience that influences subsequent development. From this perspective, there can be no pure genetic–biological or environmental–social effects on development. Everything must be conceived as the result of the continuous and bidirectional interchange between structure and function, operating at all levels of organization, over the course of development. From this perspective, there is neither biological nor environmental determinism. Rather, development is probabilistic. Genes are part of the developmental system, but they are not immune to influences from other levels of the system.

Nevertheless, different members of a species usually develop to be highly similar to one another. That is, despite the probabilistic nature of development, species-universal behaviors exist. Gottlieb (1997) presented his position as follows:

> The resolution I have opted for in this volume is to accept that certain developmental outcomes are species-typical or species-specific, adaptive, and responsive to a narrow class of stimulation in the absence of prior exposure to these configurations (independent of frank learning but not independent of experience, broadly defined). (p. 144)

As we view it, from a developmental systems perspective, one cannot speak of innateness without considering genetics, environment broadly defined, and the continuous interaction of these factors.

Genes, of course, interact with their environment at a variety of levels, and some researchers have proposed that we restrict "innate" to changes that arise as a result of interactions occurring within the organism. For example, we may inquire about the extent to which some neural circuits are the products of interactions strictly within the organism, seemingly uninfluenced by the external environment (M. H. Johnson, 1998; M. H. Johnson & Morton, 1991). This perspective is consistent with the idea of *constraints*, discussed briefly in chapter 2 and in greater detail in chapter 6 (Elman et al., 1996; Gelman & Williams, 1998; Spelke & Newport, 1998). When applied to the human mind, the idea of constraints is that humans (or any species) are limited in how they can make sense of their world. Such limitations, however, permit individuals to specialize. Rather than having a general-purpose mind that must be applied to every problem it encounters, individuals have specialized programs that are able to solve certain problems exceptionally well (e.g., acquiring language), but little else.

From this perspective, processing constraints enable learning rather than hinder it.

Within this perspective, there are different kinds of constraints, with different implications for what it means for something to be innate. For example, in *Rethinking Innateness*, Jeffrey Elman and his colleagues (1996) proposed three general types of constraints: representational, architectural, and chronotopic. These constraints vary in the extent to which they are genetically specified. *Representational constraints* are the most highly specified and refer to representations that are hardwired into the brain so that some types of "knowledge" are innate. For instance, several neonativist theorists have proposed that infants come into the world with (or develop very early in life) some basic ideas about the nature of objects (e.g., their solidity), mathematics (e.g., simple concepts of addition and subtraction), or grammar (see Pinker, 1997; Spelke & Newport, 1998; Wynn, 1992). Specific synapses are "preprogrammed" to process certain types of information. However, unlike earlier instinct theorists, neonativists advocating representational constraints do not propose that this innate knowledge is independent of experience. Experience and maturation of the brain (which itself is a complicated process involving the bidirectional relation between neurons, their immediate surroundings, and their own excitation) play a role in shaping this knowledge, but most proponents of representational constraints believe that infants come into the world with domain-specific neuronal representations. This does not mean that there is no development; everything develops. But, according to this perspective, children (and other animals) are born able to make sense of certain aspects of their environments (e.g., basic notions of physics, how to build webs if you're a spider) given only minimal postnatal experience.

According to Elman et al. (1996), *architectural constraints* represent an intermediate degree of genetic specificity and refer to ways in which the architecture of the brain is organized at birth. For example, neurons can have different functions (some being excitatory, others inhibitory) or can vary in the amount of activation required for them to fire. At a somewhat higher level, neurons in a particular part of the brain might be more or less densely packed or have many or few synapses with other local neurons. And at a higher level yet, different areas of the brain are connected with other areas of the brain, affecting global organization. Architectural constraints limit the type and manner of processing the brain can perform, not because the brain comes equipped with innate representations (e.g., what grammar is), but because certain neurons or locations of the brain can process only certain types of information and pass it along to certain other parts of the brain. Thus, architectural constraints imply limits on what is processed, as do representational constraints. Unlike representational constraints, however, architectural constraints permit (or require) a high degree of learning to occur. For example, the architecture of a particular area of the brain may be best suited for processing a certain type of information (e.g., language), but the architecture will change as a function of the quantity and quality of the information it receives. Structures in the brain are not preformed to "know" grammar, for instance, but are biased toward processing information about language and develop a grammar as a result of interactions with the world (Elman et al., 1996).

Developmental neuropsychologist Mark Johnson (2000) provided a set of hypotheses about how architectural constraints may work. Different areas of the brain have different sets of inputs and outputs, balances of neurotransmitters, synaptic densities, and patterns of connections, which result in biases in information processing. Such biases in human newborns are slight, however, but sufficient to ensure that some types of information (e.g., language) are processed more effectively by particular parts of the brain than others. Brain specialization increases over the course of development, as particular parts of the brain gradually restrict the range of stimuli that they respond to. The speed with which this information can be processed also increases. Thus, although infants' brains come equipped with "innate" biases, they are weak by nature, but specialization increases with experience. De Haan, Oliver, and Johnson (1998) provided evidence for this when they looked at developmental differences in face processing. Adults process upright and inverted faces differently, reflected by differences in reaction times and activation of different neural pathways. Six-month-old infants similarly process upright and inverted faces differently, but this is independent of whether they are looking at human or monkey faces. Adults, in contrast, process human faces differently than monkey faces. These findings suggest that cortical processing of faces becomes more specialized with age and experience, specifically for the processing of human faces.

Chronotopic constraints are the least specified and refer to limitations on the developmental timing of events. For example, certain areas of the brain develop before others. This means that early developing areas are likely to have different processing responsibilities than are later developing areas. Similarly, some areas of the brain may be most receptive to certain types of experiences (to "learning") at specified times, making it imperative that such experiences (e.g., exposure to patterned light or language) occur during this "critical period" of development. For example, children universally acquire language in about the same way and at about the same time (Pinker, 1994). If, however, for some reason children are not exposed to language until later in life, their level of language proficiency is greatly reduced. The human brain appears to be prepared to make sense of language and makes it easy for children to acquire the language that they hear around them. But such neural readiness is constrained by time. Wait too long, and the ability to acquire a fully articulated language is lost.

Gottlieb (1997) has been critical of the "constraints" perspective of innateness because, for the most part, it ignores the role of functional activity (e.g., the activation of neurons influencing their subsequent organization and operation) in development. Although we concur with Gottlieb on the importance of considering self-produced activity as a form of "experience," we believe that the idea of innate constraints has great heuristic value. Some aspects of development are less easily perturbed than others, in that they will be expressed in most members of a species given a species-typical environment. This does not contradict, we believe, the basic ideas of the developmental systems approach, but only acknowledges that some psychological outcomes are more highly predictable than others, given an "expected" environment. Deviations from the species-typical pattern can occur, and such deviations negate any notion of genetic determinism. This is true for each of the three types of constraints described by Elman and his colleagues (1996), even the highly specified representational constraints.

From our perspective, what we mean when we say something is "innate" is that individuals have evolved biases that constrain how they make sense of certain types of information. These outcomes are not inevitable, but arise as a result of the continuous and bidirectional interaction between an organism (including its self-produced activity) and its environment at all levels of organization. Genes are an important component of the developmental system, but their actions are influenced by other parts of the system. From the perspective of evolutionary developmental psychology, the information-processing biases of infants and children are continuously modified by experience (some more so than others). Thus, "innate," as we understand it, does not imply inevitability, unmodifiability, or biological determinism; it does imply that evolution has prepared infants and children for life in a human group, making the general pattern of ontogeny predictable and the adult generally well adapted for continuing the species.

EVOLUTIONARY DEVELOPMENTAL PSYCHOLOGY, BEHAVIORAL GENETICS, AND INDIVIDUAL DIFFERENCES

One of the lasting contributions of Darwin to psychology was a focus on individual differences (Charlesworth, 1992). Variation is the stuff upon which natural selection works, and differences among individuals and the origins, maintenance, and modification of such differences has been a major focus of developmental psychology (Bjorklund, 2000; Cairns, 1979). The formal, scientific study of individual differences can be traced to the polymath Francis Galton (1822–1911), a first cousin of Darwin's. After reading *On the Origin of Species*, he wrote to Darwin "your book drove away the constraint of my old superstition as if it had been a nightmare" (cited in Shipman, 1994, p. 111). Galton's words were not empty ones. He turned his immense curiosity and energy to a lifetime study of the inheritance of human mental abilities. Thus, the systematic study of human individual differences had an explicitly Darwinian beginning, a perspective that modern behavioral geneticists maintain (Plomin, DeFries, McClearn, & Rutter, 1997; Scarr, 1995a).

Social Darwinism and Other Misuses of Evolutionary Theory

Another early advocate of Darwin's theory was social scientist Herbert Spencer (1820–1903), whose influence in promoting Darwin's theory, particularly in Great Britain and the United States, was considerable. However, what Spencer is remembered for today in most circles is the *social Darwinism* movement. Whereas Darwin applied natural selection to biological life, Spencer envisioned a broader application of the concept, believing that it typified every form of existence, from the origin of the solar system to human society (Shipman, 1994). Unlike Darwin, Spencer, who coined the phrase "survival of the fittest," was convinced that evolution was progressive, always moving toward perfection. When applied to society, it resulted in the very self-serving position that both social success and failure are inherited characteristics, and it is nature's way to weed out those who are unsuccessful. Thus, Spencer argued against social programs such as poverty laws, universal education, and housing regulation that would artificially help poor people (i.e., the biologically unfit) and thus impede the progressive improvement of society. Differences in social success between races and social classes were not attributed to differences in educational or economic opportunities but to some unspecified mechanism of inheritance. Views such as these were not original to Spencer but had been advocated by people in political and economic power before; what was new was the scientific justification for the notions of inherited superiority and inferiority. Although in these pre-Mendelian days Spencer lacked a mechanism for inheritance, Darwin's theory provided the scientific "proof" that the rigid British class system was "natural" and that attempts to fiddle with it were misguided.

At the same time, Galton was advocating a similar program. In his 1869 book *Hereditary Genius*, he traced the family histories of successful men over a 200-year span. He observed that "genius" runs in families and, like Spencer, attributed the phenomenon to inheritance, not opportunity. Galton coined the term *eugenics*, a proposed science that would improve the condition of the human species by selective breeding, much as cattle are improved by ranchers.

In the United States during the early decades of the 20th century, the new field of intelligence testing was taking off, prompted in large part by Galton's earlier work (as well as the research of the French psychologists Alfred Binet [1837–1911] and Theophile Simon [1873–1961]). Lewis Terman (1871–1956) and Robert Yerkes (1876–1956), both giants of early American psychology, undertook a mass testing of 1.75 million recruits to the U.S. Army in World War I. Two forms of the test were constructed: the Army Alpha, which was a paper-and-pencil test used for recruits who could read English; and the Army Beta, which was used for recruits who were not fluent in English. Despite many difficulties in administration, the findings of the tests were considered to be reliable, perhaps because they confirmed the preconceived notions of the researchers. The results showed lower levels of intelligence of Black people and more recently arrived immigrants from southern and eastern Europe than for native-born White people or those who had lived in the country for longer periods of time (mostly of northern European descent). Figure 3.1 presents the average mental age

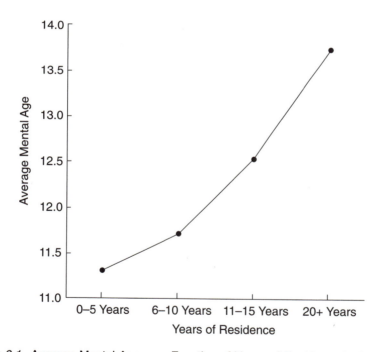

Figure 3.1. Average Mental Age as a Function of Years of Residence in the United States, from U.S. Army Alpha and Beta Tests.
Note. Data from *The Mismeasure of Man* (p. 221), by S. J. Gould, 1981, New York: Norton. Copyright 1981 by S. J. Gould. Adapted with permission.

of recruits as a function of years of residence in the United States (adapted from S. J. Gould, 1981). As can be seen, mental age varied monotonically with years of U.S. residence. However, rather than viewing these data as a consequence of the testing procedures for non-English speakers or in terms of differing degrees of cultural experience, the researchers proclaimed that the more recent immigrants to the country were innately less intelligent than the earlier immigrants (Brigham, 1923).

The theories of the IQ-testers and of Spencer and Galton extended beyond the ivory tower and influenced social policy by those who were all too willing to believe and influenced the formulation of social policy by ethnically biased legislators. In the United States, the 1924 Immigration Act limited immigration to 2% of those of the same nation that were living in the United States in 1890, when northern Europeans predominated. Sterilization laws were passed in 16 states between 1907 and 1917, making it legal to sterilize many classes of criminals as well as people with epilepsy, people with a drug addiction, people with mental retardation, and those considered insane (Shipman, 1994).

Improvements in IQ testing continued throughout the 20th century and became the primary arena for debates concerning the heritability of intelligence. The field of *behavioral genetics* emerged, which examines the similarity of characteristics among people of different degrees of genetic relatedness and provides statistical estimates of the degree to which variation in these characteristics are attributed to inheritance. In Great Britain, the major proponent of a gene-based theory of intelligence was the influential educational psychologist Sir Cyril Burt (1883–1971). In a series of studies conducted both before and after World War II, Burt demonstrated that the IQs of groups of identical twins who had been separated early in life were remarkably similar, even though they had been reared in different environments. The results, clearly supporting a strong genetic view of intelligence, were seriously questioned after his death in 1971. Burt had published three studies comparing IQ-test scores of three groups of separated identical twins. The numerical results in the three studies were exactly the same—a coincidence so unusual to be considered impossible, due either to fraud (Kamin, 1974) or, according to more recent evaluations, carelessness (Mackintosh, 1995).

Excesses of social Darwinism were also found in some segments of European society. In Germany, Haeckel became an ardent proponent of Darwin's theory and promoted it vigorously in his homeland. However, as did many researchers of his time, he saw evolution as being progressive, and for Haeckel, this meant that the different races reflected different levels of perfection, with the pure German, or Aryan, race (the *Volk*) being "imbued with power and intrinsic goodness and mystically tied to their holy German landscape" (cited in Shipman, 1994, p. 95). Evolutionary theory justified

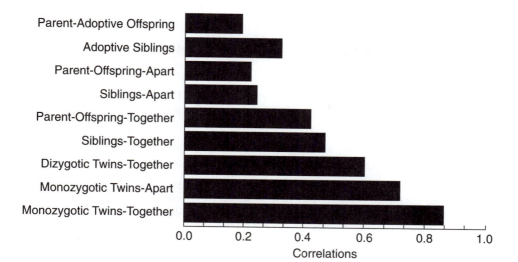

Figure 3.2. Average Correlations of Familial Studies of Intelligence.
Note. Adapted from data from "Familial Studies of Intelligence: A Review," by T. J. Bouchard, Jr. and M. McGue, 1981, *Science, 212,* pp. 1055–1059. Copyright 1981 by American Association for the Advancement of Science.

overt racism for Haeckel, and he advocated that German authorities be given the power to enhance and protect the race through eugenics (Shipman, 1994).

Given these early misappropriations of evolutionary theory to human variation, it is little wonder that scientists were reluctant to apply Darwinian principles to the study of human individual differences. Gradually, however, Darwinian thinking returned to the issues of individual differences and has produced an impressive corpus of results illustrating the heritability of a wide range of characteristics from intelligence through religiosity (see McGuffin et al., 2001; Plomin et al., 1997).[2] For example, research conducted since the uncovering of the Burt controversy has found that IQ scores between people vary predictably as a function of their degree of genetic relatedness (Bouchard & McGue, 1981; Bouchard, Lykken, McGue, Segal, & Tellegen, 1990; Plomin & Petrill, 1997), reflecting a substantial heritability of intelligence (see Figure 3.2). Theory in modern behavioral genetics is explicitly Dar-

---

[2] *Heritability* is the extent to which differences in any trait within a population can be attributed to inheritance. Heritability is expressed as a statistic that ranges from 0 (none of the differences in a trait are attributed to inheritance) to 1.0 (100% of the differences in a trait are attributed to inheritance). It reflects the proportion of variance in an observed trait that is due to genetic variability. Heritability is a population statistic, in that it describes average differences among people within a population. It does not refer to how much of any one person's intelligence (or height or personality characteristics) can be attributed to genetic factors, only what percentage of the difference in a trait within a specific population can be attributed to inheritance, on average. Heritability of a trait can vary over time as environments vary. The more homogenous environments are among people within a population, the higher heritability will be.

winian, at least among some prominent practitioners (Harris, 1998; Scarr, 1995a).

In contrast, evolutionary psychologists have generally steered away from individual differences. Rather, evolutionary psychology has primarily been concerned with species *universals*—characteristics that typify all members of a species (or all of one sex of a species) at certain times in development. To many evolutionary psychologists, individual differences are simply noise (such as whether a belly button turns inward or outward)—uninteresting variation based on underlying evolved universals (Buss, 1995). But variation is at the very core of Darwinian theory, and so any evolutionary account that views variability among individuals as simply noise is likely missing some important points. According to developmental psychologist and behavioral geneticist Sandra Scarr (1995a), evolutionary psychologists have applied Darwin's idea of natural selection to understanding human nature but have virtually ignored variation. A notable exception is Gottlieb (1983), who considered the study of individual differences to be a defining principle in the study of development.

Developmental Systems Approach and Behavioral Genetics: Different Views of Individual Differences

Child developmental psychology is concerned both with universals (developmental function) and with individual differences, and an evolutionary approach to development encounters the same criticisms as does evolutionary psychology in general with respect to how it deals with individual differences. In fact, a perspective that emphasizes the expression of epigenetic rules through interaction with the environment, as does the developmental systems perspective favored here, has been criticized for making individual differences interpretable only after the fact. There are so many interacting factors at so many different levels of development that individual differences can never be predicted with any accuracy (Scarr, 1993). This is in contradiction to what is known about the stability of many individual differences over time (Bayley, 1949; Schneider, Perner, Bullock, Stefanek, & Ziegler, 1999) and about the heritability of many psychologically important traits (Plomin et al., 1997). Behavioral genetics research provides a means of predicting and explaining individual differences by using an underlying evolutionary theory that, at one level, is substantially different from that advocated by the developmental systems approach.

Perhaps the best developed behavioral genetics approach with respect to development is Scarr's genotype → environment theory (Scarr & McCartney, 1983; Scarr, 1992, 1993; see also Plomin, DeFries, & Loehlin, 1977). In this theory, there are three types of genotype → environment effects: passive, evocative, and active. *Passive effects* occur when genetically related parents provide the rearing environment of the child. When biological parents rear a child, the effects of genetics and environment cannot be separated because the people who provide the genetic constitution for a child also provide the environment. The influence of passive effects is proposed to decline with age. *Evocative effects* occur when the child elicits responses from others that are influenced by his or her genotype. For example, people respond differently to an irritable child than to a well-tempered child, and the type of attention received by an infant who likes to cuddle is different from that received by an infant who does not want to be held. Evocative effects are proposed to be stable over development.

For our purposes in this book (and for Scarr's), active genotype → environment effects are the most important. According to this model, a child will seek environments that are compatible with his or her genotype. This is captured by the phrase "genes drive experiences." The experiences children have will shape their personalities and intellects, but it is one's genotype that determines, to a large extent, which environments are sampled (see also Bouchard et al., 1990; Harris, 1998; Plomin & Daniels, 1987). *Active effects* increase with age, as children are increasingly able to choose their own environments. Children thus develop consistent with their genotypes. Children with identical genotypes (monozygotic twins) develop similarly, and children with different genotypes develop differently.

At one level, the behavioral genetic position gives a substantial role to the environment in shaping individual differences—experience is most directly responsible for crafting important aspects of cognition and personality. But genes drive children to seek these environments (McCartney, Harris, & Bernieri, 1990; Plomin & Daniels, 1987). Scarr (1995a) contended that children will develop characteristics consistent with their genotype in all but the most impoverished environments. Similar to the developmental systems approach, a species-atypical environment will restrict children's options and thus limit their development. But differences among ordinary environments will have little impact on the expression of characteristics, and individual differences will best be predicted based on a child's genotype. This is because, according to Scarr (1992, 1993), evolution does not permit easy environmental modification of characteristics that are essential for survival. Rather, important aspects of social, emotional, and cognitive development are highly canalized and are not much influenced by the vagaries of parenting behavior. It would be a short-lived species that had an extended childhood and a narrow range of parental behaviors necessary to rear a child to adulthood.

Based on these observations, Scarr (1992) proposed the concept of *"good-enough" parenting*. Because children will develop according to their genotype in all "ordinary" environments, it is not necessary to be "superparents" to rear healthy and (re)productive members of the community. More is said about Scarr's ideas about parenting in chapter 8.

Behavioral genetics research has generally confirmed many of Scarr's observations, particularly for intelligence as measured by IQ. For example, identical twins reared apart are highly similar to one another on a wide range of traits (see Bouchard et al., 1990; Plomin et al., 1997), whereas siblings become less alike the older they get (McCartney et al., 1990; Plomin & Daniels, 1987; Scarr & Weinberg, 1976; Turkheimer & Waldron, 2000), presumably because the differences in their genotypes (they share 50% of their genes compared to the 100% for identical twins) lead them to different experiences, with these active genotype $\rightarrow$ environment effects increasing with age for many characteristics.

These findings should not be viewed as indicating that postnatal environment has little influence on children's developing intellects and personalities. Most behavioral geneticists would place the heritability of intelligence at between .5 and .6 and the heritability of personality somewhat lower (J. R. Harris, 1998; Plomin & Daniels, 1987; Plomin et al., 1997; Plomin & Petrill, 1997). This means that 50% or more of the variance for important psychological characteristics are attributable, presumably, to postnatal environmental factors. However, the bulk of environmental effects on IQ and personality remain unspecified (so-called *nonshared environmental effects*; see discussion below). A smaller proportion of the differences in IQ and personality is attributed to *shared environment effects*, a result of siblings growing up in the same family. Shared environment, the family environment that siblings share, independent of shared genes, is also discussed below.

The modest effects of shared environment on IQ are illustrated in research by Segal (2000), who developed the virtual-twin method to assess the joint contribution of genes and shared environment on IQ. *Virtual twins* were defined as unrelated children of about the same age who grew up in the same family from early infancy. For instance, two adopted siblings, or a biological child and an adopted child, fewer than 9 months apart in age would be virtual twins. In such cases, they share the same family environment, similar to that shared by monozygotic or dizygotic twins, but are genetically unrelated. Segal reported a correlation between the IQs of the virtual twins of .26—significant, but substantially less than that found between monozygotic twins (.86), dizygotic twins (.60), or full siblings (.50). This finding is consistent with the argument that shared environments have only small effects on intellectual development.

How does a behavioral genetics approach differ from the developmental systems approach? Although both take an explicitly evolutionary perspective, they have a different view of what underlies development. As advocates of a developmental systems approach have emphasized (Gottlieb, 1991a,

1998; Lickliter & Berry, 1990; Oyama, 2000a, 2000b), there is no straightforward relation between a gene and a behavior. Development proceeds via the bidirectional relation between all levels of organization, including the genetic. Moreover, the effects of genes and environment on any particular phenotypic characteristic cannot literally be partitioned, as is done via statistical techniques by behavioral geneticists. As we have stressed throughout this book, genes are part of a developmental system, with continuous and bidirectional interaction between adjacent levels within the system, ranging from the DNA through the culture.

Behavioral geneticists, in contrast, look for genetic links to behavior and have actually found genes associated with IQ, and thus, to certain aspects, associated with intelligence (see Chorney et al., 1998; McGuffin et al., 2001; Plomin & Rutter, 1998). Gottlieb (1995a) has interpreted the claims of behavioral genetics as being unidirectional, genes → behavior, rather than bidirectional, as postulated by the developmental systems approach. Of course, behavioral geneticists understand that the connection between genes and behavior is multilevel and complex; they believe that Gottlieb has misinterpreted contemporary behavioral genetics theory and exaggerated the uncertain course of development, and they balk at the label of "genetic determinists" (Burgess & Molenaar, 1995; Scarr, 1995a; Turkheimer, Goldsmith, & Gottesman, 1995). Nonetheless, we believe that Gottlieb is correct in stating that behavioral genetics research typically simplifies the notion of "the environment," in part by not explicitly acknowledging the role that self-produced function may play in early development and by failing to objectively define specific aspects of children's environments that contribute to the characteristics under study. For example, in most behavioral genetics models, nonshared environmental effects account for the greatest amount of nongenetic variance (Harris, 1998; McCartney et al., 1990; Plomin & Daniels, 1987; Scarr & Weinberg, 1976; Turkheimer & Waldron, 2000). *Nonshared environments* refers to siblings' unique experiences, those not experienced by the whole family. Such nonshared environmental effects are usually not objectively measured, but when they are, they account for only a small portion of the nongenetic variance (Turkheimer & Waldron, 2000). One interpretation of this finding is that there are other (nonshared) environmental factors that are not being measured in current behavioral genetics research.[3]

---

[3] In addition to proposals about the source of nonshared environmental effects, some theorists have proposed that shared family environment effects are larger than have been assumed. For example, Stoolmiller (1999) proposed that statistical anomalies, associated with restricted range of family environmental qualities in adoption studies, greatly underestimates the degree of shared family environment, and when this is taken into consideration, the amount of variance accounted for by shared family environment increases substantially.

As we mentioned earlier, although the developmental systems approach predicts substantial plasticity in development, most individuals develop according to a species-typical pattern because they develop in a species-typical environment. We believe that the predictive power of behavioral genetics arises from the same phenomenon. Genes associated with certain levels or types of behavior are inherited, but they will be expressed in their "expected" form only when children experience species-typical environments. Environments, including prenatal ones, that deviate from the species-typical norm will result in genes that do not get expressed "as expected." This is consistent with the proposal of Greenough and his colleagues (Greenough, Black, & Wallace, 1987; see also M. H. Johnson, 1998) of experience-expectant processes. Neurons in the developing brain are biased (in part by genetics but also in part by the influences of their own growth and the activity of surrounding neurons) to "expect" certain experiences (e.g., vertical lines for some cells in the visual cortex). These cells will thus be organized as expected given species-normal experience. However, should such experiences not be forthcoming, these cells can be organized differently or can atrophy and die. This perspective recognizes the bidirectionality of structure and function in development, even at the genetic level, and that genes will be expressed differently in different environments, but it also recognizes that most environments that have been assessed by contemporary behavioral geneticists provide "good-enough" environments for the expression of "expected" personality and cognitive characteristics. In fact, species-typical environments, so critical to the normal expression of genes, are also inherited (Lickliter & Berry, 1990; Oyama, 2000a). For this reason it is difficult to differentiate genetic from nongenetic inheritance. What are inherited are developmental systems, and variations in any part of the system, including genes (in the form of mutations) and environment (such as exposure to species-atypical sensory experience) can alter a phenotype.

Correlations versus Means

Although adherents to both behavioral genetics and the developmental systems approach are concerned with phenotypic variation in their study participants, they have tended to emphasize different aspects of this variation. Behavioral geneticists have focused primarily on the relations of traits among people of different degrees of relatedness, such as the correlations of IQs of adopted children with their biological and adopted parents. In contrast, researchers taking an explicitly epigenetic perspective have been more concerned with mean values of a trait, such as the average level of adopted children's IQs relative to those of their biological and adopted parents (see Gottlieb, 1995b). It is possible, using the same data set, to come to different conclusions concerning the genetic influence on a trait depending on whether one focuses on correlations or means. This is illustrated by the results of the transracial adoption study of Scarr and Weinberg (1976; Weinberg, Scarr, & Waldman, 1992). Black children born primarily of parents from lower-income homes were adopted by White, primarily upper-middle-class parents. The average IQ of the adopted children who were placed in middle-income homes as infants was found to be 110, 20 points higher than the average IQ of comparable children being reared in the local Black community and similar to the estimated IQs of their adopted parents. This effect is consistent with the position that genes associated with IQ are expressed differently in different environments, yielding substantially different phenotypes. Yet, the correlation between the children's IQs and their biological mothers' educational level (IQ scores were not available for the biological mothers) was significantly higher (.43) than a similar correlation with the children's adoptive parents' educational level (.29), indicating a substantial genetic influence (i.e., genotype → behavior). Skodak and Skeels (1945) had reported similar findings 30 years earlier.

The seeming discrepancy in interpretation is related to the homogeneity of environments in which children in this study were reared. The adopted children had all been placed in intellectually stimulating homes, and such stimulation from adoptive parents was responsible for their relatively high IQs in comparison with those of their biological mothers. Moreover, assuming that the IQ-influencing experiences of the home environments were similar for the adopted children, individual differences among the children would be best predicted by genetics. When environmental conditions are relatively homogeneous, as they presumably were for the adopted children in the Scarr and Weinberg study, the best predictor of children's rank order in intelligence is the rank order of the biological parents' IQs (i.e., genetics). When environments are heterogeneous, differences in environments will play a more substantial role in individual differences in intelligence. Such an argument merely acknowledges that genes and rearing environments each contribute in nonspecified ways to any phenotype and that patterns of genetic relatedness will vary as a function of the statistical variability in these factors.

Differences in environments do not have to be extreme, however, to alter patterns of genetic relations. For instance, several researchers have argued that heritability should increase with improved educational opportunities (Bronfenbrenner & Ceci, 1994; Scarr, 1992). According to this position, less-than-optimal environments for educability may have a particularly strong influence on the development of intelligence, whereas intellectually supportive environments will have little impact beyond that contributed by genetics. Support for this contention comes from a study by behavioral geneticist David Rowe and his colleagues (Rowe, Jacobson, & van der Oord,

1999). They examined the heritability of verbal IQ scores from a sample of 3,139 sibling pairs (all adolescents), including sets of monozygotic and dizygotic twins. Consistent with previous research, they reported an overall heritability of IQ of .57 and an effect of shared environment of only .13. But these patterns varied when education level of the parents was considered. For adolescents whose parents had a high school education or less, the heritability of verbal IQ was reduced to .26 and the effect of shared environment increased to .23. In other words, both genetics and shared environment accounted for about one-fourth of the variability of differences in verbal IQ for the low-education group. In contrast, for adolescents whose parents had more than a high school education, the overall heritability increased to .74 and the shared-environment effect decreased to 0. These differences are substantial and reflect the role that differences in "ordinary" environments can have in the expression of genes that influence verbal intelligence. We find these results consistent with both the developmental systems approach and that of behavioral genetics. Genetic similarity predicts a substantial portion of individual differences in many important psychological characteristics, but we believe careful examination will prove that differences in environments also interact with these genetic dispositions, as Rowe and his colleagues have demonstrated for parents' education level on verbal IQ.

The study by Rowe and his colleagues demonstrates that differences in parental education can result in shifts in the amount of variance in children's IQ scores accounted for by shared environment and genetics. However, shared environment in Rowe's model, as well as in those of other behavioral geneticists, refers only to shared postnatal environments. Siblings also share a prenatal environment, with twins sharing the same womb at the same time.

In a recent meta-analysis of IQ studies, Devlin, Daniels, and Roeder (1997) demonstrated that a model that included shared maternal (i.e., prenatal) environment as a separate factor explained the variance in IQ performance significantly better than traditional models that excluded this factor. In this model, 20% of IQ differences among twins was accounted for by shared maternal environment, whereas shared maternal environment accounted for only 5% of IQ differences among nontwin siblings. This pattern makes sense, given that twins share the womb concurrently, whereas siblings share it consecutively. Although a woman may maintain a common physiology and personal habits (e.g., in terms of diet) from one pregnancy to another, the temporal separation between offspring and the presumed differences over time in the physiological status of the mother contribute to a lessening of the similarities of IQ among nontwin siblings. Coupled with the increase in maternal effects, the amount of IQ variance attributed to genetics (i.e., the heritability of IQ) was lower than in most other studies, accounting for only 48% of the variance (including both additive and nonadditive genetic effects; it was only 34% when only additive genetic effects were considered).

It is easy to imagine the myriad factors that may contribute to the maternal effects Devlin and colleagues reported. The brain grows rapidly during the prenatal period, and although there is substantial cognitive plasticity in postnatal intelligence, it should not be surprising that patterns of neural organization can be affected prenatally by factors impinging on the fetus. Individual differences in childhood intelligence have been reported as a function of malnutrition (Lukas & Campbell, 2000; Stein, Susser, Saenger, & Marolla, 1975), exposure to drugs and other potential teratogens (Reinisch, Sanders, Mortensen, & Rubin, 1995), alcohol and cigarette consumption (Olds, Henderson, & Tatelbaum, 1994; Streissguth, Barr, Sampson, Darby, & Martin, 1989), and lead poisoning (Baghurst et al., 1992), among others. Other research has shown that the degree of bodily symmetry (i.e., the degree to which the right and left sides of the body are physically similar) correlates positively with adult IQ (Furlow, Armijo-Prewitt, Gangstead, & Thornhill, 1997); bodily asymmetry is believed to be caused principally by prenatal stressors. Thus, more symmetrical (and higher-IQ) people likely experienced less stress in the womb than less symmetrical (and lower-IQ) people, providing another source of nongenetic, environmental variation in IQ.

These findings suggest that when behavioral geneticists broaden their definition of environment, as with the inclusion of a very general measure of prenatal environment in the study by Devlin and colleagues, estimates of heritability decline as the contribution of experiential factors increases. This pattern is predicted by the developmental systems approach, with its emphasis on the continuing interaction of structure and function at all levels of organization, beginning not with birth but with conception. Such effects are also being recognized by contemporary behavioral geneticists. For example, in reference to the often unknown source of nonshared environmental effects, Molenaar, Boomsma, and Dolan (1993) suggested these influences may not be attributed to the environment, as conventionally defined, but rather "result from intrinsic variability in the output of deterministic, self-organizing developmental processes" (p. 523), a statement that clearly invokes an epigenetic analysis (see also Turkheimer, 2000).

Evolutionary Developmental Psychology and Individual Differences

Predicting individual differences in complex psychological characteristics following our proposed perspective requires assessing aspects of children's environment throughout development, recognizing that, although early experiences are important in establishing and maintaining individual differences, so are later experiences (Bjorklund, 2000; Cairns, 1979). Some genes that influence behavior are surely expressed early and, in all but extreme environments, will be similarly expressed throughout life. Other genes' expression, we argue, will be more susceptible to environmental variations and will be expressed differently over the course of development, depending on the experiences of the individual. Genetic effects will be strong when the children who are assessed inhabit relatively homogenous environments, and stability of traits will be found when environments are stable over time. That is, not only will genes get expressed as "expected" when the individual experiences a species-typical environment but, for many characteristics, genes will be expressed differently at different points in development depending on the stability of environments. The trick, of course, is to identify the characteristics that are most (and least) susceptible to modification as a result of variation in experience, as well as the environmental conditions that are most apt to produce change at different times in development.

Genetic differences among individuals may also help explain the relative effectiveness of alternative strategies in response to varied environments. For example, several researchers have noted that children from father-absent and high-stressed homes reach puberty earlier and become sexually active sooner than children from father-present and low-stressed homes (Belsky et al., 1991; Graber, Brooks-Gunn, & Warren, 1995; Kim, Smith, & Palermiti, 1997). These data have been interpreted as reflecting how differences in rearing environments can influence the adoption of different mating and parental-investment strategies (see further discussion in chapter 9). However, there is also evidence that children who develop early have inherited this characteristic, and it is not the stressful environment, per se, that is responsible for the early maturation and mating strategies but rather their inherited genes (Moffitt, Caspi, Belsky, & Silva, 1992). A third possibility is that a rapid maturation rate may have been selected in individuals who experience stressful rearing environments, so that, over many generations, rearing environments, maturation rates, and modal mating strategies have become confounded. Environments in developed countries today can be quite variable, with children growing up to have different occupations and living in different parts of the country (or world) from their parents. But social stability has been the norm in most of the world over historic time and likely during prehistoric time. Under such circumstances, it is quite reasonable, we argue, that children's rate of maturation and adoption of mating strategies have been sensitive to differences in rearing environments and to genetics. Such a hypothesis should be testable using contemporary and historical data.

We believe that individual differences cannot be ignored in an evolutionary approach to development and that aspects of modern behavioral genetics must be incorporated into such an account. In doing so, we maintain our position that development is best understood by the expression of epigenetic rules unfolding in interaction with the environment over the course of development and that genetic and environmental effects cannot literally be partitioned. All development is the product of the continuous and bidirectional interaction between structure and function at all levels of organization, making it impossible to specify "genetic" versus "environmental" effects. However, we believe that the expression of many genes that influence individual differences is robust to the perturbations of a wide range of "ordinary" environments, accounting for the impressive predictions of behavioral genetics theory for many psychological traits. We view the statistical techniques of behavioral genetics as convenient heuristics that reveal differences in the extent to which certain characteristics are modifiable by postnatal environments. Keeping in mind both the bidirectionality of structure and function at all levels of organization and the robustness of some traits to environmental fluctuations should help integrate evolutionary psychology and behavioral genetics.

CONFLICTING TRADITIONS, A COMMON GOAL

Evolutionary developmental psychology inherits a long legacy. The perspective we offer here differs from earlier developmental conceptions in that we have adopted many of the ideas of "innate" (very broadly defined) information-processing mechanisms from mainstream evolutionary psychology, a discipline that has focused almost exclusively on human adult functioning. But we have also adopted the findings and theories of mainstream developmental psychology, including comparative developmental psychology; our aim is to integrate these ideas with those of evolutionary psychology. The task may be more difficult than it appears on the surface, for there are some fundamental disagreements between "developmentalists" and "evolutionary psychologists" on some key issues. We are developmentalists first; but for us, evolution is a form of development, and the developmental systems perspective we advocate views ontogeny and phylogeny as intricately related and inseparable. Our ancestors evolved, but during their lifetimes they, too, developed. Rather than viewing ontogeny and phylogeny as processes that need to be untangled, a better solution may be to understand their integration. We make such an attempt in the following chapters.

4

THE BENEFITS OF YOUTH

The primary focus of this book is how an evolutionary perspective can foster a better understanding of human development. However, our phylogenetic predecessors also developed, and characteristics of their ontogeny played an important role in how they, as a species, evolved. Our primary concern in this chapter is to show how ontogenetic processes may have influenced the course of human evolution. We focus on aspects of developmental timing and the role that differences in such timing may have had on the physical, cognitive, and social structure of *Homo sapiens*. In particular, we examine the role that an extended juvenile period may have had on human phylogeny.

We present evidence consistent with the position that our species' prolonged juvenile period was one of several interacting factors (the other major factors being increased brain size and increased social complexity) that resulted in the modern human mind. We also argue that *Homo sapiens'* slow growth, particularly of the brain, affords greater plasticity and permits modern humans to adapt to a wide range of circumstances and to the ability to overcome the effects of early deprivation and other maladaptive environments.

DEVELOPMENTAL TIMING

In chapter 2, we provided a brief sketch of human evolution, from *Ardipithecus ramidus* and the various species of *Australopithecines*, through *Homo habilis*, *Homo erectus*, and the rest of the hominid family to modern humans. According to fossil and DNA evidence, we last shared a common ancestor with modern apes between 5 and 7 million years ago. Judging from fossil evidence, our common ancestor was likely similar in physical and social structure to modern chimpanzees. By what mechanisms did our genetic line evolve bipedality and, later, a big brain, among the other physical differences between *Homo sapiens* and *Pan troglodytes*? The mechanisms involved in such changes are genetic, of course. But only a small percentage of DNA actually "builds" things. Most DNA in the genome is inactive throughout the life of an organism. Some genes, however, regulate development by turning other genes on and off and, in this way, play a vital role both in ontogeny and in phylogeny.

In chapter 3, we introduced the concept of *heterochrony*, genetic-based differences in developmental timing (de Beer, 1958; S. J. Gould, 1977; McKinney, 1998; Shea, 1989, 2000). de Beer (1958) originally proposed eight types of heterochrony, and more recently Shea (1989) and McKinney and McNamara (1991) suggested that this number be reduced to six. For our purposes, we can collapse the various forms of heterochrony into two general types: those in which development is in some way retarded relative to the development of one's ancestors (sometimes referred to as *paedomorphosis*, *neoteny*, or *underdevelopment*) and those in which development is accelerated relative to one's ancestors (sometimes referred to as *peramorphosis*, *hypermorphosis*, or *overdevelopment*). We generally use the terms *retardation* and *acceleration* to refer to these two general forms of heterochrony.[1]

Humans as a Neotenous Species

As discussed in chapter 3, for most of this century, evolutionary biologists and anthropologists have argued that human evolution is especially characterized by a retardation of development, particularly by the form of retardation termed *neoteny* (Bolk, 1926; de Beer, 1958; Garstang, 1922; S. J. Gould, 1977; Montagu, 1989; Schwartz, 1999). Neoteny (literally "holding youth") refers to the retardation of development or, more specifically, to the retention into adulthood of ancestral embryonic or youthful stages.

How can humans be described as a "retarded" or neotenous species? Most claims made by advocates of "humans as a neotenous species" were based on the general juvenile characteristics of adult humans in comparison to adult apes. For example, humans retain into adulthood many embryonic or infantile characteristics of our distant ancestors (Bolk, 1926; S. J. Gould, 1977; Montagu, 1989). These include the shape of the head and face, late eruption of teeth, the size and orientation of the pelvis, a delicate (or gracile) skeleton, and a nonopposable big toe, among others. Using the chimpanzee (*Pan troglodytes*) as a model for what our distant ancestors may have been like (see Wrangham & Pilbeam, in press), we find many features in adult humans that resemble those of infant chimps. For example, at birth, both

---

[1] Following McKinney and McNamara (1991), the three types of *paedomorphosis*, or retardation, are *progenesis*, or earlier onset of some aspect of development; *neoteny*, or reduced rate of development; and *post-displacement*, or delayed onset of development. The three forms of *peramorphosis*, or acceleration, are *hypermorphosis*, or delayed offset of development; *acceleration*, or increased rate of development; and *pre-displacement*, or earlier onset of growth.

humans and apes (and many other species) have rounded heads that are large relative to their bodies with adult-sized eyes, round cheeks, a flat nose, short arms, and relatively little hair. Humans maintain into adulthood, to varying degrees, many of these characteristics more so than other primates.

Looked at another way, humans, compared to other primates, show little developmental change in form relative to the fetal stage. For example, the shape of the human skull goes through relatively minor changes in shape over infancy and childhood, whereas such changes are substantial for chimpanzees (see S. J. Gould, 1977). Great apes and humans look startlingly similar in infancy compared to their appearance at adulthood. These infantile characteristics are endearing to adults; those helpless babies who keep adults awake all night are so "cute," it is hard not to love them—a very adaptive characteristic, indeed. The Nobel prize-winning ethologist Konrad Lorenz proposed that caretaking behaviors in many animals are triggered by infants' immature features. These features are found not only in infants, but also in loveable cartoon characters (Mickey Mouse), dolls (Cabbage Patch Kids), and make-believe movie creatures (E.T.). Science fiction writers and UFO aficionados seem to be aware of this evolutionary trend because they typically describe futuristic humans and visiting space aliens as short, hairless creatures with large heads and big eyes. The implicit assumption here is that if human evolution continued (or if evolution occurred elsewhere in the universe to produce a more intellectually advanced creature), the result would be an even more infantile-looking adult.

Another neotenous characteristic of humans is the orientation of the vagina (see Montagu, 1989). In chimpanzee and human fetuses, the vagina slopes toward the front of the body. As chimpanzees grow, the vagina slopes more toward the back and, as a result, copulation is done most easily (and frequently) from the rear. In humans, the vagina retains its forward-sloping position, so that copulation can be done face to face. Such face-to-face copulation may have facilitated the formation of pair bonds, with both the male and female working together to help raise their offspring. This is particularly important because human children are dependent on their parents for so many years, making the establishment of stable "families" critical (see Crook, 1980; S. J. Gould, 1977; Montagu, 1989).

Groves (1989) provided evidence that neoteny may have played a role in human evolution in his careful analysis of the characteristics of hominid skeletons from *Australopithecines* through *Homo sapiens*. In comparison to their presumed ancestors, *Australopithecines* displayed relatively few juvenile traits. In contrast, Groves noted that *Homo habilis*, possibly the oldest species in the *Homo* line, possessed a number of juvenile features that differentiated it from its presumed ancestors, including: reduction in size of the molars and canines, smaller noses, more gracile cranial form, reduced facial height, and smaller jaws, among others. Many of these and other features went through further juvenilization again in *Homo erectus* and *Homo sapiens*. Further changes were observed between more closely related species, with the extinct *Homo neanderthalensis* displaying fewer juvenile features than fully modern humans (*Homo sapiens*).

The viewpoint from theorists earlier in this century that differences in developmental timing (particularly neoteny) played an important role in phylogeny, was, in part, in reaction to the long-held dogma of evolutionary theory that new species arise by the addition of features to ancestral adults. This was the central tenet of *the biogenetic law* as postulated by Ernst Haeckel in the latter part of the 19th century (see discussion of biogenetic law in chapter 3). To recapitulate, the theory held that the development of the individual goes through the same sequences as the evolutionary development of the species, with evolutionary additions being added to the adult stages of an organism. Neoteny is an obvious violation of the biogenetic law.

Does Neoteny Really Describe Human Evolution?

The idea of humans as a neotenous species has been a favored interpretation by developmental psychologists with an evolutionary perspective (Bjorklund, 1997b; Cairns, 1976; Mason, 1968a). They viewed important aspects of human morphology and behavior as the result of retarded development and saw *Homo sapiens*, in general, as a neotenous species. *Homo sapiens* retained their juvenile appearance and certain aspects of juvenile behavior (e.g., play) into adulthood, and required a much longer time than their primate cousins to reach reproductive maturity.

More recently, however, the general neotenous nature of humankind has been seriously questioned (Gibson, 1991; McKinney, 1998, 2000; McKinney & McNamara, 1991; Parker, 1996; Parker & Gibson, 1979; Parker & McKinney, 1999; Shea, 1989, 2000). For example, many aspects attributed to neoteny are due not to true neoteny, in which ancestral patterns are retarded so that the descendants do not develop as far as their ancestors did, but merely to an extension of the developmental period (Rice, 1997). This is a form of retardation, but it is not true neoteny. More critically, some aspects that have been classified as neotenous have been proposed to be due to other processes. For instance, the facial similarities of skulls of adult humans and infant apes have been proposed to be due to specific remodeling of the human face (see Shea, 1989), or even as a structural consequence of a large brain (Deacon, 1997), and not to an actual retardation of development.

In fact, some researchers have argued that acceleration is the predominant characteristic of human evolution (McKinney, 1998, 2000). For example, human brain and cognitive development clearly exceed that of other primates. This most characteristic of human qualities cannot be seen as anything other than acceleration of development beyond that of our ancestors. Other accelerated features of human development include the early fusion of bones in the wrist and early descent of the testes (see Shea, 1989).

Although acknowledging that some aspects of human evolution may have been brought about via neoteny, anthropologist Brian Shea (1989), in perhaps the most complete evaluation of neoteny in human development, stated that "a hypothesis of general and pervasive human neoteny is clearly no longer viable. A careful analysis of human development, morphology, and life history patterns reveals little concordance with predictions of neoteny based on accepted criteria" (p. 97). However, Shea (1989) did concur with earlier neoteny theorists (S. J. Gould, 1977) about the role of neoteny in influencing the relative size of the human brain and skull (see discussion below). However, the adult human brain itself shows no immature, or neotenic, features, but rather has more synaptic connections and fissures (as opposed to the smooth surfaces characteristic of fetal brains) of any mammal (see Gibson, 1991; Preuss, 2001).

We must concur with the critics of neoteny; humans' evolution cannot be characterized as one of general retardation. Rather, human evolution reflects a mosaic of changes in ontogenetic trends, some notable for their accelerated character and others for their retarded character (see Shea, 2000). In fact, we see aspects of retarded and accelerated development as often being different sides of the same coin. For example, we argue that our extended period of immaturity, coupled with the retention of fetal rate of brain growth in postnatal life—both examples of retardation—were necessary components for the expansion of the brain and cognition beyond that of our ancient ancestors—examples of acceleration.

EXTENDED GROWTH OF THE HUMAN BRAIN

The human species' most outstanding trait is intelligence. Here, we have clearly developed beyond our evolutionary ancestors; our development is not retarded to some earlier embryonic or infantile state (Byrne, 1995; Parker & Gibson, 1979). Compared with mammals as a group, human brains are far larger than expected for their body size (Jerison, 1973; Rilling & Insel, 1999). This is a trend seen among primates in general, but it is particularly exaggerated in humans. Brain size must be evaluated relative to body size, simply because large bodies require large brains to regulate them. Jerison (1973) developed the *encephalization quotient* to evaluate the expected brain weight/body weight ratio for animals within a family.[2] For

---

[2]Computing the relation between brain and body weight is more complicated than it may appear on the surface. When a simple ratio is computed between brain and body mass, small mammals, such as mice and rats, actually have larger ratios than larger mammals, such as humans. This is because brain size does not vary linearly relative to body size. Since the late 19th century, it has been known

example, given the typical pattern of changes in brain and body weight in mammals, brain weight should increase at a certain rate relative to increases in body weight. When a particular species' brain is larger than expected for its body weight, the encephalization quotient will be greater than 1.0. When a species' brain weight is less than expected for its body weight, the ratio will be less than 1.0. Using this technique, modern chimpanzees have an encephalization quotient of about 2.3. The encephalization quotient for modern humans, however, is more than triple this, about 7.6 (Jerison, 1973; Rilling & Insel, 1999).

Another way of viewing the brain–body weight ratio is to plot the relation between the two factors and compute a regression line that reflects what size brain an animal of a certain weight should have. (This line is essentially equivalent to an encephalization quotient of 1.0.) Species above this line have "more brain" than expected. Figure 4.1 graphs the brain–body weight relation for a selection of species, including humans. The distance a species is from the regression line can be interpreted as the residual variance in brain weight after removing body weight (Allman, 1999). Species above the regression line have "more brain" than expected for their body weight and, presumably, can devote more nerve cells to nonbodily functions. From this perspective, the residual above the regression line can be loosely thought of as the brain mass available for "intelligence" after subtracting the brain processes associated with maintaining basic bodily functions (Deacon, 1997). As can be seen, humans and porpoises have the highest positive residuals of any animal.

This pattern of enlarged brain relative to body size is also found in the fossil record for our hominid ancestors. Figure 4.2 presents average encephalization quotients for four species of hominids over the past 3.5 million years based on fossil skulls (of *Australopithecus afarensis*, *Homo habilis*, *Homo erectus*, and *Homo sapiens*) and for modern chimpanzees. As can be seen, the encephalization quotient for *Australopithecus afarensis* was only slightly greater than that of modern chimpanzees. From this point on, brain weight relative to body weight increased at a rapid rate.

---

that an animal's brain size varies as a function of its body size raised to the power of 2/3. The 2/3 exponent implies a surface-to-volume relationship, such that brain size is regularly related to a body's surface area and not to its actual size. The equation for computing the expected brain weight for vertebrates as a function of body weight is $Y = kX^a$, where Y and X are brain and body weights, respectively; k is a constant; and a is the exponent for body weight. When dealing with vertebrates, $a = 2/3$, although this value may vary for more specific groups of animals (e.g., only primates). Because the ranges of brain and body weights for different animals vary greatly, they are customarily transformed to a logarithmic scale so that $\log Y = a \log X + \log k$. This yields a linear equation of the relationship between $\log Y$ and $\log X$ with a slope of a. Log k is the intercept and has the value of $\log Y$ when $\log X = 0$. The constant k will vary as a function of which groups of animals are being examined. The encephalization quotient (EQ) is computed as a ratio of a species' actual brain weight to its expected brain weight, or EQ = actual brain weight/expected brain weight.

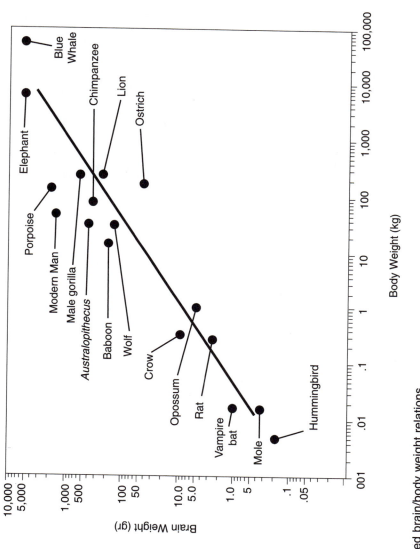

Figure 4.1. Selected brain/body weight relations.
Note. From *Evolution of the Brain and Intelligence* (p. 44), by H. J. Jerison, 1973, New York: Academic Press. Copyright 1973 by Academic Press. Adapted with permission.

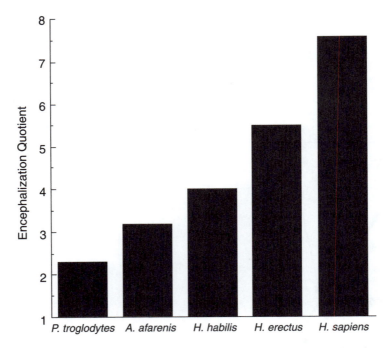

Figure 4.2. Encephalization quotients for chimpanzees (*Pan troglodytes*) and four hominid species.

Note. The data for chimpanzees are from *Evolution of the Brain and Intelligence* (p. 44), by H. J. Jerrison, 1973, New York: Academic Press. Copyright 1973 by Academic Press. Reprinted with permission. The data for hominids are from "The Brain of *Homo habilis*: A New Level of Organization in Cerebral Evolution," by P. V. Tobias, 1987, *Journal of Human Evolution, 16,* pp. 741–761. Copyright 1987 by Academic Press. Reprinted with permission.

Big Brains as a Product of Delayed Development

This enlarging of the brain was achieved, in large part, however, by maintaining the rapid rate of prenatal brain growth into postnatal life. The rate of *prenatal* brain development is remarkably similar for all primates, including humans (see Bonner, 1988). The brain develops rapidly in comparison to the overall size of the body. Brain growth slows quickly after birth for chimpanzees, macaque monkeys, and other primates, but much less so for humans. The pace of human brain development begun prenatally continues through the second year of postnatal life (see S. J. Gould, 1977). By 6 months the human brain weighs 50% of what it will in adulthood; at 2 years about 75%; at 5 years, 90%; and at 10 years, 95% (Tanner, 1978). In contrast, total body weight is about 20% of eventual adult weight at 2 years and only 50% at 10 years. This extended period of brain growth is afforded by a prolongation of the closure of the cranial sutures well into the third decade of life. Thus the brain, which grows rapidly before birth, continues its rapid development postnatally.

Increasing the time the brain grows and the number of neurons that are produced results in a larger brain (Finlay & Darlington, 1995; Finlay et al., 2001). The delay in brain maturation also results in the extension of dendritic and synaptic growth, so that the human brain has more interconnections among neurons than the brains of other primates (Gibson, 1991). However, different parts of the brain have not been equally affected in human evolution. The human neocortex has been estimated to be about 200% the size expected for an ape of comparable body size (Barton & Harvey, 2000; Deacon, 1997; Eccles, 1989; Rilling & Insel, 1999). The prefrontal cortex has been implicated in complex human cognition (Fuster, 1984; Luria, 1973); has connections to other areas of the brain, including the limbic system (Fuster, 1984); and has been hypothesized to be the locus of important inhibitory mechanisms (Harnishfeger, 1995; see Dempster & Brainerd, 1995). Most other areas of the human brain are also substantially larger than expected for an anthropoid ape, although none approach the size difference of the neocortex, and some areas actually are smaller (the olfactory cortex and some areas related to vision and motor control). Thus, although the extension of prenatal brain growth rates contributed to the overall size of the human brain (a form of retardation of development), it cannot account for the differential rate of change for different areas of the brain. Instead, different selection pressures (e.g., for language) must have played a role in shaping the size and structure of various areas of the *Homo sapiens* brain, presumably after (or during) the period when the brain was increasing in overall size (see further discussion in chapter 5).

The retention of embryonic growth rates for the brain into the first two years of postnatal life was necessitated in part by some physical limitations of human females. If a species is going to have a big brain in relation to its body, it will also, of course, have a big skull. But the skull that houses a 2-year-old human brain is far too large to pass through the birth canal of a human female. The evolutionary pressures that resulted in an enlarged brain required that gestation be extended only to the point where the infant skull will fit through the birth canal. If humans were as well developed bodily at birth as their simian cousins, their heads would never fit through the birth canal, which is limited in width because of the constraints of bipedality. The result is a physically immature infant, motorically and perceptually far behind other primate infants (see Antinucci, 1989; Gibson, 1991). However, human brain and cognitive development soon accelerate relative to their primate cousins (due, in large part, to the retention of fetal brain growth rate), all while their physical development remains delayed (Langer, 1998; McKinney, 1998).

Big Brains and "Intelligence"

Although absolute differences in brain size between species may be informative, it should not serve as a simple index of how intelligent an animal is. That is, we are not arguing, as others have (A. R. Jensen, 1980), that animals can be arranged on a linear scale in terms of intelligence (as reflected by brain size in the current discussion, or by a general intelligence factor, g). Over evolutionary time, the nervous systems of different species have evolved to solve recurring problems specific to their niche. Many species have evolved highly specific behavioral responses, or have particular sensory apparatus that is foreign to humans or higher primates (echolocation in bats; certain aspects of migratory behavior in some birds and insects). The ecological success of such animals argues for the position that each species is smart in its own way (Deacon, 1997; S. J. Gould, 1996). Furthermore, a basic tenet of evolutionary psychology is that humans possess a wide range of domain-specific abilities, which is at odds with the conception that "bigger is always better."

We concur with this central assumption of evolutionary psychology, but only to a certain degree. We find it unlikely that big brains (larger than expected for an animal's body size) contain only more domain-specific modules, although some of the neural additions of big brains are surely dedicated to domain-specific skills. We believe, rather, that large brains afford greater plasticity of learning and enhanced memory. These more general abilities can provide sufficient processing capacity in terms of greater working memory or faster speed of processing, for example, for the effective operation of more domain-specific mind/brain modules, such as those involved in the ability to detect cheaters or to perform theory-of-mind problems. Thus, we argue, species comparisons as a function of brain size can indicate plasticity of learning and memory, which are important components to the functioning of many animals, including humans, but less so for others. (The issue of the evolution of domain-general versus domain-specific cognitive abilities is discussed in chapter 5.)

Big brains, with enhanced learning and memory abilities, would not be adaptive for all animals. A strong relationship exists between brain size and longevity. Animals with bigger brains (and bigger bodies as well, of course) tend to live longer than animals with smaller brains. Long-lived animals are much more in need of bigger brains than short-lived animals. An animal that survives many years is likely to encounter a wide range of environments, some dangerous and others filled with valuable resources. It benefits such animals to have a nervous system that can readily learn and retain new information. This takes time, and, within mammals, brain size is also correlated with length of the juvenile period (discussed further below). In contrast, a short-lived animal will have too few experiences and live in too narrow a range of environments for it to benefit from the learning and memory advantages a large brain affords. Its inclusive fitness is better served by having more preprogrammed behavior patterns that depend less on "learning" (as conventionally understood; see Bogin, 1999; Deacon, 1997).

Brain size is also correlated (negatively) with litter size. Big-brained animals tend to have smaller litters and to give birth to infants at longer intervals than small-brained animals. This is the distinction of r versus K selection.[3] Small-brained species tend to produce many offspring rapidly but invest relatively little care in each one. Most insects are r-selection species, laying thousands of eggs but providing no post-hatching care. In contrast, big-brained species are more likely to have fewer offspring but to invest more care in each one (K selection). Mammals, as an order, are such animals, with primates in particular having mostly single births that are typically spaced several years apart. Following the argument we presented earlier, big-brained animals require a sustained juvenile period in order to make good use of the organ residing in their skulls, and parents must therefore invest substantially in the raising of their offspring so that such learning can occur in a relatively safe environment. And possibly, big brains are useful in choosing a mate who will also invest in the offspring. Humans are at the extreme of K selection among mammals, spacing their offspring about 4 years apart (the spacing is actually longer for orangutans) and providing both maternal and paternal support for their offspring well into the second decade of life.

THE SLOW RATE OF GROWING UP: CONSEQUENCE TO HUMAN EVOLUTION

Growing Up Slowly

One of the most important aspects of human development is our prolonged period of immaturity and dependency, and this has important implications for how we live as a species. Worldwide today, the average age that girls experience their first menstrual period (menarche) is between 12.5 and 13.5 years. However, the average age of menarche has been declining in Europe, for example, at least over the past 150 years, from between 16 to 16.5 years in 1860 to about 13.5 years in 1960 (Laslett, 1965). Moreover, for both boys and girls, there is typically a period of low fertility, thus extending the nonreproductive years even further (Bogin, 1999; Tanner,

---

[3] The terms r and K are derived from mathematical parameters used by population ecologists. In these models, r refers to the intrinsic rate of increase in a population, and K refers to the carrying capacity of the environment for any particular population (E. O. Wilson, 1975).

1978). Based on these historical data and data from traditional cultures (Hill & Hurtado, 1996; Kaplan et al., 2000), it is likely that our ancient ancestors were closer to 18–20 years of age before being fully reproductive.

The tendency toward retardation in *Homo sapiens* can be viewed as an extension of a trend observed in primates in general relative to other mammals. For example, primates live longer and mature more slowly than most other mammals of comparable body size. The gestation period is extended, with primates having substantially longer gestation periods for their birth weights than most other mammals. Similarly, puberty is typically attained in farm mammals when an animal has reached about 30% of its final adult weight. In contrast, humans and chimpanzees usually do not attain puberty until they are about 60% of their final adult weight (see S. J. Gould, 1977). Thus, humans' tendencies toward retardation, in at least some aspects of their development, seem an extension of a phenomenon apparent in primates rather than a phylogenetic innovation.

Humans' prolonged immaturity is all the more impressive when one considers the likely life expectancy of our hominid ancestors. Modern women can expect to survive childbirth, and both men and women, once they reach their teenage years, can anticipate another 60 years plus on this planet. This was not true, however, for our ancestors. Even today, the life expectancy of men and women in many developing countries barely reaches 40 years (and actually does not reach this modest level for men in some of the most impoverished nations). In the United States, life expectancy as little as 150 years ago was only 38 years for White men and 40 years for White women. High rates of infant mortality contribute significantly to these low values, so that the actual life expectancy for someone who survived past childhood in our not-too-distant past was likely beyond 50 years, similar to the age attained by some chimpanzees. Moreover, it seems likely that there have always been some "old" people in human groups, although they would have been the exception. The 3 score and 10 years promised in the Bible forecasts a rosy future, not the typical life expectancy an ancient human could realistically expect.

Thus, the 15- to 20-year wait that our hominid and hunter–gatherer forebears had before reaching reproductive maturity must have come at great expense. Many must have died of disease or fallen prey to predators before ever having a chance to pass their DNA on to the next generation and, given the risks of childbirth, many females must have died giving birth to their children, leaving their infants motherless and dependent on the kindness of strangers. Even in colonial America it has been estimated that 20% of the deaths of adult females were associated with childbirth, with one in 30 births resulting in the death of the mother. The rates were surely higher for our large-brain predecessors, dating back at least 1.5 million years ago to *Homo erectus* and possibly 2.5 million years ago to *Homo habilis*.

In hindsight, human's delayed maturation had substantial perils. Given the risks involved in our slow growth, the selective pressures for this delayed maturation must have been derived from strong compensatory advantages of the immature state, most notably increased flexibility of learning.

Human development is not only prolonged relative to other primates, but reflects stages of life history that may be unique to *Homo sapiens*. Anthropologist Barry Bogin (1997, 1999) argued that human development is comprised of five stages—infancy, childhood, juvenility, adolescence, and adulthood—two of which, childhood and adolescence, are not observed in any other species. In other mammals, infancy ends with the cessation of weaning and is followed by the juvenile period, in which the young animal is no longer dependent on its parents for survival but is not yet sexually mature (adulthood). In humans, weaning typically occurs in traditional cultures at about ages 3–4 years, but children are not able to fend for themselves until at least about age 7. (For example, they still posses "baby teeth," requiring special food preparation to receive adequate nutrition.) Human adolescence, with its characteristic growth spurt, usually begins early in the second decade and persists until reproductive maturity, which in humans is in the late teens or early 20s. (As we mentioned earlier, although girls typically have their first menstrual period at about age 13 years, there is a period of about 4.5 years of low fertility; a similar period of low fertility is found in boys.) No other species displays this rapid growth spurt before adulthood, although chimpanzees and bonobos also apparently have a post-menarche period of infertility (see Bogin, 1999).

Based on fossil evidence such as bone size and dental development, Bogin (1997, 1999) has estimated that the life stages of our australopithecine ancestors were similar to those of chimpanzees (*Pan troglodytes*): a period of infancy lasting 5 or 6 years, followed by a juvenile period, with adulthood beginning about age 12 years. According to Bogin, it is only with the beginning of the *Homo* line that a period of childhood is seen, and only in modern *Homo sapiens* is there evidence for a period of adolescence. Age of reproductive maturity apparently increased gradually in the *Homo* lineage, ranging from about 12–13 years for *Homo habilis* to 14–15 years for *Homo erectus*, to the late teens and early 20s for modern *Homo sapiens*. In addition to the emergence of childhood and adolescence, the length of juvenility and adulthood is longer in humans than in other primates and almost certainly longer than in our hominid ancestors.

Although many reasons have been proposed for the extension of the developmental period in humans and for the additions of childhood and adolescence (see Bogin, 1999), the fact of the extended prereproductive period in humans suggests that our ancestors were more successful at keeping their offspring alive than were other primates. Child mortality rates in hunter–gatherer societies are about 50% compared with 67% to nearly 90%

in other primates (Lancaster & Lancaster, 1983). Moreover, the lengthening of the developmental period, along with a threefold increase in brain volume since *A. afarensis*, suggests a substantial increase in the complexity of hominid social systems and in the ability to exploit biological and physical resources (e.g., through tool use). An extended developmental period, along with increased social play and exploratory behavior (see chapter 10), would enable the refinement of increasingly sophisticated physical, social, and cognitive competencies.

Although *Homo sapiens* have seemingly evolved many domain-specific "programs" for dealing with specific problems and with other members of their species, humans, more than any other species, depend on learning and behavioral flexibility for their success. The complexities of human societies are enormous and highly variable, and it takes an extended childhood to acquire all that must be learned to succeed. Because brain growth continues well into adolescence, neuronal connections are created and modified long after they have become fixed in other species (Jacobson, 1969). The result is a more "flexible" brain (in terms of what neural connections can be made), which means more flexible thinking and behavior. In addition, an extended youth provides the opportunity to practice complex adult roles which, because of their cultural variability and complexity, cannot be hardwired into the brain.

Archeologist Steven Mithen (1996) has recently speculated that the slow brain growth of ancient *Homo sapiens* was necessary to produce the cognitive architecture of modern humans. Consistent with contemporary assumptions of evolutionary psychology (Tooby & Cosmides, 1992), he proposed that the hominid brain was modular, with separate components for social, technical (tool use), and natural history intelligence. "Cognitive fluidity," which, Mithen claims, characterizes the modern mind, requires communication among these various modules (and general intelligence) and, he proposed, this requires an extended childhood to accomplish. To support his claim, Mithen pointed to evidence that brain development in Neanderthals, based on a discrepancy between the rate of dental and cranial development, was much faster than in modern humans (Akazawa, Muhesen, Dodo, Kondo, & Mizouguchi, 1995; Stringer, Dean, & Martin, 1990; Zollikofer, Ponce de León, Martin, & Stucki, 1995; but see Trinkaus & Tompkins, 1990 for an alternative interpretation). For example, based on dental development, a well-preserved skeleton of a Neanderthal infant was believed to be about 2 years old when it died. Yet the cranium size was equivalent to that of a modern 6-year-old child (Akazawa et al., 1995). Other fossil evidence supports the conclusion that Neanderthal brain development may have been completed substantially earlier than that of modern humans (Dean, Stringer, & Bromage, 1986; Zollikofer et al., 1995). On the basis of archeological evidence, Mithen proposed that Neanderthals demonstrated minimal cognitive fluidity, and it was only through a prolongation of child-hood that the architecture of the modern brain could develop.

Humans are not the only slow-developing and big-brained primate. As we noted earlier, chimpanzees and orangutans in particular have extended juvenile periods relative to monkeys, and they also have larger brains. In fact, in primates, the size of the adult brain is related to the length of the juvenile period; species with longer prereproductive periods, on average, have larger brains (Bonner, 1980; see Figure 4.3).

As we have noted, human development is prolonged relative to other primates when considering body size. Chimpanzees and humans have about the same adult body weight, but humans take approximately 5–7 more years to reach reproductive maturity than do chimpanzees. The picture is a bit different if, rather than body size, length of the juvenile period is predicted as a function of brain size. For mammals in general, brain size predicts rate of maturation, including maximum life expectancy (see Allman, 1999).

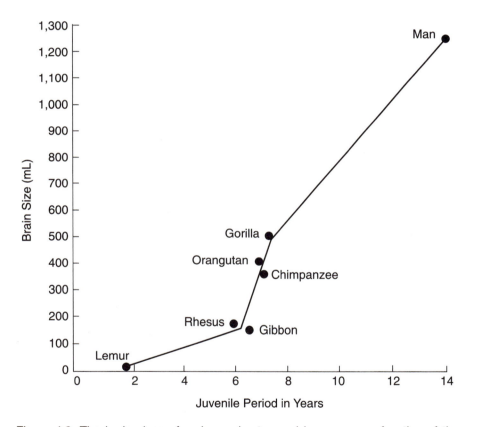

Figure 4.3. The brain sizes of various primates and humans as a function of the length of the juvenile period.
Note. From *The evolution of culture in animals* (p. 207), by J. T. Bonner, 1988, Princeton, N.J., Princeton University Press.

TABLE 4.1

Actual Maturation Stages and Predicted Maturational
Stages From Brain Weight in Humans

| | Age (Years) | |
| | Predicted From Brain Weight | Actual |
Stage		
First molar	19.3	6.4
Second molar	29.2	11.1
Wisdom teeth	37.8	20.5
Sexual maturity	44.5	16.6
Maximum life span	101.5	105.0

Note. From Evolving Brains (p. 196), by J. M. Allman, 1999, New York: Scientific American Library. Copyright 1999 by Henry Holt Co. Adapted with permission.

Among primates, brain weight predicts many aspects of physical maturation, for example, eruption of teeth and sexual maturity. Table 4.1 presents the actual and predicted age based on brain weight for several indices of maturation in humans. As can be seen, humans develop at a substantially faster rate than would be expected from brain size. (However, their maximum life expectancy is very similar to that predicted by their brain weight.)

Brain Size, Maturation Rate, and Sociality

Large brains in primates are related not only to length of the juvenile period, but also to sociality. For example, Dunbar (1992, 1995) has shown that, within primates, the relative size of the neocortex is significantly correlated with group size ($r = .76$).[4] Several theorists have argued that primates, including humans, have evolved large brains to deal with conspecifics. That is, dealing with the challenges of cooperating and competing with other members of the social group was the driving force of intellectual evolution within the primate line leading to humans (Alexander, 1989; Bjorklund & Harnishfeger, 1995; Byrne & Whiten, 1988; Geary & Flinn, 2001; Humphrey, 1976). Large brains are associated with extended juvenile periods, resulting in the argument that a prolonged childhood *and* a large brain are necessary for primates to master the complexities of their societies.

---

[4]Dunbar (1992) noted that the correlation between the size of the neocortex and home range size was also significant, suggesting the possibility that foraging demands, and not sociability, were the driving force behind increasing brain size in primates. However, when controlling for body size, the statistical relation between group size and neocortex size was virtually unchanged, whereas the relation between size of the neocortex and home range size became nonsignificant. This pattern suggests that it was sociability, rather than foraging strategies, that was the primary selective pressure for an increase in brain size in primates.

Joffe (1997) presented direct evidence for this by comparing aspects of brain size and structure with length of the prereproductive period and aspects of social complexity for 27 primates, including humans. Joffe reported that the proportion of the lifespan spent as a juvenile was positively correlated with group size and the relative size of the nonvisual neocortex. This is the part of the primate brain that is associated with complex problem solving, including memory. Joffe argued that social complexity exerted selection pressures for increased nonvisual neocortex in primates and an extension of the juvenile period.

We should be careful, however, in suggesting that big brains and an extended juvenile period were necessitated solely because of hominids' increasing social complexity. Other nonsocial factors also may have provided selection pressure for these evolutionary effects. For example, Kaplan et al. (2000) proposed that a shift to a higher quality diet was primarily responsible for humans' evolution of enhanced cognitive skills and an extended period of youth, which would be required to master the associated food-gathering skills. Kaplan and his colleagues noted that chimpanzees, for example, rely primarily on a diet of easily extracted fruit and plants with low nutrition density. Such foods, when available, can be obtained relatively easily even by juveniles. Although chimpanzees do hunt, they obtain only a small portion of their diet from vertebrate protein, which are foods of high nutrition density. Hunting, among both chimpanzees and humans, is engaged in mainly by adults (usually males) and takes considerable time to learn. Kaplan et al. (2000) examined food-gathering procedures in contemporary hunter–gatherer societies and noted that, similar to chimpanzees, young children often forage successfully for low-density, easily accessible foods, such as ripe fruit. In contrast, extracting foods of higher nutrition density, such as roots and tubers or vertebrate meat through hunting, are performed effectively only by older individuals and require many years to master.

Of course, determining the direction of causality among these various characteristics is not possible. In fact, we argue that changes in multiple factors surely acted synergistically, with changes in one factor (e.g., rate of maturation or relative brain size) both influencing and being influenced by changes in related factors (e.g., complexity of social conditions). Although one cannot discern any simple causal relation among these variables in the evolution of human intelligence, it is their confluence and the positive feedback among changes in these interacting variables that led to the evolution of *Homo sapiens*. Alternative courses of evolution were possible, of course, and *Homo sapiens* is not the inevitable product of a deterministic process involving increasingly large brains, social complexity, and delayed development. But these interacting factors were, seemingly, necessary for the evolution of *Homo sapiens* and, from a developmental perspective, delayed ontogeny appears to be the mechanism most susceptible to selection forces, and thus the linchpin in the evolutionary process that produced modern humans.

A related argument for the importance of delayed development to human evolution centers on the foundation of the human family and social structures (Allman, 1999; Crook, 1980; S. J. Gould, 1977; Hattori, 1998; Wesson, 1991). The human infant is totally dependent at birth and will remain dependent on adults for well over a decade. Pair-bonding and some division of labor (both within and between families) may have been necessary adaptations to the pressures of slow-growing offspring to increase the likelihood that children would survive to sexual maturity. The long period of dependency also meant that a male's genetic success could not be measured just by how many females he inseminated or by how many children he sired. His inclusive fitness would depend on how many of his offspring reached sexual maturity, assuring him of becoming a grandfather. To increase the odds of this happening, his help in the rearing of his children would be needed.

It has also been speculated that the juvenile features of human infants (babies are "cute") invoke positive feelings toward the infant in both males and females, and such feelings in men foster attachment to and paternal care for the infant. In fact, Hrdy (1986) suggested that the father–child bond may have been the basis of the father–mother–child bonds in human families. Noting that male primates are unique in the mammal world in the attention and care they give to infants, Hrdy suggested that perhaps this capacity preadapted "members of this order for the sort of close, long-term relationships between males and females that, under some ecological circumstances, leads to monogamy" (p. 152).

BIG BRAINS, SLOW DEVELOPMENT, AND PLASTICITY

Throughout this chapter, we have asserted that humans, more than any other species, need an extended childhood to survive and succeed. However, a childhood that extended into the fourth decade of life or beyond would do the developing human little good unless this prolonged juvenile period were accompanied by a big brain that could learn and retain the complexities of social life. As we noted earlier, within primates, there is indeed a relationship among brain size, social complexity, and length of the juvenile period (Joffe, 1997). More specifically, large brains and long juvenile periods may afford greater *plasticity*, or modifiability, of the brain. Species with larger brains (relative to body size) can make more of their early experiences, which in turn influence the maturation and size of the brain (Gottlieb, 1992).

The Plasticity of the Developing Brain

Human brains, in particular, display an extended period of growth. The human brain continues to gain weight into the third decade of life, and neurons in the associative areas of the brain are not fully myelinated until adulthood (Gibson, 1991; Yakovlev & Lecours, 1967). This slow growth provides humans with the flexibility to make many changes within their lifetimes.

Although brains grow in size, most of the growth appears to be due to increases in the size of neurons. (It has only recently been learned that new neurons continue to be produced past early infancy; E. Gould, Reeves, Graziano, & Gross, 1999.) Moreover, new connections among neurons (synapses) continue to develop throughout life. Brain (and thus behavioral and cognitive) plasticity is primarily due to the creation of new synapses (*synaptogenesis*). Although synapse formation must be under genetic influence to some degree, research over the past several decades has indicated quite clearly that experience is the primary factor in synaptogenesis and that this process is not limited to infancy, as was once believed, but continues even into old age (see Edelman, 1987; Gottlieb, 1992; Greenough, Black, & Wallace, 1987; M. H. Johnson, 1998).

Researchers have repeatedly demonstrated in laboratory animals the role of experience in the formation of synapses and the production of neurotransmitters that facilitate the conveying of messages between neurons. Animals raised in "enriched" versus "deprived" environments show different patterns of neural development, which is related to certain aspects of learning. Enriched environments usually include animals raised together in large cages that are filled with a variety of objects with which they can interact. Deprived environments often include animals raised in isolation or in smaller cages with few objects with which to interact (Hebb, 1949; Hymovitch, 1952; Turner & Greenough, 1985). A typical finding is that animals raised in enriched environments have heavier and thicker neocortexes, larger neurons with more dendrites, and more synaptic connections than animals raised in deprived environments. They also display enhanced learning abilities on a wide range of tasks. These effects are not limited to infant animals, but have also been reported for older animals (Greenough, McDonald, Parnisari, & Camel, 1986).

Results illustrating the plasticity of the animal brain as a result of experience led Quartz and Sejnowski (1997) to propose a model of *neural constructivism,* in which brain and cognitive development proceed via a dynamic interaction between the developing neural substrate and the environment. Experience changes the brain, which in turn affects what new information can be learned. The process is a constructive one, similar to the process proposed by Piaget to describe cognitive development. One interpretation that emanates from this perspective is that there are few areas in the brain that at birth are "implanted with knowledge," or what Elman and his colleagues (1996) referred to as *representation constraints* (see discussion in chapter 3). This does not preclude the very real possibility that there are processing constraints, what Elman and colleagues referred to as *architectural constraints*. Rather, experiences (many of which all normal members of a species would receive) play a critical role in shaping the brain.

Our point here is that the development of the mammalian brain is not exclusively under genetic control, a position consistent with the developmental systems approach discussed in chapters 2 and 3. The brain becomes organized from electrical and chemical activities of developing neurons and from information received through the senses as much as or more than by the unfolding of a genetic "blueprint." And *Homo sapiens*, with a large brain and an extended period of growth, should show greater plasticity over the course of its ontogeny than any species.

With respect to behavioral flexibility, Mason (1968a, 1968b) noted that the extended juvenile period in primates in general and in humans particularly, is accompanied by weaker and less persistence of primitive infantile responses (reflexive grasp, rooting, oral grasping) and a "loosening" of behavioral organization. As a result, Mason (1968a) stated,

> Developmental stages are less sharply delimited in humans than in other primates. Sensitive periods in development are more difficult to establish, there is less likelihood that the withholding of any specific experience will result in developmental arrest, and there is a much stronger tendency for behavior to reflect a blending or intermingling of different developmental stages, different response patterns, and different motivational systems. (p. 101)

Developmental scientist Robert Cairns (1976) made a similar argument, noting that the instability of individual differences and the high malleability of social behavior throughout infancy, characteristic of all social mammals, are extended in human children. From this perspective, an extended juvenile period provides not only more time to learn but, when accompanied by a reduced reliance on "hardwired" or "instinctive" behaviors, may in fact require a greater need for learning.

Although we cannot turn back the clock and observe directly heterochronic changes that produce variations in a species' morphology or behavior, some experimental evidence exists that demonstrates the effects of changes in developmental timing on social behavior across several generations. Cairns and his colleagues (Cairns, Gariepy, & Hood, 1990; Cairns, MacCombie, & Hood, 1983) noted that aggressive behavior in mice, measured by latency to attack, increased with age and experience. Cairns and colleagues selectively bred mice for latency to attack; one line was selected for low aggression and another line was selected for high aggression. Subsequent generations of low-aggression mice did not display the typical age-related increases in aggression observed in the foundational generation. Changes in the timing of development ("the progressively longer persistence of 'immature' features in the ontogeny of descendent generations"; Cairns et al., 1990, p. 59) affected the social behavior of individuals and, over several generations, altered the average value of this behavior in the genetic line. Results such as these make more plausible the hypothesis that variations in developmental timing (*heterochrony*) can influence the behavioral development of the species (Cairns et al., 1990; Gottlieb, 1987, 1992).

If plasticity is to be extended into later life, then neural circuits cannot be "fixed" early in life. Moreover, if immaturity of parts of the brain is responsible for this extended plasticity (e.g., incomplete myelination of neurons, resulting in slower and less-efficient neural transmission), then the young brain (and thus the young child) must be limited in general learning ability. Young children actually learn some things, such as language, faster than adults. The decreasing ability to acquire a second language with increasing age (much past 8 or 9 years) reflects a loss of plasticity for this skill (see chapter 6). A loss of plasticity is not always a detriment, however. As a result of genetic programming and experience, neurons become dedicated to certain functions. Such specialization, particularly when pertaining to some aspect of human life that does not change substantially over time and circumstances (e.g., which language one's social group speaks), can be greatly beneficial, and individuals are best served by a nervous system that commits neurons early in life to such basic functions. However, for a long-lived species such as humans who must contend with a large diversity of social circumstances, retaining some plasticity into adulthood is necessary (see Geary, 2001).

Despite some of the exceptional learning abilities of young children, for more general learning, the brains of infants and young children are inefficient. They process information more slowly than adults (Canfield, Smith, Brezsnyak, & Snow, 1997; Kail, 1997), which translates directly into less-efficient cognitive processing (Bjorklund & Harnishfeger, 1990; Case, 1992; Dempster, 1985). The slower processing of young children means that more of their processing is "effortful" in that it uses substantial portions of their limited mental resources (Hasher & Zacks, 1979). In contrast, more of older children's and adults' cognitive processing is automatic, in that it requires little or none of one's limited capacity. Thus, young children must exert greater effort to obtain the same results as older children. Despite the obvious disadvantages to such an inefficient system, it also has its benefits. According to Bjorklund and Green (1992),

Because little in the way of cognitive processing can be automatized early, presumably because of children's incomplete myelination, they are better prepared to adapt, cognitively, to later environments. If experiences early in life yielded automization, the child would lose the flexibility necessary for adult life. Processes automatized in response to the demands of early childhood may be useless and likely detrimental for coping with the very different cognitive demands faced by adults. Cognitive flexibility in the species is maintained by an immature nervous system that gradually permits the automization of more mental operations, increasing the likelihood that lessons learned as a young child will not interfere with the qualitatively different tasks required of the adult. (pp. 49–50)

Reversal of the Effects of Early Deprivation

Plasticity is often more easily seen in situations in which children who experienced deprivation early in life demonstrate subsequent reversibility of those effects. Although psychologists and educators earlier in this century deemed such reversibility unlikely, both human and animal work has clearly shown that reversibility is a reality (Clark & Hanisee, 1982; O'Connor et al., 2000; Skeels, 1966; Suomi & Harlow, 1972).

We start with an example from the animal literature because the notions of "critical period" and its intellectual offspring, early deprivation, are rooted there in the form of "imprinting." As we discussed in chapter 3, the most famous example of imprinting comes from the work of ethologist Konrad Lorenz (1937, 1965) with geese, in which early exposure during a "critical period" resulted in goslings' preference for specific partners in specific situations. Current evidence on imprinting, specifically, and critical periods and timing more generally, suggests that the generalizability of early effects depends on several factors (Bateson, 1981a). For example, imprinting effects are generalized only to very similar contexts. Individuals' willingness to learn different preferences beyond a critical period can be changed. In this regard, imprinting is treated as a dimension of learning and is influenced by experiences and developmental age. This view is consistent with our view that development can be represented as a transaction between individuals and their environments.

This ability to change after a critical period is further illustrated by recent studies that examined the recovery of intellectual function of children reared in stultifying orphanages in Romania who were later adopted (Kaler & Freedman, 1994; O'Connor et al., 2000; Rutter & the English and Romanian Adoptees Study Team, 1998). For example, O'Connor and his colleagues (2000) evaluated the psychometric intelligence of children at age 6 who had been reared in Romanian orphanages as a function of their age at adoption and immigration to the United Kingdom. These children were also compared with a group of UK children who were adopted by UK parents between the ages of birth and 8 months of age. Scores on the General Cognitive Index (GCI) of the McCarthy scale are presented for the various groups of children in Table 4.2. The GCI can be interpreted much as an IQ score, with a value of 100 representing the population average. Note that, despite the substantial developmental delay that all of the Romanian adoptees displayed on arrival in the United Kingdom, by age 6 years, each group had mean scores within the normal range. There was no difference in scores between the UK adoptees and the Romanian children adopted within their first 6 months of life. Scores were lower, however, for the Romanian children who had been adopted later; those who spent the most time in the institution had the lowest scores.

These results reflect the remarkable resiliency of children to the effects of early deprivation, but they also demonstrate that there are limits to intellectual plasticity. The more time children spent in the deprived environment, the less able their brains were to change, at least by age 6. (Of course, children who had spent more time in the orphanage had spent less time in their adoptive homes. Perhaps the negative effects of the early deprivation will be reversed when they spend more time in their adoptive homes.) Nonetheless, the overall impression of these findings is that, given proper stimulation, children can overcome the effects of an early negative environment. Young brains are not like tape recorders, recording everything for posterity. Rather, young brains are pliable. Were children born with more mature brains, or if development proceeded more rapidly, the mental, social, and emotional flexibility of young children would be severely compromised. This behavioral and cognitive flexibility is perhaps our species' greatest adaptive advantage, and it is afforded by the prolonged period of mental (and thus brain) inefficiency.

TABLE 4.2
Scores on the General Cognitive Index (GCI) at
Age 6 Years for UK and Romanian Children by
Age They Were Adopted

Nation	Age (Months)	GCI Score
United Kingdom	0–8	117
Romania	0–6	114
Romania	6–24	99
Romania	24–42	90

Note. From "The Effects of Global Severe Privation on Cognitive Competence: Extension and Longitudinal Follow-Up," by T. G. O'Connor, M. Rutter, C. Beckett, L. Keaveney, J. M. Kreppner, and the English and Romanian Adoptees Study Team, 2000, *Child Development, 71*, pp. 376–390. Copyright 2000 by Society for Research in Child Development. Adapted with permission.

Plasticity and Evolution

As we noted earlier, Gilbert Gottlieb (1992) has proposed that animals with larger brains relative to body size should show greater behavioral plasticity. When using exploratory behavior and general learning ability as indicators of behavioral plasticity, such a relationship is indeed found (see Gottlieb, 1992). Gottlieb has extended this argument to propose that big-brained and behaviorally plastic animals are able to adapt to environments more readily than animals with smaller brains and less behavioral plasticity, not only in ontogeny but also in phylogeny. That is, animals with larger brains should show a faster evolutionary pace than smaller-brained animals.

Evidence for this contention comes from a study by Wyles, Kunkel, and Wilson (1983), who compared the average relative brain size with the rate of anatomical changes from the fossil records for different groups of animals. Results from their study are shown in Table 4.3. To interpret the table, all one needs to know is that higher scores reflect greater relative brain size and faster rate of anatomical change, respectively. As can be seen, the relation between these two factors was almost perfect (correlation = .97). *Homo sapiens* had both the largest relative brain size and the fastest rate of anatomical change, followed in both categories by the hominoids (which include the lesser and great apes). Wyles and colleagues (1983) interpreted their findings as reflecting the importance of behavioral flexibility and innovation in evolutionary change:

> Behavioral innovation refers to the nongenetic (or genetic) origin of a new skill in a particular individual, leading it to exploit the environment in a new way . . . [the] nongenetic propagation of new skills and mobility in large populations will accelerate anatomical evolution by increasing the rate at which anatomical mutants of potentially high fitness are exposed to selection in new contexts. (p. 4396)

According to Gottlieb (1992), changes in developmental conditions activate heretofore inactive genes, which can result in behavioral and morphological modifications, which in turn can be influenced by natural selection. Although Gottlieb's model can be applied to all levels of animal life, these types of changes are most apt to be found in behaviorally flexible species, which have large brains and an extended juvenile period.

McKinney (1998) made a similar argument; he proposed that the extension of brain and cognitive development over ontogeny is an important mechanism for overcoming the limitations of morphological complexity. Although there is no single "progressive" tendency in evolution (i.e., everything gets more complex with time), there has been a trend toward greater complexity over evolutionary time (Bonner, 1988). For example, the maxi-

TABLE 4.3
Brain Size in Relation to Rate of Anatomical Evolution

Taxonomic Group	Relative Brain Size	Anatomical Rate
Homo	114	>10
Hominoids (apes)[a]	26	2.5
Songbirds	23	1.6
Other mammals	12	0.7
Other birds	4.3	0.7
Lizards	1.2	0.25
Frogs	0.9	0.23
Salamanders	0.8	0.26

[a]Including *Australopithecus* but excluding *Homo*.
Note. From "Birds, Behavior, and Anatomical Evolution" by J. S. Wyles, J. G. Kunkel, and A. C. Wilson, 1983, *Proceedings of the National Academy of Sciences USA, 80*, pp. 4394–4397. Copyright 1983 by J. G. Kunkel. Adapted with permission.

mum number of cell types in multicellular organisms has increased over geological time. However, there is an upper limit to morphological complexity, and evidence suggests that morphological change has plateaued. However, organisms that are limited in evolving greater morphological complexity to meet the demands of changing environments can respond by evolving larger and more efficient brains and the behavioral and cognitive flexibility that they afford (see also Parker & Gibson, 1979).

Some examples of how modified early environments can alter species-typical behavior that is particularly pertinent to human evolution come from observations of human-reared (*enculturated*) great apes. Great apes (mostly common chimpanzees) that have been raised by humans, much as human children, often display cognitive abilities that are more like those of children than those displayed by mother-reared animals (see Call & Tomasello, 1996). For example, the most successful of the "language-trained" chimpanzees have been enculturated (Gardner & Gardner, 1969; Savage-Rumbaugh et al., 1993). Similarly, mother-reared chimpanzees rarely imitate tool use, particularly *deferred imitation* (imitating a behavior following a significant delay). In contrast, enculturated common chimpanzees, bonobos, and orangutans have all been shown to display above-chance levels of deferred imitation of object manipulation (Bering, Bjorklund, & Ragan, 2000; Bjorklund, Bering, & Ragan, 2000; Bjorklund, Yunger, Bering, & Ragan, in press; Tomasello, Savage-Rumbaugh, & Kruger, 1993). Deferred imitation has traditionally been interpreted as requiring symbolic representation (Meltzoff, 1995; Piaget, 1962), and aspects of these apes' atypical, human-like rearing history apparently prompted the emergence of representational skills, at least in limited contexts, which are absent from their mother-reared conspecifics (see chapter 7).

It is not possible at this time to say what aspects of the apes' experiences are responsible for the change in their cognitive abilities and behavior toward more human-like thinking. One attractive candidate, however, has been joint attention strategies, whereby adults draw the attention of the young animal to a third object. An important aspect of this research is that it provides an experiential vehicle by which our hominid ancestors (using contemporary great apes as a model) could have begun to modify their cognition in the direction that resulted in *Homo sapiens*.

THE YOUNGEST SPECIES

Homo sapiens is a young species. Although our ancient predecessors failed to leave us a time capsule to verify all the facts and dates, the earliest of our kind appeared about 300,000 years ago, presumably evolving from *Homo ergaster* (or *Homo erectus*), who had been around for about 1.5 million years. Not until about 100,000 years ago, however, is evidence of fully modern anatomical humans found. Moreover, although human artifacts (tools) date back to our earliest *Homo* ancestors, solid evidence of art, which along with science and religion may be a quintessential human characteristic, is not found until about 35,000 thousand years ago. So, in the big picture of things, humans are infants in the biological world.

But humans are young in at least one other way: We are slow to grow up. This may well be, in part, a consequence of our large brain and complicated social structure. Our slow developing brains and extended childhoods give us the time and the neural plasticity to learn the complexities of human culture and to protect us from deleterious early environments.

Humans' slow growth gives us not only a juvenile body for extended years, but also juvenile behavior and thought patterns well into our teens. Juvenile features such as play, curiosity, and a love of novelty result in an exploring and fun-seeking character which, in many ways, is unique to humans, at least among adults of a species (see Bjorklund, 1997b). These youthful behavioral characteristics are the side effects of slow physical growth, play a vital role in psychological development, and may have played a critical role in human evolution.

5

CLASSIFYING COGNITION

Human cognition—our ability to think and to be aware of our thoughts—is, along with our ability to use language, perhaps the most outstanding characteristic of our species. A central tenet of evolutionary psychology is that our species' cognitive abilities are the products of natural selection. But proposing that the human mind has evolved does not mean that all contemporary forms of intelligence are adaptive and have experienced selection pressure over the millennia. For instance, reading and writing are new forms of communication, requiring new cognitive skills, that trace their roots to fewer than 10,000 years ago. Even today, a substantial portion of the Earth's population is illiterate. There has not been sufficient time for environmental pressures to select for reading ability (to say nothing of computer literacy). Other abilities associated with modern, schooled societies are similarly evolutionarily novel accomplishments.

But how can new cognitive abilities appear if they have not gone through the sieve of natural selection? One suggestion is that human intelligence consists of a single, domain-general learning mechanism that is applied to whatever problems the environment throws at it (be it tracking a saber-toothed tiger or reading about Winnie the Pooh and Tigger). As we discussed in chapter 2, this position, at its extreme, has been rejected by evolutionists and is at odds with that of the emphasis of domain-specificity most cognitive scientists favor (see Tooby & Cosmides, 1992). The human brain does not process all information equally well, but it is specialized to process some types of input (e.g., language or faces) more easily than others. Although the end result is an organism that cannot do everything expertly, it is able to do certain things (e.g., comprehending language and interpreting social cues) especially well. This does not preclude the existence of domain-general mechanisms that have undergone selection pressure, but it is unlikely that the adaptability of the modern human brain and mind can be accounted for solely by a few general mechanisms (e.g., a "general intelligence").

But humans are cognitive generalists (at least in comparison with other species) and, as we have noted, success in the modern world requires the acquisition of cognitive skills not needed by our ancestors. To handle this conundrum, we propose, as have others, that children enter the world with sets of hierarchically organized, relatively specific cognitive modules, which can be co-opted over the course of development to solve "new" problems (Geary, 1998; Geary & Bjorklund, 2000; Gelman & Williams, 1998; Pinker, 1997; Spelke & Newport, 1998). However, we also propose, as have others, that domain-general mechanisms have evolved that serve as foundations in development both for the more domain-specific mechanisms and for the creation of "new" cognitive skills (Bjorklund & Harnishfeger, 1995; Elman et al., 1996; MacDonald & Geary, 2000; Mithen, 1996).

Human cognition is not monolithic but includes various types of processes, some of which are evolutionarily quite ancient and others that may be unique to humankind. How we conceptualize thought influences our understanding of cognition, as well as its ontogeny and phylogeny. In this chapter, we discuss aspects of cognition as they relate to evolutionary developmental psychology. We examine the distinction between implicit and explicit cognition, cognitive abilities selected over the course of human evolution (biologically primary) versus those "invented" to deal with culturally created problems (biologically secondary), the immediate versus deferred benefits of children's cognition, and the domain-general versus domain-specific distinction and its significance to evolutionary psychology. In the process, we examine some evolutionarily relevant aspects of cognition and their development, and we set the stage for interpreting other important aspects of cognitive development in later chapters.

IMPLICIT VERSUS EXPLICIT COGNITION AND MEMORY

A hallmark of evolutionary psychological explication is that adaptive behavior is predicated on adaptive thought (Tooby & Cosmides, 1992). Human behavior, as well as the behavior of other species, is said to be governed by information-processing systems and "strategies" designed by evolution to solve recurrent problems faced by our ancestors. Such language connotes a self-aware being who is conscious, at some level, of the strategies he or she is executing and of the processes underlying such decisions. This bias may be especially potent for cognitive developmentalists, whose science has been strongly influenced by Piaget's theory to focus on high-level cognitive tasks, quite different from the basic-level and often unconscious processes that have characterized the study of adult cognition (Siegler & Ellis, 1996). But this is obviously not the definition of cognition and strategies that evolutionary psychologists intend. Rather, brains and minds have been shaped by natural selection to be sensitive to certain classes of information and to respond in certain ways outside of the immediate awareness of the individual. Such cognition is *implicit*. This is in contrast to the *explicit* cognition that most people think about when they think about "thought" (Karmiloff-Smith, 1992; Nelson & Bloom, 1997; Parkin, 1997; Schacter, 1992).

But explicit cognition—occurring with conscious awareness—is obviously of substantial significance to *Homo sapiens*. It is because we are conscious that we are writing this book and you are reading it. It is only self-conscious individuals who are capable of pursuits such as science, literature, art, and city governments, and consciousness may be responsible for the complex social patterns characteristic of human culture, as well as for culture itself. Although some researchers have speculated that human consciousness is a by-product of a large brain (S. J. Gould & Lewontin, 1979), we propose, as have others (e.g., Rebers, 1992), that self-awareness had strong adaptive value for a long-lived, large-brained, socially complex species with an extended juvenile period and has been selected over the course of hominid phylogeny. In this section, we examine briefly the nature of implicit and explicit cognition and memory, their development, and their possible evolution.

Implicit versus Explicit Memory

Cognitive psychologists and neuropsychologists have made two broad distinctions for the contents of memory. *Explicit*, or *declarative memory*, refers to memory with awareness and can be assessed using conventional memory tests (i.e., free recall, recognition). *Implicit*, or *nondeclarative* or *procedural memory*, in comparison, refers to memory without awareness and is reflected in memory for routinized skills (i.e., procedural memory), priming, and operant and classical conditioning. Explicit memory is usually conceived as consisting of two interacting systems: *semantic memory*, which is world knowledge, particularly of language, rules, and concepts, and *episodic memory*, which refers to a knowledge of one's past experiences or personal history, as in autobiographical memory (Tulving, 1985).

This distinction has more than heuristic value, for implicit and explicit memory appear to be governed by different brain systems, as revealed by research with people with brain damage (Schacter, 1992; Schacter, Norman, & Koutstall, 2000). For instance, the hippocampus is involved in transferring new explicit information from the short-term store (the "location" of immediate awareness) to the long-term store. People with damage to the hippocampus can acquire a new skill as a result of repeated practice but will have no awareness of ever learning such a skill. For example, neuropsychologist Brenda Milner (1964) reported the case of H. M., who had hippocampal damage. H. M. was given a mirror-drawing task over several days, in which he had to trace figures while watching his hand in a mirror. H. M.'s performance was quite poor initially but improved after several days of practice, despite the fact that he had no recollection of ever performing the task previously. The enhancement of performance as a result of practice is a reflection of implicit (procedural) memory, whereas H. M.'s failure to recall previously performing the task is a reflection of a lack of explicit memory.

It seems obvious that the memory of animals and of human infants is of the implicit type, making implicit memory both ontogenetically and phylogenetically ancient (Rebers, 1992). One interesting question concerns human infants' first display of explicit memory and whether there is any evidence of explicit memory in nonhuman animals, such as the great apes. One area of research that has shed some light on this issue is *deferred imitation*, which refers to an individual observing a model and, after some significant amount of time (5 minutes or longer), reproducing that behavior. Such imitation requires that the observed event be stored in memory and retrieved at a later time. Piaget (1962) proposed that deferred imitation is first seen in children at about 18 months of age and is a reflection of symbolic (mental representational) functioning. Recent research has shown that children as young as 9 months of age can display deferred imitation for simple behaviors (Carver & Bauer, 1999) and that preverbal toddlers can retain such memories for as long as 1 year (Bauer, Wenner, Dropik, & Wewerka, 2000; Bauer & Wewerka, 1995; Mandler & McDonough, 1995), suggesting to some that symbolic functioning is present in human infants late in the first or early in the second year of life (Mandler, 1998; Meltzoff, 1990) and that deferred imitation reflects a preverbal form of explicit memory (Bauer, 1997; Meltzoff & Moore, 1997).

Evidence for the explicit nature of deferred imitation was provided in a study of patients with anderograde amnesia who, similar to H. M., were unable to acquire new explicit information due, presumably, to hippocampal damage (McDonough, Mandler, McKee, & Squire, 1995). Patients were administered a series of explicit memory tasks, which they obviously failed. They were also given a series of deferred-imitation tasks, similar to those passed by preverbal toddlers (Bauer, 1997). If deferred imitation is a form of implicit memory, amnesic patients should perform the tasks well, just as H. M. eventually learned to copy figures while watching his hand in a mirror. If, however, deferred imitation is a form of explicit memory, the patients should perform poorly on these tasks as well. McDonough and colleagues reported that the amnesic patients did indeed fail the deferred-imitation tasks, suggesting that it reflects an explicit ability. The fact that 1-year-old infants can pass these tasks suggests "that the neurological systems underlying long-term recall are present, in at least rudimentary form, by the beginning of the second year of life" (Schneider & Bjorklund, 1998, p. 474).

If deferred imitation is a reflection of nonverbal explicit memory, it should be possible to assess such cognition in animals. This has been done with nonhuman primates. Although it is almost axiomatic that monkeys and apes imitate ("monkey see, monkey do"; the verb "to ape"), relatively little empirical evidence exists of deferred imitation in nonhuman primates (see Tomasello & Call, 1997). What evidence there is of deferred imitation in great apes comes from animals who have had significant human contact early in life. For example, observational research with rehabilitant (i.e., formerly captive) orangutans (Russon, 1996) and an enculturated (human-reared) orangutan (Miles, Mitchell, & Harper, 1996) reported evidence of deferred imitation in animals as young as 2 years, 11 months of age.

Several experimental studies, using procedures with appropriate controls similar to those used with human infants (Bauer, 1997; Meltzoff, 1995), also have been conducted (Bering et al., 2000; Bjorklund, Bering, & Ragan, 2000; Bjorklund et al., in press; Tomasello, Savage-Rumbaugh, & Kruger, 1993). In these experiments, enculturated great apes (chimpanzees, bonobos, or orangutans) interacted with target objects during a baseline phase. A human model then displayed some target behavior with the objects (e.g., putting pegs in a form board and hitting the pegs with a hammer) and, after a delay ranging from 10 minutes (Bering et al., 2000; Bjorklund et al., 2000, in press) to 24 or 48 hours (Tomasello, Savage-Rumbaugh, & Kruger, 1993), the apes were re-presented the objects (deferred phase). The incidence of imitating the modeled behavior during the deferred phase was contrasted with the apes' behavior at baseline, before they had seen the modeled actions. Tomasello, Savage-Rumbaugh, and Kruger (1993) also tested a group of mother-reared (i.e., nonenculturated) chimpanzees and bonobos and 18- and 30-month-old human children.

The deferred imitation results of the Tomasello, Savage-Rumbaugh, and Kruger study are shown in Figure 5.1. As can be seen, the enculturated apes displayed the highest level of deferred imitation, significantly greater even than that of the children. Following the interpretation of the results of McDonough and her colleagues (1995) of an inability of amnesic patients to perform explicit memory tasks, including deferred-imitation tasks, the finding of deferred imitation in great apes is consistent with the position that these animals possess symbolic representation and explicit ("conscious"), as opposed to only implicit, memory. The fact that these abilities have been demonstrated convincingly only in enculturated animals suggests that great apes have a flexible cognition that is amenable to modification toward a more "human-like" form when they experience a human-like rearing environment. As we commented in chapter 4, we believe that evidence of species-atypical cognition in a large-brained, slow-growing, genetic cousin to *Homo sapiens*, under species-atypical rearing conditions, has important implications for theories of human cognitive evolution.

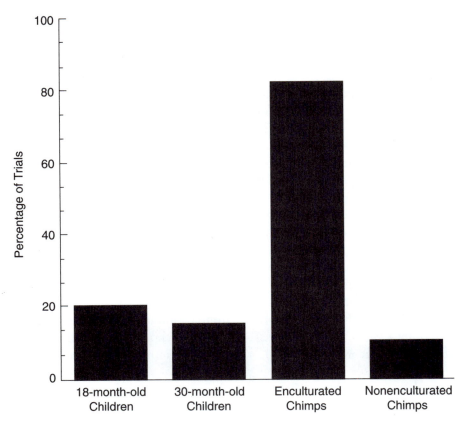

Figure 5.1. Percentage successful deferred imitation for enculturated apes, mother-reared apes, and 18- and 30-month-old children.
Note. From "Imitative Learning of Actions on Objects by Children, Chimpanzees, and Enculturated Chimpanzees," by M. Tomasello, S. Savage-Rumbaugh, and A. C. Kruger, 1993, *Child Development, 64,* pp. 1688–1705. Copyright 1993 by Society for Research in Child Development. Adapted with permission.

Implicit Cognition and Its Development

Most research in cognitive development beyond infancy has dealt with explicit cognition, particularly in the field of memory development (see Blasi & Bjorklund, 2001; Schneider & Bjorklund, in press; Schneider & Pressley, 1997). However, over the past decade or so, research into children's implicit learning and memory has increased, led, in large part, by an innovative theory of developmental psychologist Annette Karmiloff-Smith (1991, 1992). Central to Karmiloff-Smith's theory is the concept of *representational redescription,* which involves the mind re-representing knowledge it already possesses. According to Karmiloff-Smith (1991), "human development crucially involves the passage from representations that constitute knowledge in the mind to representations that acquire the status of knowledge to other parts of the mind" (p. 175). From this perspective, representational redescription is similar to Piaget's *reflective abstraction*, whereby children discover new information by examining the contents of their own mind (i.e., introspecting). Karmiloff-Smith proposed four levels of representation, the first being *implicit*, followed in development by three levels of *explicit* representation (E1, E2, and E3).

Implicit Representations

Implicit representations reflect the earliest stage and the most primitive type of representation. They correspond to special-purpose knowledge (e.g., knowledge related to physical objects, language, and cause and effect) that is modular in nature, is activated by specific environmental stimuli, and is inaccessible to other parts of the cognitive system (i.e., outside of conscious awareness). This is complex, procedural knowledge that is similar to that possessed by other "unconscious" species (e.g., a bird's knowledge about building a nest or a spider's about building a web). Implicit knowledge may still require some experience for it to be expressed properly, but it is limited in its application to a narrow range of situations and objects, and it is not easily modified. There is no redescription of implicit representations.

Consistent with Karmiloff-Smith's idea, implicit learning and memory are early developing abilities that show relatively little improvement over the course of ontogeny in comparison to explicit learning or memory (Hayes & Hennessy, 1996; Russo, Nichelli, Gibertoni, & Cornia, 1995; Vinter & Perruchet, 2000). For instance, research has demonstrated that 6- and 10-year-old children learn serial sequences of responses as well as adults, despite having no explicit (verbalizable) knowledge of what they have learned (Meulemans, Van der Linden, & Perruchet, 1998). In other research, 9- and 10-year-old children were shown pictures of 4- and 5-year-olds and asked to determine whether each picture was a former preschool classmate (Newcombe & Fox, 1994). In addition to their verbal (explicit) responses, changes in the electrical conductance of children's skin were also recorded and used as an indication of implicit recognition memory (higher skin conductance being predicted when the children saw pictures of former classmates than when they saw pictures of unknown children). As might be expected, performance was relatively poor on both the explicit and implicit tasks but greater than expected by chance, indicating that the children had some memory of their preschool classmates. Importantly, there was no difference in skin conductance between children who performed well on the explicit task and those who performed poorly, suggesting that even children whose performance on the explicit memory task was no greater than chance still "recognized," implicitly, as many of their former classmates as those children who performed well on the explicit task. This pattern of data indicates that some children "remembered" (implicitly) more than they "knew" (explicitly; see also Lie & Newcombe, 1999).

One example of young children's possession of implicit knowledge on a typically "explicit" task was provided by developmental psychologists Wendy Clements and Josef Perner (1994). False-belief tasks (Wimmer & Perner, 1983) assess theory of mind, specifically children's understanding that other people have knowledge that may be different from their own. (Theory of mind is discussed in chapter 7.) In the standard false-belief task, children watch as a treat is hidden in one location in the presence of two other people. One of those people, Maxi, then leaves the room, and the treat is moved to a new location. The child is then asked where Maxi will think the treat is hidden when he returns. Most children by age 4 pass this test, saying that Maxi will mistakenly think the treat is hidden in its original location. Children much younger than 4, however, typically say that Maxi will look in the new location, not understanding, apparently, that Maxi has different knowledge than they have.

Clements and Perner conducted a variant of this task, assessing children's implicit as well as explicit understanding of false belief. Preschool-age children were told a story about a mouse named Sam who had placed a piece of cheese in a specific location (Location A) so that he could get it later when he was hungry. While Sam was sleeping, Katie mouse moved the cheese to a new place (Location B). When Sam returned, children were asked where he would look to find his cheese. Most of the younger preschoolers erroneously said that he would look in Location B, reflecting their lack of understanding of false belief. This is the standard, or explicit, false-belief task, and it replicates previous research findings. However, Clements and Perner also recorded where children first looked, at Location A or Location B. This is an implicit task, requiring (seemingly) no conscious awareness or verbal response. Figure 5.2 shows the level of performance for children of different ages on the implicit and explicit tasks. As can be seen, only the youngest children (ages 2 years, 10 months or younger) "failed" the implicit false-belief task. By age 2 years, 11 months, children typically looked at the proper location (Location A), despite providing the incorrect verbal response (Location B). What these findings imply is that by about age 3 years, children have a well-developed implicit understanding of false belief that exceeds their explicit (verbalizable) knowledge. (This finding has been replicated by Clements, Rustin, & McCallum, 2000, and by Garnham & Ruffman, 2001; a similar finding has been reported for infants' performance on A-not-B object permanence tasks, with infants looking at the right location before searching at the right location; Ahmed & Ruffman, 1998.)

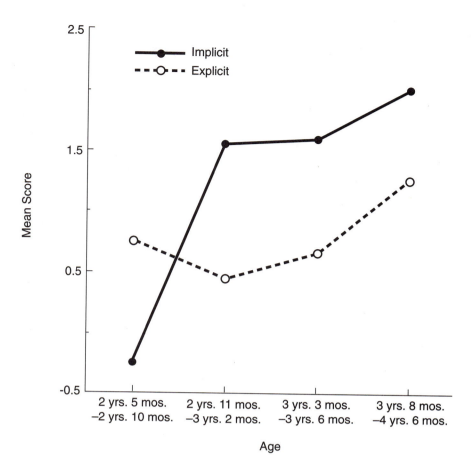

Figure 5.2. Average implicit and explicit understanding scores on false-belief task by age.
Note. From "Implicit Understanding of Belief," by W. A. Clements and J. Perner, 1994, *Cognitive Development, 9,* pp. 377–395. Copyright 1994 by Cognitive Science Society, Inc. Adapted with permission.

Explicit Representations

Redescription begins with the first level of explication (E1). Knowledge is now available to other cognitive systems but, according to Karmiloff-Smith, it is not yet conscious. For example, as young children acquire language, they are able to ignore aspects of the external stimuli and to focus on the internal meaning of language, which they use to formulate linguistic theories. Young children are very good, for example, at determining what is and is not a grammatical sentence—but they can rarely tell you why this is so. Because such knowledge is unavailable to conscious awareness, we feel more comfortable describing such cognitions as implicit. But the existence of such knowledge, which has been substantially modified by experience but is still unavailable to consciousness, suggests that much sophisticated knowledge may exist in an implicit (or E1) form before it can be realized consciously.

According to Karmiloff-Smith, knowledge requires greater explication to become conscious (E2), and greater explication yet before it can be verbalized and shared with others (E3). For example, Karmiloff-Smith (1979) described English-speaking children's correct use of the articles *a* and *the* during the preschool years. Yet, it is not until 9 or 10 years of age that they become consciously aware of them. According to Karmiloff-Smith, it is not until this time that children are able to redescribe their representations so that these representations are available not only to other cognitive systems but also to consciousness.

It is not surprising that children are born with well-developed implicit learning and memory abilities that develop rapidly. What is somewhat surprising, perhaps, is that human cognition develops beyond sophisticated implicit knowledge. Explicit memory is a phylogenetically recent innovation found only among humans and perhaps, to a lesser extent, among some great apes, and it develops slowly. Unlike age differences in implicit memory, age differences in explicit memory are substantial and do not asymptote until young adulthood (see Schneider & Bjorklund, 1998; Schneider & Pressley, 1997). Although the origins and functions of explicit memory/ representation systems may forever remain a mystery, researchers have provided interesting speculations, some of which we present below.

Evolutionary Considerations of Different Memory and Representational Systems

Humans and great apes are not the only species that need memory to survive and thrive. Learning and memory abilities, for any species, are adaptive specializations that have been shaped by the processes of natural selection to solve particular problems confronted in ancestral environments (Rozin, 1976; Sherry & Schacter, 1987). Memory is critical for finding food; avoiding predators; and identifying kin, friends, and foes and may be especially important for a species with a complex social network involving social exchanges (Cosmides & Tooby, 1992). Implicit memory theoretically exists to varying degrees, in all species with a nervous system (Rebers, 1992), but explicit memory seems to be unique to humans and perhaps to great apes raised in species-atypical conditions. Explicit memory's particular adaptive function and how it evolved are important questions for evolutionists.

One possibility was proposed by Sherry and Schacter (1987), who suggested that *functional incompatibility*, which exists when an adaptation that is dedicated to serving one function cannot serve other functions, led to the evolution of at least two memory systems. The implicit memory system (which Sherry and Schacter referred to as "Memory System I") supports the type of incremental learning involved in the acquisition of habits and skills and the development of smooth and automatic responses to invariant features of episodes. In contrast, the explicit memory system ("Memory System II"), supports rapid, one-trial learning of specific situations and the preservation of context details that characterize individual experiences. Because the implicit system is specialized to operate on invariance, and the explicit system is specialized to preserve variance, the two systems are incompatible and must be functionally separate. After all, it would be detrimental for the implicit system to preserve temporal, spatial, and other contextual details of individual episodes, because this is not its primary function. A similar argument could be made for the functional specialization of explicit memory.

Psychologist Merlin Donald (1991) provided a broader conception of changes in the cognitive/memory systems during hominid evolution. Donald proposed four stages of cognitive evolution, expressed in terms of how knowledge could have been represented in memory. Donald believed that each stage reflects a type of "culture." Donald called the earliest stage *episodic culture*, which corresponds to the type of representations possessed by contemporary great apes, particularly chimpanzees, and presumably by our common ancestor and the *australopithecines*. Apes, Donald proposed, "perceive complex, usually moving clusters and patterns of stimuli as a unit" (p. 153). Episodes, however, are bound in time and space, and chimpanzees and our hominid ancestors lived entirely in the present. This does not mean that chimpanzees have no memory for previous episodes, but rather that those memories cannot be retrieved for inspection or used to plan for the future. They function only to help an individual deal with events in the present.

The second stage in Donald's theory is called *mimetic culture* and typified the mind of *Homo erectus*. As the name implies, mimetic culture was based on imitation, initiated, most likely, from the freeing of the hands as a result of full-time bipedality. According to Donald, *mimesis* included *mimicry* (the exact copying of a behavior) but also more advanced *imitation*, in which one individual represents the actions and goals of another and attempts to reproduce the outcome achieved by another (see Tomasello, Kruger, & Ratner, 1993, and discussion of social learning in chapter 7). It seems that mimesis would clearly entail explicit cognition, in that, according to Donald, it involves "intentionality, generativity, communicativity, reference, autocueing, and the ability to model an unlimited number of objects" (p. 171).

Mimetic culture was followed by *mythic culture*, reflective of late Pleistocene *Homo sapiens*, beginning as recently as 35,000 years ago. Donald used the advent of advanced stone tools, art, evidence of religious rituals, and other signs of human culture as the beginning of mythic culture. This transition likely corresponded with the acquisition of language in a form in which narratives could be produced, and it was through language that greater communication resulted and tribal myths were invented and transmitted.

Donald referred to the final stage as *theoretic culture*, which corresponds to modern humans and the use of *external symbolic storage systems*, beginning with writing. Although this last transition was mediated by cultural and not biologic changes, it nonetheless reflects a cognitive transformation in thinking. In other words, the representational systems hominids have used over their phylogenetic history have influenced both their social and mental lives, with the latest change coming from without, as opposed to within, the organism.

Developmental psychologist Katherine Nelson (1996) has applied Donald's phylogenetic model to ontogenetic development and proposes a recapitulation theory in which representational stages in child development follow a course similar to that of our species over geological time. Donald's four stages correspond roughly with Piaget's four developmental stages: episodic representations characterizing thinking in the sensorimotor period, mimetic representations characterizing preoperations, mythic representations characterizing concrete operations, and theoretic characterizing formal operations. According to Nelson, before their first birthdays, infants are able to form episodic representations similar to those Donald proposed. The beginnings of mimetic representation can be seen early in the second year with the advent of referential pointing, pretense, deferred imitation, and self-recognition. Children are also learning language, but it is not until about 4 years of age that they are able to form narratives (corresponding to Donald's mythic culture). According to K. Nelson (1993, 1996), language is a form of mental representation (cf. Bruner, Olver, & Greenfield, 1966) that influences other forms of thought. Many of the changes that occur around age 4 (e.g., theory of mind) can be attributed to children's increasing sophistication to represent thought in terms of language. According to K. Nelson (1996), "it is human language and its potential for different ways of formulating thought that has driven and continues to drive human cognitive development on both the evolutionary and the individual scale" (p. 87). The theoretic stage in Donald's theory is attained beginning in adolescence when children develop logical abstractions, extensive use of external storage systems, and "scientific" reasoning.

K. Nelson (1996) does not suggest a simple "repeating" of phylogenetic stages over the course of ontogeny, as our brief description might imply. She acknowledged that eventually children have multiple representational systems available to them and that development in one (e.g., episodic) may not be complete before another begins (e.g., mimetic). Nonetheless, the use of a phylogenetic model as the basis for a theory of ontogenetic development is one way evolutionary accounts can inform child developmental theory.

BIOLOGICALLY PRIMARY AND SECONDARY ABILITIES

Geary (1995) postulated two general types of cognitive abilities: biologically primary abilities and biologically secondary abilities. *Biologically primary abilities* are those that have undergone selection pressure and evolved to deal with problems faced by our ancestors. Language and simple quantitative abilities, such as counting small numbers, would be examples. In contrast, *biologically secondary abilities* do not have an evolutionary history but are abilities that are instilled in children by their cultures to deal with "new" ecological problems unknown to our forebears. Reading and higher mathematics are examples. According to Geary, biologically secondary abilities are highly specialized neurocognitive systems that are supported by biologically primary abilities. They are "new" abilities only to the extent that the specific function they serve (e.g., reading) was never performed by earlier members of the *Homo* line. Their novelty, however, reflects a new application of biologically primary abilities for purposes other than their original evolution-based function.

Biologically primary and secondary abilities develop and require environmental support for their full expression. But, because of their different origins, some important differences exist between the two. First, biologically primary abilities are universal. As evolved mechanisms, they represent species-typical abilities and will be found in all biologically "normal" members of *Homo sapiens*. They also will be acquired by children in all but the most deprived environments (i.e., they are highly canalized), following a universal developmental function. The acquisition of biologically secondary abilities, in contrast, is dependent on growing up in a particular cultural context. Children in illiterate cultures, for example, do not spontaneously learn to read. Also, children will spontaneously exercise biologically primary abilities (e.g., labeling things in their environment or enumerating small quantities of objects) and, although there are sometimes individual differences in the rate and eventual level of the ability achieved (e.g., language disabilities will impair achievement), most children will attain an "expert" level of performance without adult instruction. In comparison, the variability in both the rate and end-level acquisition of biologically secondary abilities is much greater. Children must often be exhorted or coerced to practice biologically secondary skills such as reading and math. Whereas motivation for practicing biologically primary skills is endogenous, external reinforcement is often necessary for the sometimes tedious practice necessary to master biologically secondary abilities. Given this perspective, it is little wonder that reading, a supposed "language art," and higher mathematics give many children substantial difficulty.

IMMEDIATE VERSUS DEFERRED BENEFITS OF COGNITIVE ABILITIES

It is worthwhile to distinguish *immediate* from *deferred* benefits of children's cognition. Some characteristics of youth help children adapt to their environment only at that time in development, whereas other characteristics prepare them for later life (see Bjorklund, 1997b; see chapters 2 and 6). Developmental psycholinguist John Locke (1996) has addressed this issue with respect to language development. Many aspects of children's early vocal behavior may indeed lead to language, but infants do not likely engage in them for that explicit reason. For instance, by 5 months of age, many infants can recognize their name (Mandel, Jusczyk, & Pisoni, 1995), and by 8 months they can understand more than 30 words (Bates, Dale, & Thal, 1994). This stored information is of little immediate use to babies, however, for they lack the physical abilities (vocal control) and social–cognitive abilities (theory of mind, referential pointing, sense of self) to express themselves and to communicate a personal point of view. Rather than assuming that such "prelanguage" abilities have evolved only with language in mind, perhaps these abilities serve (or served for our ancestors) some other immediate function, which was co-opted for language development. For example, infants learn to recognize their mothers' voices, as well as variations in the way their mothers speak to them. Infants soon preferentially respond to their mothers' voices, which increases maternal satisfaction and, likely, strengthens attachment (Fernald, 1992). The immediate consequence for infants is that they are soothed, calmed, and cared for by their mothers. The deferred consequences are that the infants survive and learn language.

The point is that aspects of children's cognition may have immediate benefits that are different from the deferred benefits afforded by different aspects of the same cognition. Moreover, some aspects of children's cognitions may have only immediate benefits; they help children adapt to their current environment and do not prepare them for life as an adult. We gave the example in chapter 2 of the possible immediate benefits of imitation of facial gestures by newborns. Although it is possible that such imitation reflects the same cognitive mechanisms that underlie imitation in older infants (Meltzoff & Moore, 1985), its decline over the first two months of life (Fontaine, 1984) has suggested to some researchers that it may serve a unique function, such as fostering social or communicative interaction between an infant and its mother at a time when infants cannot intentionally regulate their social behavior (Bjorklund, 1987; Legerstee, 1991).

We have advocated taking a functional approach to development, but in doing so one should not assume that every aspect of children's cognition and behavior is adaptive and the product of natural selection. Nor should one assume that the similarity of early and later surface behaviors necessarily means that the early behavior "is for" the later behavior. It is worthwhile to ask what a behavior or cognitive ability is for, but the dynamic nature of development requires us to differentiate the potential immediate and deferred benefits of any behavior.

GENERAL AND SPECIFIC DOMAINS OF MIND IN DEVELOPMENT

Human intelligence, at least since the time of Charles Darwin (1871), has been viewed as a continuation of intelligence from predecessor species. Intelligence, from this view, is concerned with how interactions around tools, food, and other social actors contribute to reproductive success (Humphrey, 1976; Tomasello & Call, 1997). Evolution-minded psychologists and ethologists, further, have different opinions about the contexts in which intelligence evolved and its generality. One view holds that intelligence evolved in the context of foraging and the cognitive demands associated with locating food and the manipulations (e.g., fishing for termites and cracking nuts) necessary for making it edible (see Tomasello & Call, 1997, for a review) and that this intelligence generalized to other domains. This view, like Piaget's theory, suggests that general intelligence proceeds from sensorimotor intelligence.

Another domain-general view of intelligence holds that it evolved in the context of solving social problems and consequently affected all aspects of intelligence (Alexander, 1989; Bjorklund & Harnishfeger, 1995; Geary & Flinn, 2001; Humphrey, 1976; Jolly, 1966). This position holds that these types of social encounters shaped the evolution of primates' general intelligence. For example, in a ground-breaking paper of the social origins of intelligence, primatologist Alison Jolly (1966) presented ontogenetic and phylogenetic support for this argument from her study of lemurs. Ontogenetically, lemurs learn behaviors related to tools and food through social example. This sort of early social learning is also evident in interactions between human toddlers (Eckerman & Whitehead, 1999).

This argument for domain-general mechanisms in evolution is counter to the canonical claims of evolutionary psychology. However, as we noted in chapter 2, we believe that evidence exists that natural selection has resulted in sets of domain-specific and domain-general mechanisms in human evolution. Phylogenetic changes in domain-general mechanisms, as reflected by increases in speed of processing, inhibitory abilities, or working memory (Bjorklund & Kipp, 2001) and by the increased size of the neocortex in humans and other social primates (see chapter 4), permitted animals to hold more information in mind at a single time, for example, allowing them to execute effectively the domain-specific modules that process the information. We repeat our belief, counter to the orthodox view, that evolutionary–psychological accounts need not rely solely on domain-specific mechanisms as an explanation for human adaptive psychological functioning.[1]

Yet we believe, as others do, that much of human and primate intelligence is modularized to solve very specific problems encountered in evolutionary history. In this view, psychologists and ethologists suggest that cognition evolved as early humans learned to solve specific ecological problems and, consequently, it is primarily domain specific (Bugental, 2000; Cheyne & Seyfarth, 1990; Cosmides & Tooby, 1992; Geary, 2001).

Moreover, we reiterate our belief that domain-specific and domain-general abilities are no different from one another in terms of their ontogeny. Both types of abilities develop as a result of the bidirectional interaction among multiple levels of organization, beginning with the genes. Domain-specific abilities are not equivalent to the classic concept of "instincts," in which some complex behavior is expressed without the need of previous experience (i.e., genetic determinism). Domain-specific abilities develop following the same ontogenetic pattern as domain-general abilities, requiring a species-typical environment (that includes prenatal and internal environments as well as external environment) for their proper expression. The fact that such abilities have evolved to be sensitive to a limited range of stimuli (e.g., language) or contexts (e.g., mother–infant interactions) does not obviate the requirement that they develop following the precepts of the developmental systems approach as discussed in chapter 2.

In the remainder of this section, we discuss in greater detail the role of domain-specific and domain-general mechanisms in evolutionary developmental psychology.

Hierarchically Organized Domains of Mind

Among developmentalists, David Geary has perhaps best articulated the idea of a set of evolved, hierarchically organized, domain-specific modules that develop as children engage their physical and social worlds (Geary,

---

[1]For a philosophical argument against "massive modularity" (i.e., that the human brain/mind is completely or primarily composed of cognitive modules), see Fodor (2000).

1998, 2001; Geary & Bjorklund, 2000). As can be seen in Figure 5.3, Geary proposed two overarching domains—social and ecological—each consisting of two more specific domains (individual and group for social, and biological and physical for ecological), each of which comprise more specific domains. Geary pointed out that this list of domains is not complete (e.g., no numerical domain is listed here, which Geary believed exists), and one could argue about the organization of some of these domains (e.g., should "language" be organized within the social domain, or is it best conceptualized a separate domain?). Nonetheless, Geary's organization is consistent with the dominant perspective of evolutionary psychologists (Buss, 1995; Cosmides & Tooby, 1992; Pinker, 1997) and captures much of the developmental data.

The individual domains can be thought of as brain/mind modules that have evolved to be sensitive to relatively specific types of information (e.g., face recognition) and that function relatively independent of other modules. This conventional cognitive-science perspective, however, does not mean that the modules are fully formed at birth or with their first expression. They are "innate" to the extent that they have been shaped by endogenous (within the body) factors operating before birth, which are similar for all members of the species. (But recall that even for this definition of innateness, prenatal experience, broadly defined, plays a necessary role; see chapter 3.) However, the modules do not emerge fully formed but are skeletal in nature (Gelman & Williams, 1998; Gopnik & Meltzoff, 1997) and require the normal experiences of infancy and childhood for their expression. Children's interactions with physical objects, parents, siblings (see chapters 7 and 8), and peers (see chapters 9 and 10) are all guided by, but also in turn modify, these inherited modules, which increase in specificity with experience in a species-typical environment. As a result, the same species-general modules can produce different phenotypes, which emerge via epigenetic processes.

Although we believe that this description typifies the development of all species, this is particularly the case for humans. Our species' extended childhood has substantial costs (the likelihood of death before reproducing), and there must have been substantial benefits associated with it for it to have evolved (see chapter 4). Those benefits include the opportunity to master complex physical and (especially) social worlds. The benefits of a prolonged juvenile period would be wasted if evolved mechanisms produced rapid and unmodifiable learning. Evolved mechanisms limit the type of information children can process but make easier the processing of ecologically important data (Gelman & Williams, 1998). But *Homo sapiens*' prolonged period of immaturity requires that these mechanisms are highly plastic, that is, able to produce different phenotypes depending on the specific environmental conditions children face over the course of ontogeny.

We also believe that, although these modules are possessed by all biologically normal members of the species, they may not be identical

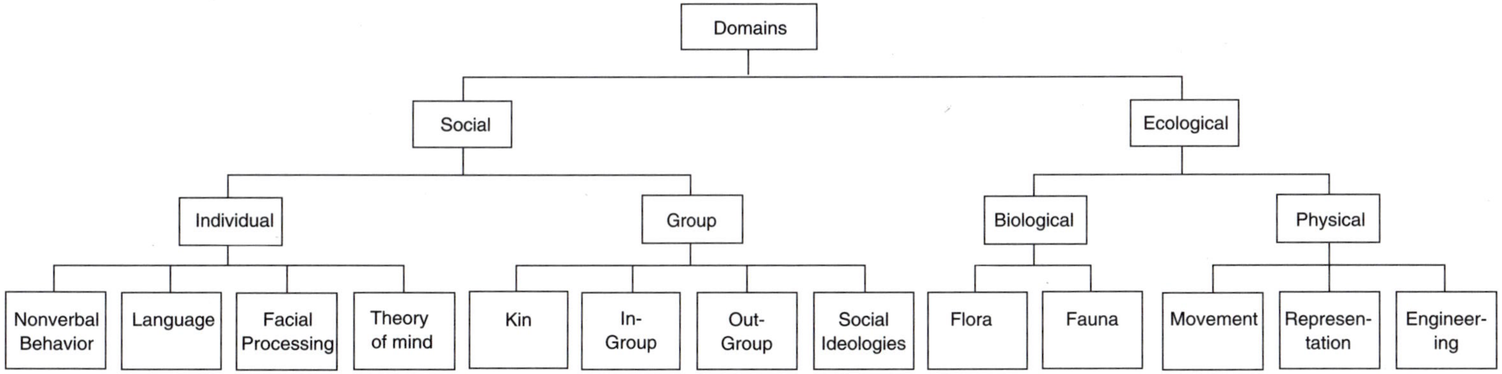

Figure 5.3. Proposed domains of mind.

Note. From *Male, Female: The Evolution of Human Sex Differences* (p. 180), by D. C. Geary, 1998, Washington, DC: American Psychological Association. Copyright 1998 by American Psychological Association. Reprinted with permission.

between individuals, an issue Geary does not specifically address. Individual differences in any of these modules may make children more or less sensitive to the target information (e.g., facial characteristics and vocal stimulation). Such endogenously based individual differences will then interact with a child's exogenous environment to yield a particular pattern of development. The process is still an epigenetic one, and the modules are still domain specific and universal. However, acknowledging such individual differences implies that behavioral-genetic analyses need not be incompatible with an epigenetically based theory of evolutionary developmental psychology (see chapter 3).

This perspective of combining epigenetic and behavioral–genetic viewpoints is consistent, we believe, with psychologist Howard Gardner's (1983, 1999) theory of multiple intelligences. Gardner originally proposed seven intelligences and has recently added an eighth: (a) linguistic, (b) logical–mathematical, (c) musical, (d) spatial, (e) bodily–kinesthetic, (f) interpersonal, (g) intrapersonal, and (h) naturalist. Most of these modules can be fit into Geary's conceptualization, except perhaps musical and logical–mathematical, the latter of which Geary noted exists outside of his hierarchic model. Like Geary, Gardner proposed that these modules are independent of one another, are potentially localized in the brain, and evolved by means of natural selection. Moreover, Gardner proposed that each intelligence has a distinctive developmental history along with a set of expert end-state performances. However, Gardner emphasizes that genetically based individual differences among these various abilities exist and each is educable. That is, although all members of the species possess each form of intelligence, individual differences can be observed early in life (best exemplified by prodigies and savants), and these various intelligences will develop to different degrees of proficiency as a function of experience over childhood.

How can one know if development is being governed by domain-specific evolved cognitive modules? Three general developmental features are predicted for evolved cognitive modules (see Exhibit 5.1). First, modules should be hierarchically organized, with the degree to which they are specialized inversely related to their level in the hierarchy. That is, the lower the module in the hierarchy, the more specialized it should be. Second, there should be sensitive periods for the acquisition of evolved skills, and children should display biases for attending to some classes of information or for performing certain activities. The sensitive period proposed for language acquisition (Locke, 1993) and infants' responsiveness to infant-directed speech (Cooper & Aslin, 1994) are well-known examples. Third, much knowledge associated with these modules should be implicit and unavailable to conscious awareness. Furthermore, the lower a module is in the hierarchy, the less under the influence of intentional control it will be.

EXHIBIT 5.1
Predicted Developmental Features of Evolved Cognitive Modules.

Hierarchical Organization
1. The modules shown in Figure 5.3 (p. 130, this volume) are very likely to be comprised of a hierarchy of submodules. For instance, the language system includes specialized systems for the comprehension and production of speech. These, in turn, are supported by sensory and motor systems for processing and articulating language sounds, among others.
2. The degree to which submodules are specialized is likely to be inversely related to their level in the hierarchy. The most basic submodules are likely to be highly specialized, designed to process a restricted range of stimuli such as specific language sounds. Modules at the highest level receive information from many lower level modules and may show moderate to high levels of flexibility in the range of stimuli that can be processed and in their output. An example is the apparently infinite number of utterances that can be generated by the language production system.

Sensitive Periods and Child-Initiated Activity
1. If cognitive modules emerge by means of an epigenetic process, then the development of the associated neural and cognitive systems will be dependent on exposure to domain-specific information (e.g., language sounds). Experiences interact with the inherent skeletal features of these modules to produce the phenotypic competencies.
2. A bias in children's activities and the types of information to which they attend is expected. These activities are expected to correspond to the domains shown in Figure 5.3 (e.g., orienting to people, exploring the environment) and provide the experiences that are an intimate feature of the epigenetic development of the modules.
3. Sensitivity to environmental input is expected to be time-limited to some degree; that is, sensitive periods are expected. The length of the sensitive period may be related to the submodules' level in the hierarchy. Because the functioning of higher level submodules may depend on input from lower level modules, the sensitive period for lower level modules is expected to be shorter and occur earlier in life than that for higher level modules. In short, the length of the sensitive period may be directly related to the complexity of the information processed by the module.

Implicit Knowledge
1. The skeletal structure of evolved modules reflects the organization of the underlying neural systems and the types of information to which these systems respond.
2. Knowledge is built into the organization of these cognitive and neural systems, that is, they respond to appropriate ecological information and produce intelligent responses to this information. The functioning of many of these systems is likely to be automatic and largely outside of the realm of conscious control.
3. The degree to which the functioning of modules is relatively automatic and the associated knowledge implicit may be inversely related to the modules' level in the hierarchy. Lower level modules are likely to be characterized by automatic processing (assuming adequate attention to this information) and a high degree of implicit knowledge. Individuals are more likely to gain explicit awareness of the output of higher level modules, and the functioning of these modules may be more open to top-down control by the individual.

Note. From "Evolutionary Developmental Psychology," by D. C. Geary and D. F. Bjorklund, 2000, *Child Development, 71,* p. 58. Copyright 2000 by Society for Research in Child Development. Reprinted with permission.

Domain-General Mechanisms in Evolutionary Developmental Psychology

Although one of the hallmarks of evolutionary explication is domain specificity, we also believe that there is evidence that domain-general mechanisms play an important role in human functioning and development and have been modified by natural selection over human evolution. Cognitive developmentalists have long argued about the nature of children's cognition, and we think it is fair to say that contemporary thinking holds that cognition is a multifaceted phenomenon, consisting of both domain-specific and domain-general mechanisms (see Bjorklund, 2000). From this perspective, domain-general mechanisms influence performance on a wide range of tasks but do not necessarily account for all of the variance in task performance. Rather, they account for a significant proportion of variance of task performance and developmental differences, with the remainder of the variance being attributed to other, presumably domain-specific factors. Candidates for domain-general mechanisms, accounting for both individual and developmental differences, include speed of processing (Kail, 1997; Kail & Salthouse, 1994); Piagetian-like cognitive structures (Case, 1992; Pascual-Leone & Johnson, 1999; Piaget & Inhelder, 1973); working memory (L. T. Miller & Vernon, 1996; Swanson, 1999); inhibitory mechanisms (Bjorklund & Harnishfeger, 1990; Dempster, 1992); attention resources (Cowan, Nugent, Elliott, Ponomarev, & Saults, 1999); and general intelligence, or g (Daniel, 1997; A. R. Jensen, 1998). In fact, evidence exists that cognitive tasks that load heavily on the g factor of psychometrically measured intelligence, despite having very different surface content, activate the same specific neural area in the frontal cortex of the adult human brain (Duncan et al., 2000). This gives us the somewhat paradoxical finding that an aspect of general intelligence is modularized.

We examine research in cognitive performance as a function of speed of processing to illustrate how a domain-general ability affects children's cognition. In general, young children need more time to execute cognitive processes than do older children. This is seen on a host of tasks, ranging from memory span (Case, Kurland, & Goldberg, 1982) to mental addition (Kail, 1991), with performance on these tasks being strongly related to the speed with which children can process the information. Developmental psychologist Robert Kail (1991, 1997; Kail & Salthouse, 1994) has performed a series of experiments reporting a similar developmental pattern of changes in speed of processing across a wide range of tasks, including mental rotation, visual search, name retrieval, and mental addition. In Kail's studies, participants ranging in age from 6 to about 21 years were given a series of reaction time tasks. For example, in a name-retrieval task, the participants were shown pairs of pictures and asked to determine whether they were physically identical or had the same name (e.g., different examples of a car, one a convertible and the other a sedan). In mental-rotation experiments, participants were shown a pair of letters in different orientations and were to determine as quickly as possible whether the two letters were identical or mirror images of each other. To do this, participants had to mentally rotate one letter into the same orientation as the other. Patterns of responses over these two, and several other, tasks were highly similar, with reaction times becoming faster with age (see also Hale, Fry, & Jessie, 1993; L. T. Miller & Vernon, 1997). Furthermore, despite the substantial differences in the task requirements, all showed essentially the same age-related pattern of changes in reaction times. Recent research has even shown a similar (although not identical) age trend in reaction time over the first year of life (Canfield et al., 1997) and for children between 22 and 32 months of age (Zelazo, Kearsley, & Stack, 1995).

Based on this and related evidence, Kail and Salthouse (1994) proposed that speed of processing is a *cognitive primitive*, or a basic aspect of the human cognitive architecture. Differences in this capacity change with age in regular ways and influence much cognitive functioning by limiting how much (or how quickly) information can be handled at any one time. Yet, age differences in speed of processing do not account for all age differences in cognitive performance. Figure 5.4 presents Kail and Salthouse's (1994) model of the relation between age, speed of processing, and cognitive performance. As

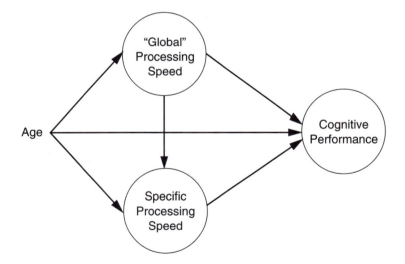

Figure 5.4. Kail and Salthouse's model showing possible relations among age, global processing speed, specific processing speed, and cognitive performance.
Note. From "Processing Speed as a Mental Capacity," by R. V. Kail and T. A. Salthouse, 1994, *Acta Psychologica, 86,* pp. 199–225. Copyright 1994 by Elsevier. Reprinted with permission.

can be seen, factors associated with age may directly influence cognitive functioning independent of processing speed, or age may affect task-specific processing speed. So although global speed of processing does not account for all of the variance in children's cognitive performance, Kail and Salthouse (1994) proposed that it is perhaps the most important single factor underlying cognitive development.

Domain-General Mechanisms as Foundational to Domain-Specific Mechanisms

Our proposal is that a certain level of domain-general resources (speed of processing, working memory) is necessary for the execution of more domain-specific mechanisms. For example, research into children's theory of mind has frequently used variants of the false-belief task. In one version of the false-belief task (Perner, Leekam, & Wimmer, 1987), children are shown a familiar box, such as an M&Ms box, and asked what they think is in it. They quickly identify it, in this case saying M&Ms. The box is then opened and, to their surprise, it contains something other than M&Ms, pencils for example. Children are then asked what a friend of theirs, who has not seen the contents of the box, would think was in it. Most children 3.5 years of age or younger say "pencils," apparently not understanding that their friend has different knowledge than they do. Children also are asked what they initially thought was in the box. Again, most children ages 3.5 or younger say "pencils," seemingly forgetting their original response. This last phenomenon has been called *representational change*. It is not until about age 4 that children typically answer these problems correctly. In other tasks, children must learn to inhibit a desired response (don't pick the window with the treat), because when they do, another person gets that treat and they get nothing (J. Russell, Mauthner, Sharpe, & Tidswell, 1991). As in more standard false-belief tasks, children much younger than age 4 have difficulty performing tasks such as this (Peskin, 1992).

Correct performance on false-belief tasks requires that children be able to keep several thoughts in mind (what I know and want, what the other person knows and wants) and sometimes to inhibit a prepotent response ("Don't tell what I know or what I want"). The individual must possess sufficient general-purpose information-processing resources before any domain-specific mechanisms can operate effectively. For example, there is evidence that children's abilities to inhibit responses, which are involved in executive-control processes, are significantly related to performance on false-belief tasks in preschool children (see Bjorklund & Kipp, 2001; Perner & Lang, 2000; Perner, Stummer, & Lang, 1999).

Alternatively, a single symbolic or representational ability may underlie several different social–cognitive mechanisms. For example, Gopnik and

Astington (1988) reported similar developmental trends in children between ages 3 and 5 years for performance on false-belief tasks, representational change (memory for one's initial belief on false-belief tasks), and appearance/reality distinction tasks (Flavell, Green, & Flavell, 1986). These latter tasks involve children realizing that something can look like one thing (e.g., a rock) but actually be another (e.g., a sponge). The consistency of patterns on these three seemingly different tasks suggests that a single, domain-general mechanism underlies performance of children's representational abilities. How "general" such a mechanism is can be debated, however. For example, a single mechanism may guide only children's representational abilities. This is not the same level of mechanism that might affect all aspects of cognition, as speed of processing or working memory may, but more general than a mechanism that is specific to theory of mind or making appearance/reality distinctions.

In general, we believe that many domain-specific abilities require a certain level of domain-general cognitive resources before they can be executed. For example, to effectively deceive another, one must hold in mind one's interpretation of the current situation, that of one's competitor, and perhaps inhibit a desired response (not gazing toward a cache of favorite food). This requires substantial mental resources, and it is likely that animals could not execute domain-specific social modules involved in deception, cheater detection, or theory of mind until a certain level of such resources was available. From this perspective, basic, domain-general cognitive mechanisms, such as inhibition, played a permissive role for more advanced domain-specific mechanisms (Bjorklund & Kipp, 2001).

Domain-General Mechanisms Interacting With Domain-Specific Mechanisms

In other situations, developmental changes in domain-general abilities may interact with domain-specific abilities. For example, language acquisition has been proposed to be governed by a host of domain-specific abilities, with children possessing a universal grammar and a language acquisition device to compare the language they hear with the grammatical theory they were born with (see Locke, 1993; Pinker, 1994). We concur with the general position that domain-specific mechanisms can foster children's language learning. However, this does not preclude the possibility that this process is aided further by developmental changes in general information-processing mechanisms. More specifically, several theorists have proposed that young children's limited working memories limit how much language they can process, thus simplifying the corpus they must deal with (Elman, 1994; Newport, 1991). Children start talking using single morphemes, gradually increasing the complexity of language they can speak. Both Newport (1991) and Elman (1994) performed computer simulations, restricting the amount of information the simulations could process at any one time (equivalent to restricting how much children can hold in working memory). They each reported that aspects of language were more easily acquired when the input was limited (either by presenting a reduced corpus or by limiting the working memory of the system). These researchers concluded that young children's limited working-memory capacity restricts how much language information they can process, which simplifies what is analyzed, making the task of language acquisition easier. This research is discussed in greater detail in chapter 6.

Evidence of the selection pressure of domain-general abilities can be found in comparative studies of brain size in primates. The size of the neocortex, relative to the size of the body, varies in primates as a function of the size (and thus presumably the complexity) of the social group (Dunbar, 1992, 1995; see chapter 4). Dunbar (1992) and others (e.g., Geary & Flinn, 2001; Tomasello & Call, 1997) have argued that as social complexity increased, animals required larger brains to deal successfully with the social demands of group living. There is also evidence that size of the nonvisual neocortex, group size, and length of the juvenile period are significantly correlated in primates, suggesting that an extended juvenile period is required for a large-brained species to master the intricacies of a complex social group (Joffe, 1997; see chapter 4). Although it is possible that the increased neocortex size observed in social primates could be the result of an increased number of domain-specific modules, we believe that it is more parsimonious to propose that most of the increased brain size can be attributed to enhanced general processing mechanisms that may have permitted the evolution or execution of more domain-specific mechanisms, particularly in the social realm (see Bjorklund & Harnishfeger, 1995; Bjorklund & Kipp, 2001). Consistent with this perspective is the interpretation of Finlay, Darlington, and Nicastro (2001) that most of the increase in brain size over mammalian evolution (including hominids) is not attributed to changes in specific areas of the brain associated with particular functions but rather to a more general increase in (nearly) all areas of the brain associated with delaying the "neuronal birthdays" (when precursor nerve cells stop dividing symmetrically and begin their migration within the neural tube). Thus, it is not tenable to assume that all features of the human brain and mind are inherently prespecified (see Geary & Huffman, 2001). Nonetheless, comparisons between the brains of monkeys, chimpanzees, and humans reveal significant differences in the microcircuitry of many areas (e.g., in the visual cortexes), suggesting that specific brain areas and cognitive functions have undergone selective pressure (Preuss, 2001). This pattern of mosaic brain evolution, with many changes in specific areas of the brain reflecting selective adaptation, has been inferred to occur within mammalian orders (including primates; de

Winter & Oxnard, 2001). In other words, even if much of the increase in brain size over hominid evolution can be attributed to a general mechanism associated with the delaying neuronal birthdays, this does not preclude the later specialization (through the mechanism of natural selection) of uncommitted cortex and the production of brain modules that are relatively domain specific in their application (Geary, 2001).

Comparative Evolutionary Developmental Cognition: The Evolution of "Piagetian Intelligence"

The most successful domain-general theory of cognitive development is that of Jean Piaget (1962, 1965; Piaget & Inhelder, 1973). Piaget argued for a monolithic cognitive developmental function. Children's thinking was relatively homogenous at any point in time, with cognition in one realm being related to cognition in other realms. Development through infancy, for example, was governed by the acquisition of certain forms of understanding, reflected in the six substages of the sensorimotor period. Underlying the transition from sensorimotor to preoperational intelligence was the emergence of the *semiotic* (or symbolic) *function*. This was represented by a variety of forms, including imagery, deferred imitation, symbolic play, and language, but the "cause" of each of these specific skills could be located in the emergence of a single, domain-general symbolic ability. Piaget himself acknowledged that development was not always as integrated as he proposed (his concept of décalage), but even in these cases, development followed a predictable course, and Piaget did not waiver from his emphasis of a single, homogeneous mind that changed uniformly over the course of ontogeny.

Piaget, of course, has been much criticized for his view on the homogeneity of cognitive function (see Brainerd, 1978), as well as for his proposal that mental representation is not available to children until the final substage of the sensorimotor period (Meltzoff, 1995). Although we believe that much of this criticism is apt and that various types of cognition exist, each with its own developmental function, we believe that the type of cognition that Piaget was studying is indeed relatively homogeneous at any point in time (other forms of cognition display greater heterogeneity of function) and reflects a form of domain-general cognition. Several researchers interested in animal cognition have adopted Piaget's model of development, and a particularly impressive database has been obtained for primates (Gibson, 1990; Jolly, 1966; Langer, 1998; 2000; Parker, 1996; Parker & McKinney, 1999). The extensive database Piaget and his followers accumulated in a variety of domains, including physical knowledge, logical mathematical knowledge, and social knowledge, renders his model rich for comparative studies of primate cognitive development and the possible existence of asynchronies in development in nonhuman as compared to human species.

One interpretation that has emerged from comparative cognitive developmental research performed from a Piagetian perspective is that cognitive abilities have been added over phylogenetic time. Anthropologist Sue Taylor Parker and paleontologist Michael McKinney (Parker, 1996; Parker & McKinney, 1999) recently have presented an overview of these findings, and they relate, we believe, to the evolution of domain-general abilities in human phylogeny.

Important for an evolutionary interpretation is the relationship that various nonhuman primate species have to humans. Figure 5.5 presents the phylogenetic relations among extant primates with approximate ages when the various lineages diverged. New World (South American) and Old World (African and Asian) monkeys last shared a common ancestor approximately 35–40 million years ago, and Old World monkeys and apes split approximately 25–30 million years ago. Lesser apes (gibbons, siamangs) and great apes diverged about 20 million years ago. Orangutans separated from the rest of the great apes approximately 12–15 millions years ago, and gorillas split off from the common ancestors of chimpanzees and humans approximately 10 million years ago. Modern-day chimpanzees, bonobos (pygmy chimpanzees), and humans last shared a common ancestor about 5–7 million years ago. *Homo sapiens* are the only remaining species of hominids, which included Australopithecines (*Australopithecus afarensis*) and extinct members of the *Homo* genus (*Homo habilis*, *Homo erectus*, *Homo neanderthalensis*; see chapter 2). By evaluating which aspects of cognitive development various

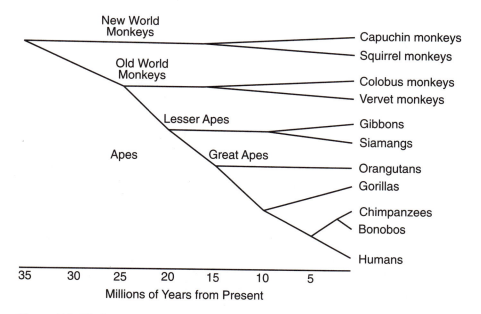

Figure 5.5. Phylogenetic relations among extant primates.

living species share with one another (particularly with humans) as a function of their last shared common ancestor, we can infer the evolutionary progression of cognitive development. That is, for any given cognitive content, we can examine the highest level a species attains and the timing (onset and offset) of its ontogeny and, based on its phylogenetic proximity to *Homo sapiens*, make some educated guesses about when (and perhaps how) these abilities evolved.

Most of the primate research has focused on achievements through Piaget's sensorimotor period, although some aspects of early preoperational thought, which requires symbolic representation, also have been investigated (see S. T. Parker & McKinney, 1999). Table 5.1 provides a description of sensorimotor intelligence through Piaget's substages. Some of the cognitive contents of the sensorimotor period (object concept, physical causality) in a variety of primate species have been examined extensively, whereas research on other topics (spatial knowledge) has received scant attention from primatologists. There also has been extensive research on non-Piagetian topics, such as theory of mind, some of which we discuss in chapter 7.

For illustration purposes, Table 5.2 presents the average age of attainment (in months) in humans, great apes, and monkeys of Substages 2 through 6 of object permanence, following Piaget's formulation (based on data presented in Parker & McKinney, 1999). As can be seen, humans and great apes reach Substage 6 (invisible displacements), whereas monkeys attain only Substage 5 (A-not-B task). Humans and chimpanzees progress through the substages at comparable rates, whereas gorillas go through the stages a bit faster, and monkeys faster yet.

Table 5.3 presents Parker and McKinney's (1999) determination of the age of attainment of Substages 2 (primary circular reactions) through 6 (mental combinations) of Piaget's sensorimotor period for humans, great apes, and monkeys. First, all primates tested progress through the same sequence of sensorimotor stages. Second, note that there are species differences in the highest stages attained, with only humans and the other great apes attaining Substage 6, and both Old and New World monkeys not progressing past Substage 4 (coordination of secondary circular reactions). Third, there are differences in rate of development. Although humans and the other great apes attain the first four substages at about the same age, humans reach Substages 5 and 6 sooner than chimpanzees, gorillas, and orangutans. Similar to the development of object permanence, macaques develop relatively rapidly through the first four substages, although capuchins' development is a bit slower for sensorimotor intelligence than for object permanence. Although not shown in the tables here, Parker and McKinney (1999) reported that, unlike human children, monkeys and apes display different schedules of development for different sensorimotor contents. One interpretation of this pattern is that the cognition of monkeys

TABLE 5.1
Characteristics of the Sensorimotor Substages in Piaget's Theory

Substage and approximate age range	Major characteristics
Substage 1 (basic reflexes): birth to 1 month	Cognition is limited to inherited reflex patterns.
Substage 2 (primary circular reactions): 1 to 4 months	Infant acquires first adaptations, extends basic reflexes; reflex is activated by chance (for example, thumb comes in contact with lips, activating sucking reflex); infant attempts to reproduce reflex, resulting in acquisition of new, noninherited behavior (thumb sucking) that can be activated at the infant's discretion; initial occurrence of behavior is by chance, so need follows action.
Substage 3 (secondary circular reactions): 4 to 8 months	By chance, infant causes interesting event in environment to occur (for example, infant kicks mattress, causing mobile over crib to move) and tries to re-create event; beginning of control of objects and events external to infant; as in previous stage, initial occurrence of interesting event is by chance, so need follows action.
Substage 4 (coordination of secondary circular reactions): 8 to 12 months	Infant uses two previously acquired schemes in coordination with each other to achieve a goal; first sign of need preceding action, or goal-directed behavior, is exhibited.
Substage 5 (tertiary circular reactions): 12 to 18 months	Infant discovers new means through active experimentation, develops new techniques to solve problems; goal-directed behavior is exhibited, but entire problem-solving process is conducted by overt trial and error; intelligence is still limited to child's actions on objects.
Substage 6 (mental combinations): 18 to 24 months	Infant shows first signs of symbolic functioning; infant is able to represent events in environment in terms of symbols (for example, language, imagery); problem solving can now be covert.

Note. From *Children's Thinking: Developmental Function and Individual Differences* (3rd ed.; p. 81), by D. F. Bjorklund, 2000, Belmont, CA: Wadsworth. Copyright 2000 by Wadsworth. Reprinted with permission of Wadsworth, an imprint of the Wadsworth Group, a division of Thompson Learning. Fax (800) 730-2215.

TABLE 5.2
Developmental Sensorimotor Substages in Object Permanence in Monkeys, Great Apes, and Humans

Species	Age, in Months				
	Stage 2	Stage 3	Stage 4	Stage 5	Stage 6
Human	2	4–8.5	7–8	7–10	11–21
Chimpanzee	3	6	9	12	18
Gorilla	1.5	3.5	7.5	7–9	9–10
Orangutan	?	?	?	?	?
Macaque	0.75	1–1.5	1.75	3.5	none or partial
Capuchin	1	2–3	3–7.5	8–12	none
Squirrel monkey	?	?	?	?	none

Note. From *Origins of Intelligence: The Evolution of Cognitive Development in Monkeys, Apes, and Humans* (p. 42), by S. T. Parker and M. L. McKinney, 1999, Baltimore: Johns Hopkins University Press. Copyright 1999 by Johns Hopkins University Press. Adapted with permission.

TABLE 5.3
Parker and McKinney's (1999) Determination of Age of Attainment of Substages 2–6 of Piaget's Sensorimotor Period for Humans, Great Apes, and Monkeys

Species	Age, in Months				
	Stage 2	Stage 3	Stage 4	Stage 5	Stage 6
Human	2	3	8	12	18
Chimpanzee	2–5.1	5–6	6–11.5	11.5–36	?
Gorilla	1.5	3–4.5	4.5–10	39	39
Orangutan	1.5–2	2.5–6	6–11.5	11.5–36	24–48
Macaque	0.5–3	1–4	2.5–5	none	none
Capuchin	1	9–12	12	none	none

Note. From *Origins of Intelligence: The Evolution of Cognitive Development in Monkeys, Apes, and Humans* (pp. 73–76), by S. T. Parker and M. L. McKinney, 1999, Baltimore: Johns Hopkins University Press. Copyright 1999 by Johns Hopkins University Press. Adapted with permission.

and apes is not integrated; different abilities develop at different rates. The pattern shown by human children, in contrast, reflects synchronous development and is consistent with Piaget's explanation that a single symbolic function underlies cognitive development.

Necessary cautions must be taken when interpreting these findings, of course. There are limited data for many sensorimotor contents for many species, and sometimes only human-reared apes attain many of Substage 6 accomplishments, which presumably require mental representation (see Call & Tomasello, 1996). Some researchers have attributed apes' performance on these tasks to lower level (nonrepresentational) processes (Heyes, 1998; Povinelli, Bering, & Giambrone, 2000). Nonetheless, Parker and McKinney's compilation and interpretation of the primate findings provide a valuable database for making inferences about human cognitive evolution.

Parker and McKinney's interpretation of these findings revolves around the concept of heterochrony, discussed in chapter 4. *Heterochrony* refers to differential timing of ontogeny that affects the eventual phenotype of an organism. As discussed briefly in chapter 4, Parker and McKinney's claim is that the cognitive developmental data on primates indicate a *recapitulation* of cognitive evolution, such that hominid phylogenetic changes are best interpreted as reflecting additions to the adult stages of ancestors (*terminal extensions*; see Langer, 1998, 2000; McKinney, 1998, 2000; S. T. Parker, 1996; S. T. Parker & McKinney, 1999). The basic argument is that primates traverse the same sequence of sensorimotor stages, with apes attaining higher stages (evolutionarily new stages) than monkeys, and humans attaining yet higher (and newer) stages than gorillas, orangutans, and chimpanzees (those reflected by preoperations and beyond). These extensions are afforded by an extended juvenile period (in apes relative to monkeys, and in humans relative to apes), which permits time for brain development to continue beyond the levels attained by the ancestors.

We believe that these arguments make sense, and we basically concur with the interpretations of Parker, McKinney, and their colleagues on the evolution of Piagetian-like intelligence. However, we believe that intelligence is multifaceted and that a Piagetian-like, domain-general intelligence exists alongside other more domain-specific abilities, which have their own phylogenetic history. Although monkeys, apes, and humans share a common ancestor and primitive cognitive abilities, each group also surely has evolved species-specific cognitive adaptations, suited to their particular ecological niche.

Cognitive primatologists have generally asked how apes and monkeys are like us, using humans as the "standard," a reasonable approach when the goal is to understand human cognitive evolution. However, apes may have evolved different cognitive abilities that are not very human-like since we last shared a common ancestor with them, as *Homo sapiens* surely have evolved their own set of cognitive specializations. Our point is that human cognition and cognitive development include both domain-general and domain-specific abilities, each of which evolved and is compatible with an evolutionary developmental psychological perspective.

The Co-Evolution of Domain-General and Domain-Specific Abilities

Parker and McKinney (1999) proposed that ape cognition involves the use of various combinations of skills in different contexts, which requires cognitive flexibility, in contrast to the modular position advocated by contemporary evolutionary psychologists. Over the course of evolution, the genetic line that led to *Homo sapiens* simply expanded this flexible intelligence, adding concrete operational abilities in *Homo erectus* and formal operational abilities in *Homo sapiens*. Mithen (1996) has made a similar claim, proposing that hominids evolved powerful, domain-specific modules to deal with their natural and social worlds, but it was not until the emergence of modern humans about 100,000 years ago that *Homo sapiens* were able to integrate the information-processing abilities of these modules to produce a general-purpose intelligence. In both cases, it appears that a domain-general mechanism is proposed as the necessary addition for the emergence of the modern human mind.

Under what conditions should domain-general versus domain-specific abilities evolve? Domain-specific mechanisms will be favored when environments remain relatively stable, with individuals facing recurrent problems generation after generation. For *Homo sapiens*, characteristics of social organization, such as male–male competition, female care for and provisioning of young, kin recognition, reciprocal social interactions, and aspects of mate selection, likely remained relatively stable over evolutionary time. As a result, we should expect domain-specific mechanisms to have evolved to deal with these classes of problems, and evolutionary psychologists have devoted much time to demonstrating exactly this (Buss, 1995; Tooby & Cosmides, 1992). In contrast, domain-general mechanisms will be favored when environments are unstable and the nature of the problems individuals face varies over generations. Under these circumstances, flexible, decontextualized problem-solving routines would be most adaptive (MacDonald & Geary, 2000). For example, we noted in chapter 2 that the environment in which humans evolved was characterized by frequent and noncyclic changes in climate (Potts, 2000). This would have resulted in unpredictable changes in habitat, requiring individuals to be able to respond to situations unlike any their recent ancestors faced. It is exactly in such situations that flexible, domain-general mechanisms would be favored.

Given the evidence from human cognition and cognitive development indicating the existence of domain-general and domain-specific abilities, data from comparative cognitive development, and evidence of the conditions under which our species evolved, both domain-general and domain-specific abilities should be considered as plausible components of the evolved human mind. We believe that such a stance, although perhaps at odds with the canonical interpretations of evolutionary psychology and the developmental systems approach, best reflects the extant evidence and can be interpreted within a framework that considers that cognitive abilities evolved in the dynamic interaction between developing organisms and their environments.

COGNITION, DEVELOPMENT, AND EVOLUTION

Human cognition and its development are complicated phenomena. People do not possess a single "mind" that functions similarly across diverse situations. Via the process of natural selection operating over geologic time, the human brain has become specialized for processing particular types of information. Consistent with the tenets of evolutionary psychology, we believe that many information-processing programs can be described as domain-specific mechanisms designed to handle specific problems faced repeatedly by our ancestors in the environment of evolutionary adaptedness. However, these programs are not preformed, but become manifest only with experience over the course of development.

Research into human cognition and cognitive development has clearly shown that the human mind is multifaceted, with different types of cognition displaying different developmental functions. Much human cognition is implicit, unavailable to consciousness, and reaches mature levels relatively early, but other cognition is explicit, slow developing, and may be unique to *Homo sapiens*.

Evolution has prepared humans with a unique set of cognitive skills built on a foundation that we share with our primate cousins. Human cognition is highly flexible, with many important cultural skills being co-opted from a set of biologically primary abilities that have undergone natural selection. These skills, both those designed by natural selection and those that emerge only in a specific cultural context (biologically secondary skills), do not arise fully formed but have developmental histories, which will determine the eventual cognitive phenotype of an individual. In the next two chapters, we examine some of the specific cognitive contents that, we believe, children have been prepared by evolution to deal with, and for which an evolutionary developmental perspective can provide us with greater insight into child development.

6

PREPARED TO LEARN

As we have seen, evolutionary developmental psychological theories propose that humans evolved both domain-specific (modular) and domain-general information-processing mechanisms for adapting to the myriad problems our ancestors faced. Solutions to some problems are likely very old and shared with our primate relatives; others may be unique to our species, although built on abilities possessed by our simian ancestors. The majority perspective, however, both among evolutionary psychologists in general and among child developmental psychologists who have taken an explicitly evolutionary perspective, is that most evolved abilities are domain specific in nature.

The implication of this position, one with which we largely agree, is that the human mind is not a general-purpose problem solver. Rather, there are constraints on learning (Gelman & Williams, 1998), meaning that human thought cannot wander to whatever topic it might desire and cannot solve with equal facility any problem presented to it simply by exerting sufficient effort. Such constraints should be viewed not only for what they do not allow, however, but also for what they permit. Limitations on the range of information that an animal can make sense of permit species to be specialists in a limited number of domains. Bats and dolphins use sonar to navigate around their very different worlds. We can never know what it is like to be a bat; instead we are specialized to process language, human social relations, as well as life in a three-dimensional, gravity-dependent world. This perspective—that infants come into the world with knowledge or processing constraints for particular domains—is often referred to as *neonativism* or *structural-constraint theory*.

At the extreme, neonativist arguments hold that little, if anything, develops; instead, infants are born with innate knowledge for interpreting the world (see Spelke, 1991). For example, this approach proposes that knowledge of language, objects, and space is innate, requiring only the proper environmental stimuli for their expression (see Spelke & Newport, 1998). Such knowledge is proposed to be domain specific and to impose constraints on how that information is processed and thus what is learned (thus the name *structural-constraint theory*). Stated more positively, these constraints make it easier for the child to learn certain classes of information, such as language (Gelman & Williams, 1998). This is consistent with the idea of *representational innateness* that was discussed in chapter 3.

Most neonativists, however, advocate a more moderate position concerning what type of knowledge is innate; they argue that constraints within a domain may be more general in nature, consistent with the idea of *architectural innateness* as discussed in chapter 3 (Gopnik & Meltzoff, 1997; Wellman & Gelman, 1998). For example, developmental psychologists Alison Gopnik and Andrew Meltzoff (1997) proposed a "starting-state" nativism, in which children enter the world with sets of rules for operating on particular representations. However, the rules and the representations they generate are altered by experience. Thus, domain-specific mechanisms constrain what and how information can be processed, but the content (e.g., knowledge of objects) and the processes for operating on that content vary with experience. This level of flexibility is clearly necessary given the diverse situations into which humans are born.

In this chapter we sample some of the research pertinent to the idea that human infants and children come into the world prepared to acquire some information more readily than other. We first examine briefly the research on prepared learning, then examine research on infants' understanding of their physical world. We follow this with sections on young children's quantitative–mathematical and spatial abilities. We then devote substantial space to various evolutionary-friendly views of language acquisition and conclude the chapter by examining the role of immaturity in cognitive development.

PREPARED LEARNING

The classic demonstration of prepared learning is conditioned taste aversion, termed the *Garcia effect* after its discoverer psychologist John Garcia (Garcia, Ervin, & Keolling, 1966; Garcia & Keolling, 1966). In a series of experiments, Garcia and his colleagues gave rats novel foods and, several hours later, exposed them to radiation, which made them sick. Despite the fact that the nausea occurred several hours after they ate the food, the rats later avoided the food they had eaten shortly before getting sick. It was not as easy to condition the rats to avoid nonfood stimuli (flashing lights and sounds) that the researchers paired with the nausea; it was the association of sickness and food (especially novel food, as later research showed) that was conditioned, despite the long delay between consumption and illness. Garcia argued that, counter to the conventional wisdom of the day, animals are prepared to make associations between nausea and food and that the rules of learning are not independent of the stimuli used.

For humans, the argument for prepared learning of fears extends back more than 30 years (Marks, 1969; Seligman, 1971). According to Seligman (1971), fear responses are more easily made and are more difficult to extinguish for biologically relevant stimuli—stimuli that may have represented danger to our ancestors—than for biologically irrelevant stimuli. Learning is still necessary (e.g., there is not an innate fear of snakes or spiders), but emotional responses are more easily conditioned to objects and situations that would have represented threats to our ancestors than to other stimuli. In support of this, clinical phobias are more likely to be found for evolutionarily-significant stimuli, such as spiders, snakes, heights, and closed spaces, than for other contemporary objects and events that reflect greater danger for modern humans, such as automobiles, knives, or guns (Nesse, 1990). Moreover, in laboratory work, adults conditioned with fear-relevant stimuli display greater resistance to extinction than adults conditioned with fear-irrelevant stimuli (Cook, Hodes, & Lang, 1986; Öhman, 1986; but see McNally, 1987, for exceptions).

Research with monkeys suggests that such "prepared fears" are not limited to humans. Psychologists Michael Cook, Susan Mineka, and their colleagues performed a series of studies examining rhesus monkeys' acquisition of fears to biologically relevant stimuli versus biologically irrelevant ones (M. Cook & Mineka, 1989, 1990; M. Cook, Mineka, Wolkenstein, & Laitsch, 1985; Mineka, 1992; Mineka, Davidson, Cook, & Keir, 1984). In their studies, laboratory-raised monkeys showed no initial fear of snakes (realistic toy snakes were always used as stimuli). Wild-raised monkeys, however, do display fear of real snakes. Cook and Mineka showed laboratory monkeys videos of other monkeys displaying fear reactions to snakes versus fear reactions to "neutral" stimuli such as flowers and rabbits. (A "fear of flowers" video clip was created by combining shots of a monkey's fear response—actually to a snake—with shots of a flower.) The observer monkeys later showed fear of snakes but not of the neutral stimuli. These and related data suggest that the animals were prepared to make fear associations with snakes, which would have been adaptive when living in an environment in which snakes were a major danger. Note that the fear response was not just to "animate objects." Fear reactions were not acquired from watching videos of monkeys reacting fearfully to rabbits.

Although fear responses are proposed to be acquired in infancy and childhood, there has been no published research on humans, to our knowledge, that has looked at the acquisition of children's fear responses as a function of the evolutionary significance of the stimuli.[1] Rather, develop-

---

[1] We are aware of one unpublished conference presentation in which infants displayed fear responses to toy snakes and artificial stimuli that had lifelike qualities when their mothers also responded fearfully (Walden & Passaretti, 1998). The authors interpreted their findings as consistent with

mental psychologists have looked at perceptual and cognitive biases of infants and young children, examining the extent to which (and how) children are prepared to acquire certain classes of ecologically important information (see Gelman & Williams, 1998; Gopnik & Meltzoff, 1997; Mandler, 1998; Spelke & Newport, 1998), and it is to some of this research that we now turn.

What Infants Know Without Being Told (Much)

Cognitive anthropologist Pascul Boyer (1998) proposed that the human mind is equipped with an *intuitive ontology*, which is a set of expectations about the kinds of things that are to be found in the world (see also Geary, 1998, 2001). Boyer proposed that there are five intuitive ontological categories: person, animal, plant, artifact (manufactured objects), and natural object (non-manufactured and nonliving objects, such as mountains and rivers). In addition to these ontological categories, Boyer proposed that humans come into the world with (or develop very early) three classes of intuitive expectations: those related to physical objects, to biological entities, and theory of mind.

We examine two of these classes of intuitive expectations in this book, infants' expectations related to physical objects in this section and theory of mind in the next chapter. Space limitations prevent us from discussing infants' and children's understanding of biological entities (see Wellman & Gelman, 1998). First, we look at the perceptual biases of infants and how they might have played a function in the environment of evolutionary adaptedness.

Perceptual Preferences in Infancy

Not long ago well-informed people believed that infants enter the world unable to perceive sights and sounds. When I (DB) was teaching my first child development class as a graduate student in the early 1970s, I mentioned that newborn infants can "see," meaning that they can tell the difference between two visual displays. A middle-age woman informed me that I was wrong; infants cannot see. She had had four children, and her obstetrician and pediatrician had both told her that her babies were functionally blind at birth and that they "learned" to see over their first month of life. We know now that infants can see (and hear, smell, taste, and "feel") at birth, that there are some stimuli that they prefer to attend to, and that perceptual learning associated with ecologically relevant stimuli

biological preparedness, but the absence of a truly neutral stimulus requires that these findings be interpreted cautiously.

occurs early in life. Although working sensory systems are adaptive for all organisms, an evolutionary developmental psychological approach holds that certain aspects of children's developing sensory systems should be both well suited to their current environment and helpful to them in gaining information that will be useful to them in the near future. We briefly examine evidence from several developing sensory systems consistent with this argument.

Audition

At birth, infants can discriminate sound, and they display certain preferences. For example, they tend to prefer the voices of women, particularly their mother's voice, to those of men (DeCasper & Fifer, 1980; Spence & Freeman, 1996). They also prefer to listen to speech spoken in their mother tongue than in a foreign language (Mehler et al., 1988). The apparent reason for these biases can be found in prenatal auditory experience. Hearing develops before birth, and the voice a fetus is most likely to hear is that of its mother. In a classic study by developmental psychologists Anthony DeCasper and Melanie Spence (1986), mothers read one of two stories to their unborn child during the last six weeks of pregnancy. At birth, infants were fitted with headphones and a pacifier and could control what they heard (the story their mother had read to them in utero or another story) by altering their sucking rate. Neonates varied their sucking rate to hear the stories their mothers had read to them, reflecting prenatal learning but also the establishment of a bias (to attend to what was heard before birth) that may promote attachment and thus survival. Other aspects of infants' auditory abilities that, on the surface, appear to be specifically related to language acquisition are discussed later in the chapter.[2]

Olfaction

Newborns can discriminate among a wide variety of odors (Steiner, 1979), and they develop preferences for certain odors within the first week. For example, in one experiment (Macfarlane, 1975), babies were placed between two breast pads—one from the baby's mother and one from another woman. Infants showed no preference in terms of head turning at 2 days of age, but by Day 6 they turned to their own mother's pad more often than to the pad of another woman. Other research has shown that by 4

---

[2] Infants have also been found to show a preference for well-formed, as opposed to poorly formed, music (Krumhansl & Jusczyk, 1990) and to be able to discriminate out-of-tune versus in-tune sequences of music from different musical traditions (Lynch, Eilers, Oller, & Urbano, 1990). It is interesting to speculate on the evolutionary and biological roots of music (Gardner, 1983) and to propose that human infants are predisposed to acquire musical systems, just as they are seemingly predisposed to acquire language. However, such speculation is beyond the scope of this book.

days of age, infants develop a preference for the odor of milk versus the odor of amniotic fluid (which they had been living in for 9 months; Marlier, Schaal, & Soussignan, 1998), and that bottle-fed, 2-week-old infants prefer the breast odor of a lactating female to that of a nonlactating female (Makin & Porter, 1989). These findings suggest that infants are particularly sensitive to the odor of breast milk and develop a preference for it (and particularly for the odor of their mother's milk) within the first days of life.

Vision

More research has been performed on infant vision than on any other sense. Vision is relatively poor at birth but develops rapidly over the first 6 months (Kellman & Banks, 1998). The lens is not very flexible, so that focusing is difficult. Stimuli that are most likely to be "in focus" for a newborn will be about 20 cm (8 in.) from its eyes (Haynes, White, & Held, 1965), about the distance between the faces of an infant and its mother during nursing. Much (but not all) of an infant's looking behavior during the first two months is controlled by subcortical areas of the brain, reflecting "automatic" as opposed to "purposeful" processes (M. H. Johnson, 1998). By 3 months of age, infants show greater control of their looking behavior, and they display decided preferences for what they like to look at. Among the preferences for physical stimulus features include movement (Haith, 1966), vertical symmetry (i.e., stimuli that are similar on their right- and left-hand sides; Bornstein, Ferdinandsen, & Gross, 1981), and curvilinearity (Ruff & Birch, 1974). These are all characteristics of the human face, and early research showed that infants generally prefer facelike to nonfacelike stimuli (Fantz, 1961).

One somewhat surprising visual preference found in young infants is that for attractive faces (Langlois et al., 1987). Langlois and her colleagues presented pairs of photographs of attractive and less attractive female faces, as judged by groups of college students, to groups of infants ranging between 2–6 months of age. Infants of all ages spent more time looking at the more attractive than at the less attractive faces. This preference was unrelated to how attractive an infant's mother was judged to be. More recent research has shown that this preference extends across sex, race, and age of the modeled face; 6-month-olds consistently showed a preference for attractive faces of both men and women, Black and White adult females, and 3-month-old infants, despite the fact that the infants had little or no experience with some of these classes of faces (Langlois, Ritter, Roggman, & Vaughn, 1991).

One possible explanation for these and related findings is that attractive faces might have more of the physical stimulus characteristics that attract infant attention than do unattractive faces. For example, symmetry is a sign of physical (Gangestad & Thornhill, 1997) and psychological (Shackelford & Larsen, 1997) health and, among adults, facial symmetry is perhaps the most potent factor in determining attractiveness (Gangestad, Thornhill, & Yeo, 1994). An alternative possibility is that infants are born with a schema of a human face, and that attractive faces match that schema better than unattractive ones. Evidence indicating that preference for attractive faces involves more than simply symmetry comes from work with newborns, who displayed a preference for (i.e., looked longer at) attractive faces positioned upright (versus less attractive faces) but not for the same faces presented upside down (Slater, Quinn, Hayes, & Brown, 2000). Such a finding suggests that if infants are born with a schema, or prototype, for a human face, it includes information about its orientation.

Some evidence indicates that attractive faces are those that reflect the "average" of a range of facial features (e.g., average distance between the eyes, average nose length; Langlois & Roggman, 1990; Rhodes & Tremewan, 1996). This would make sense from an evolutionary perspective. If attention to faces is important in forming social relations in early infancy, and thus aids survival, the best bet would be to have an attentional bias toward average features (even if the combination of these average features results in an attractive, and thus "above-average," face). More recent research has found a relationship between attractiveness and averageness not only for human faces, but for dogs, birds, and wristwatches (Halberstadt & Rhodes, 2000). This suggests the possibility that the preference for average faces is built, in part, on a mechanism for averageness in general rather than on a domain-specific mechanism exclusively for faces.

The work of Langlois and her colleagues suggests that by at least age 2 months, infants are influenced by individual differences in the features of human faces. This work also implies that young infants know something about faces, with little in the way of experience necessary for them to express this knowledge. Such knowledge, however, is implicit, and does not imply self-awareness. In retrospect, such findings should not be surprising, for there is likely no single visual stimulus that is of greater significance to a human infant than that of the face of another member of its own species.

Research has demonstrated that infants show a preference for face-like as opposed to nonfacelike stimuli from the first weeks of life. Working with 1-month-old infants, Johnson and his colleagues showed infants differ-ent head-shaped stimuli, moving each stimulus across the babies' line of vision (Johnson, Dziurawiec, Ellis, & Morton, 1991; Johnson & Morton, 1991; Morton & Johnson, 1991). They then measured the extent to which the infants followed each moving stimulus (a) with their eyes and (b) by turning their heads. They found significantly greater eye and head movement to facelike than to nonfacelike stimuli for infants ranging in age from several minutes to 5 weeks. These data suggest that infants are born with some idea of "faceness" and will visually track face-like stimuli more than nonfacelike stimuli. More recent research has found visual preferences for face-like stimuli for newborns (Mondloch et al., 1999).

Morton and Johnson (1991) developed a two-process model to explain infant face preference. An initial process involves primarily subcortical neural pathways. This system is responsible for human newborns' preference for the human face, but because of limited sensory capabilities, infants are not able to learn about specific features of faces until about 2 months of age. Beginning about this time, the second process, which is under the control of cortical brain areas, begins to take over. The functioning of this system depends on cortical maturation and experience with faces during the first two months of life, as infants begin to build a representation that enables them "to discriminate the human face from other stimuli and especially from faces of other species" (p. 178). Recent research has shown, consistent with the model, that 3- and 4-month-old infants are able to use cues from the faces of animals (e.g., dogs and cats) to identify not only specific individuals, but also an animal's species (Quinn & Eimas, 1996). That is, they are able to acquire a category for "dogs" versus "cats" based on facial features.

Other research has found that newborns look longer at the faces of their mothers than those of other women, suggesting an ability to discriminate among faces at birth (Bushnell, Sai, & Mullin, 1989).[3] In other research, 12- to 36-hour-old infants varied their rate of sucking more to see a picture of their mother's face than they did to see a picture of another woman's face (Walton, Bower, & Bower, 1992). These data suggest not only that newborns can tell the difference between faces and nonfaces, but also that they learn a preference for their mothers' faces within the first day of life (see also Walton & Bower, 1993).

Although the research findings are not 100% consistent, they agree that even newborns are predisposed to be attentive to human faces and that learning about faces develops early. This makes sense from an evolutionary perspective; it is also consistent with the speculation of pioneering infant researcher Robert Fantz (1961), who wrote more than 40 years ago that

---

[3] When using infant looking time to determine whether infants can discriminate between two stimuli, it is sufficient to demonstrate that they consistently look more at one stimulus than another. Most research using visual preferences with infants has found that looking time is greater for a novel versus a familiar stimulus. This is not always the case, however, especially for young infants and for infants learning about new classes of stimuli. Although there is debate about what various patterns of looking behavior indicate (Bahrick & Perkins, 1995; Bogartz & Shinskey, 1998), to demonstrate that infants can tell the difference between two stimuli, it does not matter whether they prefer the novel or the familiar.

infants' preferences for facelike patterns may "play an important role in the development of behavior by focusing attention to stimuli that will later have adaptive significance" (p. 72).

Physical Objects

As adults, we take many things for granted about physical objects: Only one can occupy a space at a time; unsupported objects will fall; one object cannot pass through another; objects continue to exist even when they are out of our immediate perception. Do infants enter the world with such knowledge, or must they discover it through experience? If the latter, are these early developing abilities, or ones that take time before being mastered? Neonativists argue the former, proposing that infants have innate knowledge about objects, or are biased to acquire such knowledge very early in life.

Core Knowledge

Writing about young infants' knowledge of continuity and solidity, developmental psychologist and neonativist Elizabeth Spelke (1991) proposed that such knowledge

> may derive from universal, early-developing capacities to represent and reason about the physical world. These capacities may emerge in all infants whose early growth and experience fall within some normal range. They may enable children to infer how any material body will move in any situation. (pp. 160–161)

Spelke (1994; Spelke, Breinlinger, Macomber, & Jacobson, 1992) proposed that there are three core principles of innate knowledge about objects that infants possess: *continuity*, the idea that an object moves from one location to another in a continuous path and cannot be in the same place as another object; *cohesion*, the idea that objects have boundaries and that their components stay connected to one another; and *contact*, the idea that one object must contact another object to make it move. Infants' understanding of these three principles develops with experience, but, according to Spelke, babies possess *core knowledge* (representational innateness) about these domains from birth.

The first of these three domains to be investigated in infants was that of continuity. Figure 6.1 shows a solid rectangle with two bars extending from its top and bottom. Adults infer that the rectangle is occluding a solid bar, although no solid bar is actually seen. What do infants think?

Researchers use changes in infants' visual attention as an indication of what they know. For instance, if a baby is repeatedly shown a stimulus, his

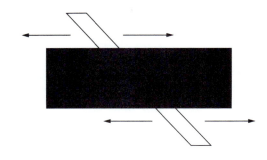

Figure 6.1. Example of Gestalt continuation.

or her attention to it (looking time) will eventually decrease (habituation). If the baby is then shown a new stimulus and recognizes it as different from the earlier stimulus, his or her attention to it will increase (dishabituation). With respect to the concept of continuity, infants who are habituated to the stimulus shown in Figure 6.1 and then shown only a solid bar in the same orientation as the partial bars in the original stimulus, should increase their attention to the new stimulus *only if they believe it is a novel stimulus*. This would indicate that they had not inferred that the two partial bars in the original stimulus were connected. If, however, when shown the solid bar they show little interest in it, the inference is that they are treating it like an "old" stimulus. But it is not literally the old stimulus; it is physically different from the habituated stimulus. If they treat it like an old stimulus, it would be because they have inferred that the rectangle in the original stimulus was occluding a solid bar, which reflects the Gestalt concept of continuity.

Results with infants following procedures similar to this indicate that by 4 months of age they treat the solid bar as if it were an old stimulus, but only for moving stimuli (as shown by the arrows in Figure 6.1) and not for stationary stimuli (S. P. Johnson & Aslin, 1996; Kellman & Spelke, 1983). Similar patterns of result have been reported for the core principles of cohesion and contact. For example, by 6 months of age, infants will look longer when a stationary block of wood moves before it has been touched by another block of wood moving in its direction (the contact principle of innate knowledge; Kotovsky & Baillargeon, 1998), and they will look longer when one object appears to travel along two different paths (the cohesion principle; Spelke, Vishon, & von Hofsten, 1995).

Spelke (1985) speculated that the pattern of data reported for the continuity experiments indicates that infants "know," at some level, that objects are continuous in space; they are born with the notions of the persistence, coherence, and unity of objects. Other research has questioned this position, however, revealing that 2-month-old (Johnson & Aslin, 1995) and 4-month-old (Eizenman & Bertenthal, 1998) infants will show evidence of inferring object unity in some but not other situations and that newborns *increase* their attention to the solid bar (Slater et al., 1990). These data indicate that infants are apparently not born with this knowledge but that it develops very early in life.

In related research, developmental psychologist Renée Baillargeon, Kotovsky, and Needham (1995) have investigated infants' understanding of the idea that an object must be supported or it will fall. In this research, infants are shown possible and impossible events reflecting the idea of support (see Figure 6.2). A gloved hand pushed a box that sat atop a platform from left to right. In the possible event, the box stopped while firmly situated on the platform. In the impossible event, the box was pushed until only 15% of it rested on the platform. If the impossible event violated infants' expectations (i.e., surprised them), they should attend to the event longer than if it did not violate what they thought should happen (Why shouldn't it remain where it is?). Three-month-old infants in these experiments were not surprised and did not look any longer at the impossible than the possible event. So long as the box maintained some contact with the platform, they acted as if they expected it to remain on the platform. Beginning around age 4.5 months, the amount of contact between the box and the platform became important, and by age 6.5 months, infants expected that the box would fall unless a significant portion of it was in contact with the platform.

Figure 6.2. Example of possible and impossible events for object support.
Note. From "How Do Infants Learn About the Physical World?" by R. Baillargeon, 1994, *Current Directions in Psychological Science, 3,* pp. 133–140. Copyright 1994 by Blackwell. Reprinted with permission.

Baillargeon interpreted these findings as reflecting young infants' gradually developing understanding of the role of support. Infants' concepts get progressively revised with experience; initially they believe that any contact between two objects is enough for one to support the other, and eventually they develop, quite early in life, an adult-like concept of support. Rather than postulating that infants have innate knowledge about the nature of objects, Baillargeon proposed that infants' possess innate processes, and these processes, in interaction with their experiences with objects, develop rapidly over infancy. This is a type of starting-state nativism as proposed by Gopnik and Meltzoff (1997) discussed earlier. According to Baillargeon (1994; Baillargeon et al., 1995), infants initially form a preliminary all-or-none concept about a phenomenon. Only with experience does this concept get modified and result in an adult-like understanding.

It is worth noting that in experiments assessing young children's understanding of solidity and support using search tasks (the child must retrieve the object in one of several potential locations), 2-year-olds generally fail to display knowledge of these concepts, unlike their younger 4- and 6-month-old counterparts (Berthier, DeBois, Poirier, Novak, & Clifton, 2000; Hood, Carey, & Prasada, 2000). This is at first perplexing, that infants should have more sophisticated knowledge of the physical world than 2-year-olds. However, it is likely that the nature of the task reflects differences in the nature of the knowledge. The tasks used with older children seem to require explicit understanding (i.e., available to conscious awareness), whereas the looking-time tasks used with infants require only implicit knowledge. It is likely, we propose, that this implicit knowledge develops earlier than explicit knowledge. As such, postulating that infants who look longer at an "impossible" than at a "possible" event have the same type of knowledge that older children have for the phenomenon under question (here, support) is likely unwarranted. What young infants appear to possess, or develop early, is implicit knowledge, which likely cannot be used as flexibly as explicit knowledge (see research by Clements & Perner, 1994, for a similar distinction for children's solutions to false-belief tasks, discussed in chapter 5).

Development of Object Permanence

Perhaps the most studied and most familiar aspect of infants' object knowledge is the concept of *object permanence*. This concept was introduced to developmental psychology by Piaget (1952) and refers to the idea that objects continue to exist independent of one's perception of them. It is typically tested via search tasks, in which an infant must retrieve a hidden object. Research by Piaget, replicated repeatedly across cultures, produces the following developmental pattern. Much before 8 months of age, children do not search for an object under a cloth or cup directly in front of them unless they are moving in the direction of the object when it was hidden. Beginning at about 8 months, infants will retrieve a hidden object but will fail the A-not-B task, in which an object is hidden by an experimenter and retrieved by an infant repeatedly at one location (A) but then hidden, while a child watches, at a second location (B). Not until about age 12 months will infants reliably retrieve the object on the B trials. Finally, not until about age 18 months will infants solve the *invisible displacement* task, in which they watch as an object is hidden under one obstacle (a cloth), but when they remove the cloth they see a second obstacle (a cup); most infants younger than 1.5 years of age will not look under the cup to find the missing object.

Despite the robust nature of these findings (Uzgiris & Hunt, 1975), researchers questioned whether performance on these tasks adequately reflects what infants really know about the permanence of objects. The search requirement of the task may greatly limit the sophistication that young infants are able to display. Researchers taking a neonativist perspective developed techniques using infants' looking time, following the same logic described above for assessing infants' knowledge of core principles of objects, to assess what young infants might understand about the permanence of objects. For example, in a study by Baillargeon (1987), infants watched as a screen was repeatedly rotated from being flat in a box with its leading edge facing the infant, then rising continuously through an arc until it rested in the box with its leading edge farthest away from the infant. Once habituated to this event (i.e., once infants became bored and stopped watching it), they were shown a colorful wooden block with a clown face painted on it, placed to the rear of the flat screen. In the impossible-event condition, the screen was rotated upward (exactly as in the habituation trials) and continued its downward rotation until it lay flat, an impossible event if the object were real in time and space (and achieved by trickery in the experiment). The screen was then rotated upward again, and the wooden block reappeared. If infants have the concept of object permanence, such a series of events (the continuous movement of the screen, despite the presence of an obstacle) should have been impossible. If the wooden object were real in time and space, the screen should have stopped when it reached it. This, in fact, is what infants did see on some trials (the possible event), with the screen stopping where it should have, given that there was an object on the other side. If the infants believed that the wooden block continued to exist, they should have shown surprise and increased looking time at the impossible event relative to the possible event. Infants as young as 3.5 months of age looked significantly longer at the impossible event than at the possible event.

This and related results suggest that infants much younger than proposed by Piaget understand at least the rudiments of object permanence.

Research using the more sensitive looking-time procedures, rather than the more difficult (from the point of view of the infant) search procedures, suggests that infants either enter the world with or develop very early the idea that objects have permanency in time and space.

How Much Do Infants Really Know?

For years, psychologists (and parents) have marveled at infants' rapid changes in cognition from birth to about 2 years of age. In the past several decades, however, neonativists have informed us that the changes are more apparent than real. Infants from a very early age, perhaps from birth, know much about the nature of objects. Experience is necessary to elucidate this knowledge but, according to some, the core knowledge is there.

Perhaps we should not be too surprised that infants are prepared to quickly acquire information about physical objects in the world. The accomplishments we have described in these sections are also made by most, if not all, land animals (vertebrates, at least; Hauser, 2000). Not only humans need to understand continuity, contact, and cohesion. Any relatively large animal living in the world of other animals and objects needs to understand these things, and often more. For instance, all mammals tested on object permanence tasks seemingly pass them, at least at the earliest stages (i.e., recovering a hidden object; Dore & Dumas, 1987). (Fewer species, however, including chimpanzees, gorillas, and, some argue, dogs and monkeys [Dore & Dumas, 1987; Parker & McKinney, 1999], pass the highest level of Piaget's object-permanence scale [invisible displacement].) Thus, in general, we should not be startled that human infants acquire these abilities; what the neonativists are claiming, however, is that infants are specially prepared to acquire this information. What underlies their performance are relatively well-developed domain-specific mechanisms, rather than domain-general mechanisms as proposed by Piaget, and these mechanisms have evolved over millions of years to make life in a world of objects easier to master. There is nothing special about *Homo sapiens* in the acquisition of these abilities (except, perhaps, for the most advanced level of object permanence, which seemingly reflects symbolic representation). All animal species facing the demands of a world with gravity, and with solid, moving objects that sometimes bump into one another, should be prepared similarly by evolution.

We concur with the neonativists that human infants are well-prepared for learning information pertinent to the physical world in which they are born. But do domain-specific skills necessarily underlie these abilities? As we emphasized in the previous chapter, we believe that an evolutionary developmental account must include room for both domain-specific and domain-general mechanisms. With respect to the research findings reviewed here, it seems clear to us that some abilities are quite likely governed by domain-specific mechanisms. For example, infants' attention to and recognition of faces may involve several domain-specific mechanisms, such as those suggested by Morton and Johnson (1991), as well as mechanisms that direct infants to attend to "more attractive" versus "less attractive" faces (Langlois et al., 1987). Such skills are likely limited to face identification and recognition and do not generalize to other aspects of cognition (but recall Halberstadt and Rhodes', 2000, evidence for a general preference for "averageness" extending beyond faces). Similarly, infants' knowledge of the characteristics of objects might be restricted to certain classes of physical relations. For instance, despite their substantial knowledge of the permanence, continuity, and solidity of objects, young infants apparently know nothing about gravity and inertia (Spelke, 1991).

Other aspects of young infants' cognitions might be more domain general in nature. For example, Sophian (1997) has suggested that innate structures are more like tools that might be best suited for some contents (e.g., understanding objects) but can also be used for other purposes (see also Gelman & Williams, 1998). Along these lines, Mandler (1992, 1998, 2000) proposed that infants possess certain general learning or processing abilities that permit them to redescribe their perceptions (presumably including both the sensory and motor representations of the traditional sensorimotor stage) into conceptual form (cf. Karmiloff-Smith, 1992). From this position, it is not necessary to posit substantial amounts of innate knowledge. According to Mandler (1992),

> We need only to grant that infants are born with a capacity to abstract certain kinds of information from perceptual displays that they will process, and to redescribe them into conceptual form . . . I have suggested that this process is operative from at least a few months of age, which would allow concepts to develop in tandem with sensorimotor development, rather than having a later onset. (pp. 226–227)

Others have suggested that infants' seemingly advanced knowledge of the properties of objects requires only a better understanding of the underlying perceptual and memory processes (Bogartz, Shinskey, & Speaker, 1997; Rivera, Wakeley, & Langer, 1999), that there is nothing "special" or "prepared" about such learning.

We believe that infants are prepared by evolution for a world consisting of objects and animate beings. We do not believe, however, that evolution has solved the problem by using only a single "trick." Rather, we propose that infants' understanding of objects and the physical laws relating to them are multiply determined. They might possess some innate knowledge (representational constraints) as Spelke proposed, some innate processing mechanisms (architectural constraints) as suggested by Gopnik and Meltzoff, and at other times their behavior might be better explained by more domain-general perceptual and memory mechanisms rather than by highly constrained domain-specific processes. And most likely, infants' brains are structured so that domain-specific constraints (both representational and architectural) and domain-general mechanisms interact. We emphasize that "experience," some of it surely prenatal (DeCasper & Spence, 1986), is always a necessary ingredient in the recipe. Everything develops, and development involves the continuous and bidirectional interaction at multiple levels of organization, both endogenous and exogenous to the child.

INTUITIVE MATHEMATICS

Many of the cultural accomplishments *Homo sapiens* have made since the invention of civilization can be attributed to their ability to deal with numbers and the relation between numbers. The tendency for people to quantify objects and events is ubiquitous. Although advanced mathematics, such as algebra and calculus, is acquired only through formal education, all cultures have notational systems and techniques of adding and subtracting at least small quantities. Children over the world seem to acquire some basic numerical concepts at about the same time, independent of formal training (Geary, 1994). This pattern has led some researchers to propose that humans' quantitative abilities are not simply the product of increased general intelligence; rather, mathematical cognition has been selected over the course of evolution and is governed by relatively domain-specific mechanisms (Geary, 1995, 1996).

David Geary (1995, 1996) has proposed that several quantitative skills are good candidates for biologically primary abilities (see Exhibit 6.1). Recall from chapter 5 Geary's proposal of biologically primary and secondary abilities. *Biologically primary abilities* are those that have been selected over the course of evolution to deal with problems our ancestors faced. These abilities are universal and develop in a predictable pattern for all but the most deprived individuals. Children will spontaneously exercise these skills (i.e., their use is intrinsically motivated), and most will achieve expert-level functioning. In other words, although some children may acquire the skills before others, individual differences in eventual attainment will be small. *Biologically secondary abilities*, in contrast, are built on the primary abilities and are the product of culture, today often of formal schooling. Keep in mind, however, that despite the claim of primacy for the biologically primary abilities, they still develop and require experience before they can be used effectively.

Potential Biologically Primary Mathematical Abilities

- *Numerosity*—The ability to accurately determine the quantity of small sets of items or events without counting. In humans, accurate numerosity judgments are typically limited to sets of four or fewer items.
- *Ordinality*—A basic understanding of "more than" and "less than" and, later, an understanding of specific ordinal relationships. For example, understanding that 4 > 3, 3 > 2, and 2 > 1. For humans, the limits of this system are not clear, but it is probably limited to quantities < 5.
- *Counting*—Early in development there appears to be a preverbal counting system that can be used for the enumeration of sets up to three, perhaps four, items. With the advent of language and the learning of number words, there appears to be a pan-cultural understanding that serial-ordered number words can be used for counting, measurement, and simple arithmetic.
- *Simple Arithmetic*—Early in development there appears to be a sensitivity to increases (addition) and decreases (subtraction) in the quantity of small sets. This system appears to be limited to the addition or subtraction of items within sets of three, perhaps four, items.

Note. From "Reflections of Evolution and Culture in Children's Cognition: Implications for Mathematical Development and Instruction," by D. C. Geary, 1995, *American Psychologist, 50*, p. 36. Copyright 1995 by American Psychological Association. Reprinted with permission.

Numerosity

The first biologically primary ability proposed by Geary is *numerosity*, which refers to the ability to determine quickly the number of items in a set without counting. When referring to "determining the number of items in a set," this does not necessarily mean understanding what "two" versus "three" means but rather being able to discriminate consistently between the number of items within small arrays. This is seemingly a basic ability, based possibly more on simple perceptual skills than on abstract quantitative cognition, and perhaps we should not be surprised that it is found early in life. For example, it has been observed in many mammal and bird species (Davis & Pérusse, 1988), including cats, chimpanzees, and an African grey parrot.

However, there is evidence that numerosity judgments are made very early in humans and that they involve more than simple perceptual abilities. For example, infants within the first week of life can discriminate between arrays containing up to three (and sometimes four) items (Antell & Keating, 1983; Starkey, Spelke, & Gelman, 1990; van Loosbroek & Smitsman, 1990). Moreover, infants can make these judgments when arrays are stationary or in motion, when sets consist of the same as opposed to different types of items, and when contrasts must be made intermodally. Infants' ability to make numerosity judgments between two different sensory modalities may reflect not merely more general perceptual processing but also a degree of abstraction. This was demonstrated in a series of studies in which 6- to 9-month-old infants were simultaneously shown arrays of two or three objects and heard either two or three drum beats. Infants looked significantly longer at the visual array corresponding to the number of drum beats (Starkey et al., 1983, 1990; see also footnote 3).

This finding is particularly interesting because 3-year-old children have difficulty making audiovisual matches, that is, selecting visual arrays consisting of the same number of discrete auditory signals (Mix, Huttenlocher, & Levine, 1996). The lack of continuity between the performance of 6-month-old infants and 3-year-old children may reflect how this knowledge is represented and accessed. The infants' knowledge was represented implicitly and assessed via a looking-time procedure. In contrast, the preschool children's knowledge was assessed via an explicit task. It is likely that cross-modal judgments early in life can be made only through implicit cognition and that such knowledge does not become available to conscious awareness until much later in development. (This is similar to the discrepancy discussed earlier between the knowledge of solidity shown between 4-month-old and 2-year-old children; Berthier et al., 2000; Hood et al., 2000.) Recall the results of Clements and Perner's (1994) study discussed in chapter 5 on implicit versus explicit performance on a false-belief task. In that study, children "passed" the false-belief task earlier when a relatively passive, implicit measure (looking time) was used as opposed to a verbal (explicit) measure. Consistent with the Clements and Perner findings and our discussion of implicit and explicit cognition in chapter 5 is the suggestion that intermodal (and thus abstract) numerosity judgments may be available to infants and young children only implicitly, and not until much later is such knowledge available to conscious awareness.

Ordinality

Ordinality refers to a basic understanding of more-than and less-than relationships and seems to develop later in infancy, after an understanding of numerosity has been attained. In one study, 16-month-old infants were conditioned to touch the side of a screen containing either the smaller or larger array of dots (Strauss & Curtis, 1984). For example, an infant might be repeatedly shown arrays of three and four dots and reinforced consistently for touching the array with three dots. After training was completed, infants were presented arrays with different numbers of dots, in the present example, two versus three dots. If they had learned merely to respond to the absolute number of dots in an array, they should continue to point to the array with three dots in the transfer phase; however, if they instead had learned an ordinal relationship (select the array with the smaller number of dots), they should touch the array consisting of two dots. Infants did the latter, suggesting they had learned an ordinal relationship.

Research with mammals and some birds has convincingly shown that they are able to learn the concept of ordinality (Boysen, 1993; Gallistel, 1990; Pepperberg, 1994). However, such evidence has typically been derived only after many hours, in some cases months and years, of training. The extensive training required for nonhuman animals to display a concept as basic as ordinality suggests that it may not be a very salient aspect of the environment for them. However, a seminaturalistic study has shown that rhesus monkeys do understand the more-than and less-than relationships when small numbers are involved (Hauser, Carey, & Hauser, 2000). In this study, rhesus monkeys (*Macaca mulatta*) living on an island had been habituated to humans. The monkeys watched (one at a time) as researchers placed pieces of apples, one at a time, under two distinctive opaque containers. The number of pieces of apple placed in the two containers differed, ranging from one piece in one container and none in the other, to eight pieces in one and three in the other. The researchers then walked away and noted which container the monkey first approached. When the number of apple pieces in the two boxes varied by only one, and the total number of apple pieces in the most numerous box did not exceed four, the monkeys consistently approached the box with more pieces in it. That is, for contrasts of 0 versus 1, 1 versus 2, 2 versus 3, and 3 versus 4, the monkeys consistently approached the box with the larger cache of apples. When the number of pieces placed in the boxes was larger, the monkeys were less likely to consistently "approach the larger" first. These findings suggest that monkeys do have a "natural" understanding of ordinality, at least for quantities up to three or four, and that extensive training is not necessary for them to demonstrate this ability. These findings also demonstrate that human's ordinality skills are evolutionarily quite old and not unique to our species.

Counting

Young children around the world enumerate small sets of items, using variants of the number of words (e.g., "one," "two," "three") available in their culture. Although such counting is not observed until children are able to talk, several theorists have proposed that this ability is based on implicit, skeletal principles evident in infancy (Geary, 1994; Gelman & Gallistel, 1978). Although many children can count almost as soon as they can talk, they do not typically practice "mature" counting (i.e., similar to that practiced by adults in their community) until late in the preschool years. According to Gelman and Gallistel (1978), counting involves five principles:

1. *The one-one principle*: Each item in an array is associated with one and only one number name (e.g., "two").
2. *The stable-order principle*: Number names must be in a stable, repeatable order.
3. *The cardinal principle*: The final number in a series represents the quantity of the set.
4. *The abstraction principle*: The first three principles can be applied to any array or collection of entities, physical (e.g., chairs, jelly beans) or nonphysical (e.g., minds in a room, ideas).
5. *The order-irrelevant principle*: The order in which things are counted is irrelevant.

Gelman and Gallistel referred to the first three principles as the "how-to" principles of counting and proposed that children as young as 2.5 years of age demonstrate knowledge of them under some circumstances. For example, children as young as 3 years of age will use the one-to-one principle in counting arrays of five or less, and most will use a stable number sequence. However, children sometimes use an idiosyncratic list of number words (e.g., "one, two, six"), but they use this list consistently (see Geary, 1994; Gelman & Gallistel, 1978).

One technique that has been useful in determining what features young children believe are necessary for proper counting is to have them observe a puppet count an array of objects and ask them whether the puppet was correct or not. Using this procedure with 3-, 4-, and 5-year-olds, Briars and Siegler (1984) concluded that children's understanding that one-one correspondence and stable order were necessary for accurate counting increased over the preschool years (30%, 90%, and 100% for the 3-, 4-, and 5-year-olds, respectively). However, 60% of the 5-year-olds also viewed other features as essential, such as beginning to count at an end rather than in the middle of an array and pointing to each object only once. In other words, young children learn the critical features of counting by 4 years of age but infer, from watching others, additional features that are characteristic of, but not necessary for, proper counting.

Simple Arithmetic

Geary's proposal that addition and subtraction of small quantities reflect a biologically primary ability has received support from research with infants. In an initial study, developmental psychologist Karen Wynn (1992) showed 5-month-old infants a sequence of possible and impossible events that involved the addition or subtraction of elements. For example, infants sat in front of a stage and watched as one object was placed on it. A screen was then raised, covering the object. This was followed by a hand placing a second object behind the screen. The hand then left empty. The arithmetic logic here is that there are now two objects behind the screen, that is, 1 + 1 = 2. In the possible event, the screen was then lowered, and the two objects that the infants had previously seen placed on the stage were indeed there. For the impossible event, when the screen was lowered, only one object was there. Following the logic we described earlier in this chapter that infants should increase their looking time when a result violates their expectation, we would assume that infants would look longer at the impossible than the possible event only if they have some basic conception of addition. This was, in fact, Wynn's finding, and it has been replicated several times (T. S. Simon, Hespos, & Rochat, 1995; Uller, Carey, Huntley-Fenner, & Klatt, 1999; see Wakeley, Rivera, & Langer, 2000, for an exception).

As with the research evidence of object permanence in young infants discussed earlier, there have been alternative interpretations of Wynn's results. Such patterns may be the product of more general perceptual or attention mechanisms and not the product of an "arithmetic module" per se (Haith & Benson, 1998; T. J. Simon, 1997). For example, rather than reflecting infants' abstract understanding of integers (i.e., there should be 1 or 2 objects behind the screen), performance on such tasks may be based on representations of the actual objects (e.g., ♥ versus ♥ ♥), suggesting that decisions are based more on perceptual than conceptual relations (Uller et al., 1999; see Mandler, 2000).

There have been at least two sets of studies demonstrating simple arithmetic abilities in nonhuman primates. One study used a paradigm similar to Wynn's to assess understanding of subtraction in free-living rhesus monkeys habituated to humans (Sulkowski & Hauser, 2001). Individual monkeys were shown food objects placed on two side-by-side stages. Screens were then raised, hiding the food. In some conditions, food objects were then removed while the monkey watched, altering the number of objects behind the screen. The humans then backed away, and the subject was allowed to approach the stages. If monkeys can subtract, they should be able to keep track of how many food objects are behind the two screens and approach first the stage with the greatest number of food items. This is exactly what the monkeys did, so long as the number of items initially placed behind the screen did not exceed three.

The other set of studies demonstrating arithmetic abilities in nonhuman primates was performed with an enculturated chimpanzee. For example, comparative psychologist Sally Boysen and her colleagues taught the chimpanzee Sheba the Arabic numerals 1–8. When a number was shown on a video screen (e.g., 2), Sheba would have to point to an array containing two objects to receive a reward (see Boysen, 1993). In one set of experiments (Boysen & Berntson, 1989), one to four orange slices were placed at two of three sites in the laboratory. Sheba's task was to inspect the sites and return to a home base and select the Arabic numeral that corresponded to the sum of the orange slices. So, for example, if one site had two orange slices, a second had three, and the third had none, Sheba would have to select the Arabic numeral 5 to be correct. This she did on the very first trial, requiring virtually no training. In a second experiment, the arrays of oranges were replaced by Arabic numerals. So instead of finding two and three orange slices at two sites as in the example above, she would find only the numerals 2 and 3. Sheba performed significantly above chance on this task beginning with the first experimental session. Sheba's simple arithmetic performance is comparable to the simple counting strategies observed for 3- and 4-year-old children (Starkey & Gelman, 1982) and suggests that simple, humanlike addition abilities can be used by at least some chimpanzees.

Basic quantitative abilities are important not only for humans but for many species. Humans, however, have made especially great use of mathematics, and Geary's claim that there are a handful of biologically primary quantitative abilities finds support in the infancy and child development literatures. We are also intrigued by the evidence that some monkeys and chimpanzees can develop counting and simple arithmetic skills, suggesting that the origins of humanlike mathematics may extend back to our simian ancestors. We should emphasize that despite the primacy of these abilities, they do not appear, *de novo*, fully formed when first seen. Rather, they develop over infancy and childhood and, interestingly, there may be a discontinuity in their development, with early expression of these skills being implicit in nature, whereas the more mature forms are expressions of explicit cognition. Despite the unresolved issues, we believe that the evidence is strong that humans are prepared to deal with quantitative relations, that such preparation exceeds that shown by other species, and that it is displayed early in ontogeny.

SPATIAL COGNITION

The abilities to locate objects in space and to navigate through space are critical not only for humans but for most animals. In fact, humans' spatial abilities are greatly inferior to those of many other animals. The migratory behavior of many birds (and insects such as the monarch butterfly, fish such as Pacific salmon, and reptiles such as the loggerhead turtle) makes humans' spatial memory and navigational skills look truly puny (Gallistel, 1990). Although the wayfinding behavior of these species can (often) be described appropriately as inflexible, the spatial memory of the Clark's nutcracker cannot be. Clark's nutcrackers store seeds in the fall and return months later to retrieve them. Observations of these birds have demonstrated that they can store up to 33,000 seeds in more than 6,600 locations and recover most of them months later (Hauser, 2000). When looking at the variety of impressive spatial abilities displayed by a broad range of species, it seems wholly reasonable to assume that natural selection has worked to prepare animals with relatively specific mechanisms to find their way through space and to remember spatial locations, specific to the ecological needs of each species. *Homo sapiens* should be no exception.

Development of Spatial Abilities in Children

Spatial abilities in humans clearly develop with age. But infants seem to represent spatial location early in life and to use the location of an object, rather than other characteristics such as shape or color, to define that object. In one set of experiments (Quinn, 1994), 3-month-old infants were habituated to visual displays in which a dot was presented in one of several locations above a solid bar. (Other infants were habituated to dots consistently presented below a solid bar.) After habituating, infants were presented with two new stimuli: One was a dot presented in a new location but still above the bar, and the other was a dot presented below the bar. Both stimuli were "new" to the extent that neither had been shown to the infant previously. However, the stimulus with the dot above the bar was similar to the other above-the-bar stimuli the infants had seen during the habituation trials. If infants categorize spatial location, they should look longer at the stimulus with the dot below the bar than the one with the dot above the bar. This is what the infants did, indicating that they were able to abstract a spatial (geometric) relation.

In other research, 5-month-old infants watched as an object was hidden in a specific location in a sandbox. After a 10-second delay, the object was dug out. This was repeated four times. On the fifth occasion, instead of digging up the object at the location where it had been hidden, the experimenter dug out the object from a different location (as near as 6 inches from where the first had been hidden). Infants looked significantly longer when the object was retrieved from a location at which it had not been hidden, indicating that these infants had coded the specific location of the objects (Newcombe, Huttenlocher, & Learmonth, 1999). In a final experiment in this report, infants were not surprised (i.e., did not look longer) when a different object was retrieved from the sand. This finding suggests that spatiotemporal characteristics may play a central role in defining objects to young infants, with shape and color being unimportant.

The finding of infants being able to "solve" this problem is particularly interesting, given that a similar task, in which children must retrieve an object hidden in a sandbox after they moved to the other end of the box, is typically not solved above chance levels until 21 months of age

(Newcombe, Huttenlocher, Drummey, & Wiley, 1998). Perhaps the act of moving around the sandbox makes the task too demanding for children's limited mental resources or disrupts their memory, accounting for the discrepancy. Another possibility refers to the difference between the implicit and explicit natures of the two tasks. Similar to the argument we made with respect to judgments of numerosity, children may understand spatial relations at first only implicitly, with explicit understanding, as reflected by an overt search task, being displayed only later in ontogeny.

Despite its early appearance and its saliency in identifying objects, spatial cognition in infancy does not simply "mature" but develops as a result of experience. For example, depth perception can be assessed by how infants behave on the visual cliff, an apparatus that consists of a glass-topped table with a board across its center. On one side of the board the infant sees a checkerboard pattern directly under the glass. This is referred to as the "shallow" side. On the other side the checkerboard pattern is several feet below the glass; this side is referred to as the "deep side." In the original research using this apparatus, infants rarely crawled over the deep side to their calling mothers, suggesting that once infants can crawl, they display fear, indicating that little learning about depth is necessary (Walk & Gibson, 1961). However, more recent research has found that the tendency to show fear on the visual cliff is related to the extent of previous locomotor experience (Bertenthal, Campos, & Barrett, 1984; Bertenthal, Campos, & Kermoian, 1994). Infants with more locomotor experience are more likely to show fear than their less experienced peers, suggesting that such experience is related to depth perception. In explaining these results Bertenthal and his colleagues (1994) suggest that the

> active control of locomotion, unlike passive locomotion, demands continuous updating of one's orientation relative to the spatial layout. This information is provided through multimodal sources, such as visual and vestibular coding of angular acceleration. With locomotor experience, changes in angular acceleration detected by the visual system are mapped onto analogous changes detected by the vestibular system. Fear or avoidance ensues when the expected mapping between visual and vestibular information is violated. (p. 142)

Classic research with kittens (Held & Hein, 1963) demonstrated that self-propelled movement is critical in the development of depth perception. Kittens were raised in a darkened room and had visual experience only during training and testing. Pairs of kittens of the same age were used, one given normal visual-motor experience and the other given visual experience without associated motor experience. To do this, a special training apparatus was developed. The *active* kitten was harnessed to walk around the brightly decorated track while the *passive* kitten was placed in a gondola to have identical visual experience without the associated motor feedback. That is, although both kittens had the same visual experience, and both had motor experience in their darkened room, only the active kitten had visual experience concomitant with movement. After training, both kittens were tested on the visual cliff to determine if they showed a preference for sides. The passive kittens showed no preference for sides, indicating a lack of depth perception. By contrast, the active kittens consistently chose the "shallow" side, indicating that they could perceive depth. In research with humans, 8.5-month-old infants were divided into three groups: (a) prelocomotive (i.e., not yet crawling); (b) prelocomotive but with experience using a walker; and (c) locomotive (i.e., crawling; Kermoian & Campos, 1988). The infants were then tested with a series of tasks in which they had to retrieve an object hidden under a cloth. Infants with locomotive experience, either by crawling or in a walker, displayed more advanced performance on the object-retrieval task than did the noncrawlers. Also, there was no difference in performance between the crawlers and the walkers, suggesting that it was the locomotor experience and not maturation that was responsible for the advanced spatial memory. Other research has similarly shown advantages on some visual–spatial tasks associated with self-produced locomotion, although these effects tend to be limited to specific abilities (Arterberry, Yonas, & Bensen, 1989; Bai & Bertenthal, 1992).

The infant literature is consistent with the position that humans are born with an ability to code spatial relations and that they may use spatial location as a primary cue in defining objects. That is, where an object is may be the defining feature of that object. However, spatial abilities improve with age and develop in relation to infants' self-propelled movements. Spatial skills can get very complex, and substantial development occurs in these abilities during childhood. One interesting finding in spatial abilities concerns sex differences that have been reported consistently beginning during the preschool years (Halpern, 1996; Maccoby & Jacklin, 1974). From an evolutionary psychological perspective, males and females have somewhat different self-interests and have evolved different strategies to best deal with the problems they have faced. Evolutionary psychologists Irwin Silverman and Marion Eals (1992) have interpreted the sex differences observed in males and females from an evolutionary perspective, proposing that the pattern of differences makes sense given the division of labor that has been assumed for ancient men and women.

Sex Differences in Spatial Abilities

There is a division of labor among males and females in traditional cultures that likely reflects the way our ancestors lived in the Pleistocene. Although men engage in some gathering, the tasks of finding and gathering edible fruits and vegetables, typically found close to the home base, is primarily the job of women. In contrast, hunting is nearly exclusively in the male domain, with men often traveling far from the home base over a series of days in pursuit of game. Given this division of labor, spatial skills that permitted men to mentally manipulate space (cognitive maps of geographic locales) would be to their benefit because of the advantage this would provide in navigating over large areas.[4] Relatedly, males' greater use of projectile weapons, in the service of male–male competition, may also have contributed to a male advantage in some aspects of spatial cognition (e.g., the ability to track objects moving in space; Geary, 1998). In contrast, women may be most advantaged with spatial skills that permit them to recall the location of specific caches of fruit or edible plants and for making fine perceptual discriminations, such as between ripe and unripe berries. In general, sex differences, beginning in childhood, have been found in the predicted direction in these areas.

For example, males, beginning in the preschool years (Levine, Huttenlocher, Taylor, & Langrock, 1999) tend to perform better than females on tasks that involve manipulating spatial relations (Baenninger & Newcombe, 1995; Casey, 1996), finding their way through physical (or virtual) environments (Moffat, Hampson, & Hatzipantelis, 1998; Silverman et al., 2000), making maps (Matthews, 1987), and using maps (Dabbs, Chang, Strong, & Milun, 1998; Gibbs & Wilson, 1999; Money, Alexander, & Walker, 1965). In one study, kindergarten, second-grade, and fifth-grade children walked repeatedly through a large model town consisting of buildings, streets, railroad tracks, and trees. Following the walks, the children were asked to re-create the layout from memory. Performance increased with age, with boys performing better than girls (Herman & Siegel, 1978). This pattern is what one would expect if males were prepared by natural selection for spatial skills related to navigating large spaces.

In comparison to these spatial–analytic skills, there is a consistent female advantage for tasks that assess object-location memory. For example, females are better than males at remembering the locations of each item in an array of objects, with this ability increasing with age for both sexes (Eals & Silverman, 1994; Silverman & Eals, 1992). In several of the studies by Silverman and Eals, participants ranging in age from 8.5 years to college adults were shown an array consisting of a variety of objects. Sometime later, they were shown a larger array and were to cross out all items that

---

[4]Sex differences in spatial ability favoring males have been found in nonhuman mammalian species in which males have a larger home range during mating season than do females; the reverse pattern is observed in species in which females are more active in spatial search (see Sherry, 2000). Spatial ability in these species is also correlated with the size of the hippocampus. Thus, the pattern of sex differences observed in humans is seemingly tied to ancestral lifestyle differences and is analogous to that of other species.

were not in the original array. At all ages tested, females performed better than males. When asked about the correct location of items (e.g., where in an array a particular item was originally located), females again performed better than males, although differences were not significant until adolescence (see also Kail & Siegel, 1977). This pattern is consistent with Silverman and Eals' theory that natural selection has favored such detailed spatial–perceptual skills in females. (These results are also reminiscent of the "Where's the mayonnaise?" phenomenon, frequently observed in some homes, in which men have a difficult time finding a specific item, such as the mayonnaise jar, when it is among a large collection of somewhat similar objects in the refrigerator.)

Although our interpretation of these results is similar to that of Silverman and Eals (1992)—that there have been differential selection pressures on hominid males and females for the development of spatial skills and that this is reflected in the abilities of modern men and women—it does not address the proximal mechanisms responsible for these differences. One possibility is that differences in brain organization, mediated relatively directly by the action of different genes, or indirectly through differences in hormones, are responsible for these effects. An alternative possibility is that sex differences in locomotor play between boys and girls may be responsible, in part, for some of the sex differences observed in spatial cognition (Bjorklund & Brown, 1998; see also chapter 10 for a discussion of sex differences in play style). In other research, Newcombe and her colleagues have found a significant relation between adult females' spatial abilities and the extent to which they engage in everyday tasks involving high-spatial content (Newcombe, Bandura, & Taylor, 1983). Similarly, sex differences in children's abilities to form cognitive maps of large-scale environments is associated with sex differences in experience with the environment. For example, across a wide range of cultures, boys tend to have a larger exploration range than girls and, as a result, may acquire more information about their physical surroundings than girls (Matthews, 1992). Support for this comes from a study that equalized the experience preschool girls and boys had for their physical environment and found no sex difference in spatial–orientation task performance (Hazen, 1982).

In one study with preschool children, researchers found sex differences in visual–spatial play activities, with boys engaging in more spatial activities than girls (Connor & Serbin, 1977). The amount of boys' visual–spatial play correlated significantly with their performance on the Block Design subtest of the Wechsler Intelligence Scale for Children (Wechsler, 1974) and the Preschool Embedded Figures Test (Karp & Konstadt, 1971), suggesting that sex differences in play activity are responsible in part for boys' generally greater spatial skills. However, other studies have found smaller differences in spatial play between preschool boys and girls and no significant relation between spatial play and spatial cognition (Caldera, O'Brien, Truglio, Alvarez, & Huston, 1999). Thus, the relation between physical play, spatial cognition, and sex differences is not a simple one. It would be interesting to examine more specific relations between locomotor activity and spatial cognition to see how these effects, if they exist, vary with age. For example, are there sex differences in the specific types of physical play that boys and girls engage in, and do some of these types (e.g., play involving eye-hand coordination) predict spatial abilities better than others?

Clearly spatial abilities improve with age and are largely a result of the specific experiences children have, both in infancy and in later childhood. The results of a meta-analysis support this contention, finding significant relations between children's spatial cognition and locomotor experiences, with these effects being greater during early childhood than during the infancy and toddler periods (Yan, Thomas, & Downing, 1998). That sex differences in spatial abilities may also be partially attributed to differences in locomotor experience is provocative, but obviously not the entire answer. Psychologist M. Beth Casey (1996; Casey, Nuttall, & Peraris, 1999) has suggested that a combination of genetic and experiential factors likely contribute to the typically observed sex differences, with most males being biased from an early age toward activities involving spatial cognition, such as block building, carpentry, and throwing and catching objects; this experience eventually leads to enhanced spatial skills. Geary (1998) has made a similar proposal, stating that sex hormones influence boys' and girls' tendencies to explore environments, which in turn affects the development of the brain. Hormones, however, also directly affect the organization of the brain, which in turn influences spatial cognition. The bulk of the evidence, we believe, argues strongly that humans are prepared to process spatial information and that males and females, because of selection pressures experienced by their ancestors, develop, on average, different strengths in spatial skills. Although we are confident that both the universal and sex-differentiated skills are mediated by inherited (genetic) differences, proximal factors, such as locomotor and play experiences, contribute significantly to these patterns, consistent with the argument that development proceeds as the result of the continuous and bidirectional relationship between multiple levels of organization over time, and is not the simple product of either "genes" or "environment."

LANGUAGE ACQUISITION

The area in which the most ink has been spilled concerning *Homo sapiens'* special preparation for a cognitive ability is that of language. In fact, the modern neonativist perspective can trace its roots to the pioneering work of linguist Noam Chomsky (1957), who challenged the behaviorist–empiricist perspective popular at the time and proposed that children were born with a "mental organ" specially designed for the acquisition of language. We will not attempt to review the extensive literature that has arisen around this issue. (An excellent, thorough, and readable account of the contemporary neonativist perspective on language acquisition can be found in Steven Pinker's 1994 book, *The Language Instinct: How the Mind Creates Language*.) We focus here on presenting the basic argument of the neonativists that humans are specially prepared for language, summarizing briefly some of the research that we think is especially compelling in support of this position, particularly evidence for a sensitive period for language acquisition. We also present arguments and data suggesting that social factors and domain-general mechanisms may play more of a role than most neonativists propose.

Neonativist Perspective

Universal Grammar

The most basic claim of neonativists is that children are born with a *universal grammar*, which is some primitive knowledge about the structure, or syntax, of language. In other words, children have a "theory of syntax" in their brains that consists of the most basic grammatical rules that typify all languages. Children also possess a *language acquisition device*, which is the mental organ that compares the language input that children hear around them with their innate theory (universal grammar), makes modifications, and, eventually, permits children to understand and speak their mother tongue (see Hoff 2001; Pinker, 1994).

Infants, of course, are not born knowing any language. Rather, they possess a set of principles and parameters that guide their interpretation of speech. *Parameters* refers to aspects of grammar that vary across language. For instance, although all languages have syntactic subjects, some languages require that proper sentences state the subject (as in English, "I love you"), whereas in other languages the subject can be expressed as part of the verb, eliminating the necessity of stating the subject (as in Spanish, "Te amo"). The presence of a subject is common to all languages, but how that subject is expressed is not, and children must learn whether their language requires the subject explicitly stated (as in English) or not (as in Spanish). Consistent with the theory of universal grammar, the early sentences of English-speaking children are often similar to those of languages that do not require an explicit subject, as in statements such as "Going store" or "drink milk" (Bloom, Lightbown, & Hood, 1975). Other examples consistent with a universal grammar include the facts that all languages have (a) vocabularies divided into different part-of-speech categories including verbs and nouns; (b) movement of grammatical categories, such as the subject–auxiliary inversion used to transform declarative sentences in English into questions (e.g., "I am going" transformed to "Am I going?"); (c) words organized into phrases that follow a similar underlying rule system (called the X-bar system); and (d) prefixes or suffixes for verbs and nouns (e.g., in English, adding "s" to make a noun plural; Hoff, 2001; Pinker, 1994). The similar way and rate that children acquire many of the same grammatical forms over the world is further evidence for a universal grammar.

Another source of support for a universal grammar comes from situations in which children "invent" new languages when exposed to poorly structured verbal codes. Linguist Derek Bickerton (1990) studied groups of people who spoke different languages but who were brought together, usually to work (e.g., on Hawaiian pineapple plantations), and when they were not given the opportunity to participate in the majority culture or to use the language of their homeland, they developed a new verbal communication system. These makeshift systems are termed *pidgins*, and they combine several languages at a rudimentary level and are used to convey necessary information within the group and between the group and its "hosts." Word order is often highly variable, and there is little in the way of a grammatical system.

Children can rapidly modify pidgins into true languages, called *creoles*. Creoles are grammatically structured languages, possessing their own syntax, that reflect the integration of the majority language (usually that of the host) with the variety of other languages spoken by members of the community. (Most of Bickerton's work was based on his own studies of Hawaiian Creole, although there is historical evidence of the formation of creoles in other parts of the world.) In a single generation, children in a pidgin-speaking community transform that pidgin into a "real" language. There is no simple explanation for this phenomenon, but children's creation of a creole suggests that they possess an innate grammar and use it to "correct" the fragmented pidgin spoken by their parents and convert it into a true (and new) language.

Recent evidence consistent with Bickerton's hypothesis that children create true language from the pidgins they hear comes from a study of several cohorts of children creating Nicaraguan Sign Language (Senghas & Coppola, 2001). Deaf education in Nicaraguan began only in the 1970s, with deaf children before this time having little contact with one another. There was also no recognized form of Nicaraguan Sign Language for children to be taught. Such a language began to emerge, however, in the Managua school for the deaf and was well developed by the late 1990s. Senghas and Coppola (2001) tested deaf Nicaraguan signers who had first been exposed to Nicaraguan Sign Language as early as 1978 (Cohort 1) or as late as 1990 (Cohort 2) and examined changes in early linguistic structures (specifically spatial modulations) in the newly emerged sign language. They found that Nicaraguan Sign Language was systematically modified from one cohort of children to the next, with children aged 10 years and younger generating most of the changes. In other words, sequence of children created a new sign language from the incomplete forms used by their predecessors.

Bickerton is one of many theorists who has speculated on the evolution of language. Although we will not go into his theory (or those of others) here, we note his belief that pidgins, as well as the "language" used by signing apes (e.g., Washoe the chimpanzee; Gardner & Gardner, 1969), feral children (e.g., the "wild boy of Aveyron"; Itard, 1962), and the speech characteristic of toddlers, reflect what he termed *protolanguage,* a type of speech that may have characterized our hominid ancestors. According to Bickerton (1990), this protolanguage was a maplike representation of the world with hierarchical categories and simple infrastructure. What it lacked was syntax. Its presence in pidgins, young children, and possibly apes under some circumstances, reflects its primitiveness and greater resistance to neural damage and may give us a glimpse of how our prelanguage ancestors communicated.

Sensitive Periods and Language Acquisition

Consistent with the neonativists' perspective of language acquisition (but not necessarily inconsistent with other viewpoints) is the idea that there are sensitive periods in the acquisition of language. A child's brain is prepared to make sense of language input early in life, and this developmental window closes (or at least becomes partially shut) with time. Evidence for sensitive periods in language acquisition comes from at least four sources (Locke, 1993): children who are not exposed to significant amounts of language until later in life; second-language learning; deaf children learning sign language; and recovery of function from brain damage.

The first class of evidence is the most controversial. Data come from so-called feral children, who have been discovered late in childhood apparently living to that point without significant human contact (Itard, 1962), and from socially deprived children who, because of pathological parents, are isolated and hear little language directed to them. When such children are not identified until adolescence, they fail to achieve a level of language proficiency comparable to most 3-year-olds, even after intensive education. The best documented case is that of Genie, a child who was discovered at 13 years of age and had been living chained to her bed, virtually ignored by the rest of her family (Curtiss, 1977). Genie later developed substantial social and problem-solving skills and learned many vocabulary items. However, her language was similar to the protolanguage described by Bickerton, essentially lacking the grammatical structure that makes language unique.

When socially deprived children have been discovered and rehabilitated before adolescence, the likelihood they will develop full language is greater. For example, Koluchova (1976) documented a case of monozygotic twins who experienced abuse, neglect, and malnutrition between ages 11 months and 7 years 2 months. When discovered at age 7, neither twin had any language. The children were placed in foster care and received educational enrichment. When the children were tested at age 11, they had IQs of 93 and 95 and apparently normal language proficiency. Although evidence from socially deprived children is consistent with the sensitive-period hypothesis, children who are deprived for extended periods of time also likely have more general intellectual retardation (or their retardation as infants may have been the impetus for their isolation), making any definitive statement about the role of developmental timing impossible.

The largest body of evidence for a sensitive period comes from studies involving second-language learning, with eventual proficiency in a foreign tongue being related to the age of first exposure. This is nicely illustrated in research by Johnson and Newport (1989), who tested native Chinese and Korean speakers who had emigrated to the United States and learned English as a second language. The age of arrival to the United States varied for participants in the study between 3 and 39 years, and they had lived in their new country between 3 and 26 years at the time of the study (all were educated adults). Participants were given sets of English sentences and asked to determine whether each was grammatically correct. People who had emigrated to the United States between the ages of 3 and 7 years scored comparably to native English speakers. Performance on this test declined steadily as a function of age of arrival in the United States (*not* the number of years residing in the United States) from 8–10 years to adulthood. Although it might not be surprising that immigrants maintain an accent when they do not learn a second language until their teenage years, these data demonstrate that their syntax rarely reaches the level of a native speaker, even for people who have resided in a country for several decades.

Perhaps some of the most compelling evidence in support of the sensitive-period hypothesis comes from differences in proficiency in sign language among deaf people as a function of when they were first introduced to signing. Unlike the research with second-language learning, sign language is the first language of the deaf, and many deaf children born to hearing parents are not exposed to formal sign language until later in life. Newport (1990), using a design similar to that used in the Johnson and Newport (1989) study of second-language learning discussed earlier, reported that the grammatical proficiency of deaf adults learning American Sign Language was related to age of first exposure and not to the number of years they had been using the language.

Finally, evidence from recovery of function from brain damage reveals that the age at which language areas of the brain are damaged or removed is related to eventual language competence (Witelson, 1987). When damage occurs early in life, particularly prior to adolescence, the prognosis for full (or nearly full) recovery is good. The older a person is when the brain damage occurs, the less language recovery, on average, one is likely to find.

Related to issues of developmental timing and language acquisition is research that has identified differences in brain organization between bilingual individuals who acquired their second language in early childhood versus in adolescence or adulthood (Kim, Relkin, Lee, & Hirsch, 1997). Using brain-imaging techniques, patterns of brain activity were taken of "early" and "late" bilinguals while they silently recited brief descriptions of events from the previous day. For the early bilinguals, the same areas of their brains "lit up" regardless of whether they were using their first or second language. In contrast, different parts of the brains were activated for people who had learned their second language in adolescence or later when using their first versus their second language. Both the early and late bilinguals in this study reported comparable fluency in their second languages, so differences in language ability between these two groups is not likely the cause of this difference. Rather, these results suggest that different parts of the brain, and thus different cognitive processes, likely are involved when a second language is learned early in childhood rather than later.

Alternative Perspectives

We believe that there is compelling evidence that human children are prepared to acquire language and that this ability is sensitive to maturationally paced developmental changes. However, the neonativists' claim that language acquisition is based solely on domain-specific mechanisms has not gone unchallenged. For example, given the species-specific nature of language acquisition, human infants should possess some specializations in processing the sounds of language. This is demonstrated in research by Eimas and his colleagues (1971), who showed that 1-month-old human infants can discriminate between speech phonemes, such as /p/ and /b/. However, this ability is not restricted to humans but is found in chinchillas and Japanese quails as well (Kluender, Diehl, & Killeen, 1987; Kuhl & Miller, 1975), suggesting that the perceptual abilities underlying human language are phylogenetically ancient (Doupe & Kuhl, 1999). In other research, human newborns were able to discriminate between sentences spoken in Japanese and Dutch, but only when the sentences were played forward not when they were played backward (Ramus, Hauser, Miller, Morris, & Mehler, 2000). Given the different rhythms of these languages, the results are consistent with the idea that human infants are prepared to be sensitive to the specific rhythmic patterns found in languages. Although the data support this contention, this same pattern of results was found for cotton top tamarins (*Saguinus oedipus oedipus*), a South American monkey species. These data suggest that some of the apparent preparations human infants possess are shared with other species and may not reflect special mechanisms for language acquisition but rather represent more general characteristics of the primate or mammalian auditory system.

Consistent with our contention that human cognition involves both domain-specific and domain-general mechanisms working in concert for most phenomena, in the following section we present evidence and arguments that children's biological preparation for language acquisition may be facilitated by factors that involve greater environmental support and domain-general mechanisms than neonativists (and evolutionary psychologists) typically propose.

Social-Pragmatist Perspectives

No one suggests that the language a child speaks is "innate." The claim of the neonativists is that children are specially prepared to acquire whatever language they hear around them as they grow up. All human languages share certain structural aspects, and children's universal grammar acts to discern how the particular language that surrounds it is organized. Thus, the social environment dictates which language a child speaks, but does not influence greatly how or whether children acquire language. It was behaviorists such as B. F. Skinner who proposed that children learn language via traditional techniques of association and reinforcement. Although the empiricist position of language acquisition was defeated soundly by Chomsky more than 40 years ago, many psychologists continued to believe that aspects of the social environment facilitate language acquisition and that perhaps infants and their parents were prepared by evolution to take advantage of this social information. In support of this, there is evidence that individual differences in how mothers speak to their children are related to subsequent aspects of children's language development (Hoff-Ginsburg, 1985, 2000; Moerk, 1986).

Developmental psychologist Jerome Bruner (1983) perhaps best reflects this *social-pragmatist* position. Bruner and others (Carpenter, Nagell, & Tomasello, 1998; Tomasello, 1992) purported that the language environment that infants and young children hear is well organized, making the job of acquiring language easier. Bruner contended that language is carefully presented to children in a way that enhances their ability to learn language. He proposed that adults possess a device that responds to infants and young children by automatically altering speech to a more understandable form, which he called the *language acquisition support system*. According to this social–pragmatic view of language acquisition, "children's initial skills of linguistic communication are a natural outgrowth of their emerging understanding of other persons as intentional agents" (Carpenter et al., 1998, p. 126). Language is viewed as a powerful social–cognitive tool, used to manipulate other people's attention. It is based on more primitive social processes, such as shared joint attention, which makes language possible.

Bruner and others, most notably development psycholinguist Catherine Snow (1972), have noted that adults speak to infants and young children differently than they speak to other adults and that children seem to expect it. *Infant-directed (I-D) speech* involves the use of high-pitched tones, exaggerated modulations, simplified forms of adult words, many questions, and many repetitions (see Hoff, 2001). Such speech is used not only by mothers talking to their infants, but also to varying degrees by fathers, older siblings, and strangers in the supermarket, and may represent some innate language transmittal mechanism found in adults.

Systematic examination indicates that, compared to adult-directed (A-D) speech, I-D speech has a wider range of acoustic frequencies, a higher mean frequency, and a greater incidence of rising frequency contours (Fernald & Mazzie, 1991). Although there are cultural differences in how much adults speak to infants, these acoustic patterns of I-D speech have been found for American, British, French, Italian, German, Latvian, Comanche, Mandarin Chinese, Japanese, Sinhala, Russian, and Swedish mothers (see Fernald, 1992; Fisher & Tokuro, 1996; Kuhl et al., 1997). Not all cultures use the same exaggerated style of I-D speech as American mothers (and fathers) frequently do, but as the list of cultures mentioned suggests, some aspects of infant-directed speech might be universal (Fernald, 1992; Kuhl et al., 1997). This pattern has also been found with the mothers of deaf infants, who sign more slowly, use more repetitions, and greater exaggerations of movements to their infants than to their deaf adult friends (Masataka, 1996).

One possible reason that adults use I-D speech is that infants are more responsive to it than to A-D speech. For example, using conditioning paradigms, infants as young as 1 month will turn their heads to hear a tape playing I-D rather than A-D speech (Cooper & Aslin, 1990, 1994). In other research, hearing infants, who had never been exposed to sign language, were more responsive to the greater exaggerations found in infant-directed sign language than to adult-directed sign language, suggesting that this effect is not modality specific (Masataka, 1998). Other research has reported that infants can discriminate between words more easily when they are spoken in I-D rather than A-D speech (Karzon, 1985; Moore, Spence, & Katz, 1997) and that mothers use different intonations when they want to get their babies' attention (rising) than when they want to maintain their attention (up and down; Stern, Spieker, & MacKain, 1982). These and other findings suggest that infants are prepared to attend to and process certain types of language, and it is no coincidence that the singsong-y speech babies best understand is the same kind of speech that adults and children seem compelled to produce in the presence of an infant.

It seems unlikely that speaking to infants and children in I-D speech is necessary for language acquisition, but there is some anecdotal evidence that A-D speech heard in a nonsocial context does not lead to language learning. For example, children apparently do not learn language from watching television, which is not personally directed to them, has no social give-and-take, and consists of mainly adult-directed speech. In one case, Dutch children who regularly watched German television often failed to understand the programs and did not achieve appreciable control of German (reported in Snow et al., 1976). In another case, a hearing child of deaf parents was confined to his home because of poor health and learned sign language from his parents but could neither speak nor understand spoken English by age 3, despite frequently watching television (Moskowitz, 1978).

One proposal holds that the significance of I-D speech is not so much in subsequent language acquisition but in emotional development. Fernald (1992) has suggested that by using specific tones, the mother regulates her infant's emotions, behavior, and attention and also conveys her own emotional state to the infant. Fernald divides I-D speech into four different acoustical patterns, all of which are used by British, American, German, French, and Italian mothers when talking to their 12-month-old babies. These patterns are used to convey the mother's approval, express prohibition, ask for attention, and provide comfort to the infant. Such nonverbal communication is important in developing secure mother–infant attachment and, according to John Locke (1994), "Spoken language piggybacks on this open channel, taking advantage of mother–infant attachment by embedding new information in the same stream of cues" (p. 441). Fernald's claim is that the evolutionary origins of language may stem from mothers attempting to regulate the emotions of their infants, something that I-D speech continues to do today (see also Trainor, Austin, & Desjardins, 2000).

Language Development Without Domain-Specific Mechanisms

There is a well-regarded group of psycholinguists who believe that language acquisition can be explained by postulating only domain-general mechanisms (Bates, 1999; Elman et al., 1996). Jeffrey Elman and his colleagues propose that one does not need to hypothesize the presence of "innate knowledge" as reflected by universal grammar (representational constraints; see chapter 3) to explain language acquisition; rather, infants are born with sets of architectural constraints—information-processing limitations or biases in the brain—that require specific experience for their expression. Language acquisition is still innate from this perspective but requires more information from the environment than neonativists purport.

For example, developmental psycholinguist Elizabeth Bates and her colleagues have demonstrated a strong relation between vocabulary size and syntactic development (Bates, 1999; Bates & Goodman, 1997). Once children acquire between 300 and 400 words, there is a rapid increase in the complexity of their syntax. This pattern has been found in longitudinal and cross-sectional studies, in late talkers and early talkers, in children with Down syndrome and Williams syndrome, in children with language disorders, in people with brain damage, and for a variety of languages. The relation between vocabulary size and syntactic complexity is stronger than the relation between age and syntactic complexity, making it unlikely that age alone is the common underlying factor (see also Robinson & Mervis, 1998). Bates argued that these data suggest that there is no need for a domain-specific "grammar module." Rather, as vocabulary increases, grammar emerges from a need to organize the growing corpus. According to Bates and her colleagues, vocabulary and grammar are two sides of the same coin, developing together, with changes in one influencing changes in the other. Rather than proposing innate modules, Bates, Elman, and their colleagues proposed a connectionist model of language acquisition. Such an approach still hypothesizes that there are innate constraints that influence how language items are processed (architectural constraints), but it avoids the idea of a specific grammar module that has a ready-made theory of syntax that children will discover just by being exposed to language. Instead, through adjustments that children make as they acquire words (changing weights in the connectionist model), grammar emerges without the need of specific, innate structures.

One line of research consistent with Bates's and Elman's position that there are domain-general accounts of language acquisition that are also evolutionary friendly argues that aspects of children's more general information-processing abilities interact with their developing language skills. Specifically, aspects of children's immature cognition may facilitate language acquisition. Newport (1991) and Elman (1994) have independently suggested that limitations in young children's working memories facilitate language acquisition. (This research was introduced briefly in chapter 5.) Both theorists proposed that children initially perceive and store only component parts of complex stimuli. They start with single morphemes (usually a single syllable) and gradually increase the complexity and the number of units they can control. This results in a simplified corpus that actually makes the job of analyzing language easier. With success and time, maturationally paced abilities gradually increase, as does language learning. Newport proposed the "less is more" hypothesis, suggesting that young children's limitations serve a positive function for language acquisition. Elman used the metaphor "the importance of starting small" to get across the same idea.

Both Newport (1991) and Elman (1994) performed computer simulations, restricting the amount of information the simulations could process at any one time (equivalent to restricting how much children can hold in working memory). For example, in Newport's simulation, the input filter became less restricted over repeated trials, similar to the effect that maturation has on the size of the short-term store in developing children. The restricted filter resulted in the loss of data for morphology (the smallest units of word meanings), making initial learning worse than when a less restricted input filter was used. (This accounts for the faster initial learning of a second language that adults show relative to children.) However, there was greater loss at the whole-word level than at the level of morphology (indicating that prefixes and suffixes were often retained). Importantly, the restricted filter resulted in an improvement in the signal-to-noise ratio (i.e., the ratio of relevant linguistic information to irrelevant background information). There was also greater loss of data from accidental co-occurrences than from systematic co-occurrences of form and meaning. This means that with the less restricted filter (reflecting a larger short-term store), many language-irrelevant associations were retained, which impeded rather than facilitated language learning. Newport (1991) concluded that

> overall, then, a learning mechanism with a restricted input filter more successfully acquired a morphology; the same learning mechanism with a less restricted filter, or with no filter at all, entertains too many alternative analyses and cannot uniquely determine which is the better one. (p. 127)

Elman (1994) reached a similar conclusion using a very different connectionist computer simulation. These researchers concluded that young children's limited working-memory capacity restricts how much language information can be processed, which simplifies what is analyzed, making the task of language acquisition easier. Preliminary empirical support for the "importance of starting small" position comes from evidence that adults learn an artificial grammar faster when presented with smaller units of the language (Kersten & Earles, 2001). (See Bjorklund & Schwartz, 1996, for a discussion of these ideas applied to remediation of language disabilities in children.)

Consistent with the starting-small hypothesis is evidence that adults simplify the speech they present to infants and young children (i.e., use infant-directed speech). As we noted previously, adults around the world talk to children using highly repetitive and greatly simplified speech. As children's language competencies increase, so too does the complexity of language addressed to them. According to Bjorklund and Schwartz (1996),

"Such modified language, accompanied with young children's limited information-processing abilities, results in children receiving a much reduced body of linguistic evidence from which to extract the phonological, syntactic, and semantic rules of their mother tongue" (p. 26).

Language acquisition has been perhaps the most investigated topic from a neonativist perspective in the cognitive developmental literature. Given its long history, it has also had the time to produce counter hypotheses. We believe that the evidence that infants are specially prepared to acquire language is incontrovertible. Despite the confidence we have in this conclusion, it is also clear to us that there are multiple interacting factors that influence language acquisition. Consistent with the developmental systems approach, children's genetic dispositions interact with experiences to produce a particular pattern of development. For example, recall that children can hear while still in utero and are born having a preference for the language their mothers speak (Mehler et al., 1988). Although we do not necessarily advocate a domain-general, connectionist model as proposed by Elman, Bates, and their colleagues, we do believe that more general aspects of children's developing information-processing abilities interact with more domain-specific mechanisms to affect the course of language acquisition. We are particularly intrigued with the idea that certain aspects of children's immature cognition may influence positively the ontogeny of other aspects of their thinking. It is to this issue, the potential adaptive nature of cognitive immaturity, that we now turn.

ADAPTIVE NATURE OF COGNITIVE IMMATURITY

The major theme of this chapter is that human infants and children have been prepared by evolution to acquire some types of information more readily than others—that there are enabling constraints that make the process of learning certain classes of information easier than others. Consistent with this contention is the idea that some information will be most easily acquired at particular times in ontogeny. This is related to the concept of critical, or sensitive, periods. A related hypothesis is that some aspects of developmental immaturity, rather than simply being limitations that the child must overcome on the way to adulthood, make certain forms of cognitive development easier, much as a limited working-memory capacity is proposed to be related to greater ease of language acquisition.

We discussed in chapter 2 the potential adaptive role of immaturity in development (Bjorklund, 1997b; Bjorklund & Green, 1992; Wellman, in press). In that chapter, we provided specific examples relating to children's play and neonatal imitation. In this section, we discuss theory and research suggesting that certain aspects of children's immature cognitions are actually adaptations that facilitate overall cognitive development.

Species-Atypical Experience and Subsequent Development

Developmental psychobiologists Gerald Turkewitz and Patricia Kenny (1982) proposed that the immaturity of sensory and motor systems may play an adaptive role early in development. The limited motor capacities of altricial animals (those that are physically immature and helpless at birth and need substantial parental care) prevent them from wandering far from the mother, thus enhancing their chances of survival. Of greater interest, however, was their proposal that sensory limitations of many young animals are adaptive in that they reduce the amount of information infants have to deal with, which facilitates their constructing a simplified and comprehensible world.

In chapter 2, we presented research demonstrating how varying the prenatal environment of precocial birds can alter species-typical behavior (Gottlieb, 1976). Related to this is research showing how presenting animals with "experiences" outside of the species-typical range early in ontogeny can disrupt development. For example, Lickliter (1990) removed part of the eggshell of bobwhite quails two to three days before hatching and provided visual experience (patterned light) to these animals. Of course, quail chicks would not normally see patterned light until after hatching. Following hatching, the quail chicks were placed in a circular tub, with the maternal call of a quail coming from a speaker on one side of the tub and that of a chicken coming from a speaker on the opposite side. The findings from this study are shown in Figure 6.3. A group of control animals that had the eggshell removed but did not receive any additional visual experience displayed the species-typical pattern, approaching the maternal call of their own species on most occasions. In contrast, most of the animals exposed to light showed no preference or approached the maternal call of a chicken. It is worth noting that these animals did show enhanced visual discrimination abilities relative to control animals (a facilitory effect of early, species-atypical experience), but at a cost to auditory discrimination abilities. Other research, using ducks, quails, and rats as subjects, has demonstrated that providing young animals with stimulation that is outside the species norm has negative consequences for development (Gottlieb, Tomlinson, & Radell, 1989; Kenny & Turkewitz, 1986; Lickliter & Hellewell, 1992; Lickliter & Lewkowitz,1995; McBride & Lickliter, 1994; Spear, 1984).

Neonatologist Heidelise Als (1995) has suggested that premature human infants have experiences similar in nature to those of Lickliter's bobwhite quails. Als suggested that the stimulation that premature infants often receive in hospitals disrupts brain development (particularly of the frontal

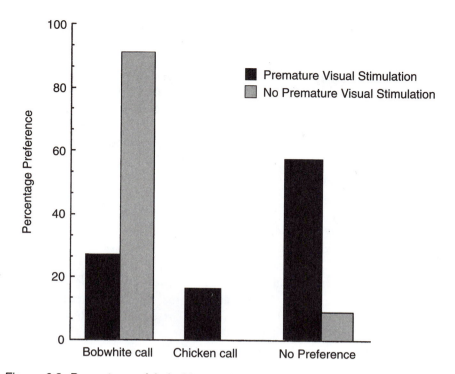

Figure 6.3. Percentage of bobwhite quail chicks that approached the bobwhite maternal call, the chicken maternal call, or showed no preference as a function of whether they received premature visual stimulation.
Note. Adapted from "Premature Visual Stimulation Accelerates Intersensory Functioning in Bobwhite Quail Neonates," by R. Lickliter, 1990, *Developmental Psychobiology, 23,* pp. 15–27. Copyright 1990 by Wiley. Adapted with permission.

cortex) during sensitive periods, frequently causing impairments resulting in speech problems, eye-hand coordination difficulties, impulsivity, attention deficits, and lowered IQ. However, these deficits are often accompanied by accelerated development or enhanced abilities in other areas, such as mathematics. Als interpreted these findings in premature infants much as Lickliter did for bobwhite quail. Stimulation outside the species-typical range can have unforeseen consequences on brain and behavior development. Als (1995) wrote,

> Social contexts evolved in the course of human phylogeny are surprisingly fine-tuned in specificity to provide good-enough environments for the human cortex to unfold, initially intrauterinely, then extrauterinely ... With the advances in medical technology, that is, material culture, even very immature nervous systems exist and develop outside the womb. However, the social contexts of traditional special care nurseries bring with them less than adequate support for immature nervous systems

... leading to maladaptations and disabilities, yet also to accelerations and extraordinary abilities. (p. 462)

The findings presented above indicate that extra prenatal stimulation in one sensory system can affect adversely later learning in another system. Related to this issue is the question of whether an early learning experience in infancy can interfere with later learning.

Several studies have found that rats who began conditioning early in life actually produced lower levels of eventual learning, compared to rats who began training at a later age (Rudy, Vogt, & Hyson, 1984; Spear & Hyatt, 1993). Similar evidence of a detrimental effect of early learning on later learning was provided by comparative psychologist Harry Harlow (1959), who was studying object-discrimination learning in rhesus monkeys. Monkeys who began discrimination training at 155 days of age or younger actually performed more poorly on the learning tasks later in life than animals who did not begin training until 190 days of age or older. This was true despite the fact that the "early" animals had more experience on the task than the older animals. Performance on these problems, beginning at age 120 days, is shown in Figure 6.4 as a function of the age at which monkeys began training. Harlow (1959) concluded that

> there is a tendency to think of learning or training as intrinsically good and necessarily valuable to the organism. It is entirely possible, however, that training can either be helpful or harmful, depending upon the nature of the training and the organism's stage of development. (p. 472)

"Readiness" and Early Education

Formal education is a novelty for *Homo sapiens*, a cultural invention only thousands of years old at most, and only in the past century has a majority of humanity received even moderate levels of schooling. Yet, within some cultures, children begin formal education during the preschool years. Moreover, some of the discoveries relating to infant and fetus learning have resulted in movements to push formal education back to the crib and, in some cases, to the womb (Doman, 1984). The evolutionary developmental perspective taken here, however, suggests that imposing formal education on children before they are biologically "ready" or "prepared" to learn will have few positive consequences and may have some negative effects. Surprisingly, there has been little systematic research on this socially important issue. There are a few notable exceptions, however.

In one study, researchers Marion Hyson, Kathryn Hirsh-Pasek, and Leslie Rescorla (1990; Rescorla, Hyson, & Hirsh-Pasek, 1991) compared the effects of early high-academic versus low-academic preschool programs on middle-class children's school-related behavior in kindergarten. Children

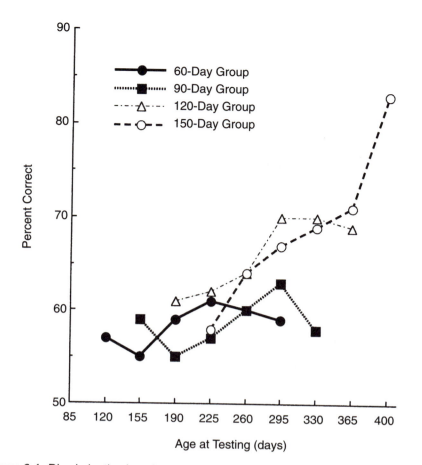

Figure 6.4. Discrimination learning set performance of rhesus monkeys as a function of age at which testing was begun.
Note. From "The Development of Learning in the Rhesus Monkey," by H. Harlow, 1959, *American Scientist,* December, pp. 459–479. Copyright 1959 by American Scientist. Reprinted with permission.

were given tests of academic skills, social competence, emotional well-being, and creativity at the end of their prekindergarten program and again toward the end of kindergarten. High-academic prekindergarten programs stressed adult-directed instruction, whereas low-academic programs did not but instead followed a "developmentally appropriate" curriculum (Bredekamp & Copple, 1997). There were no significant differences in academic ability either at the end of prekindergarten or kindergarten between the children who attended the high-academic and low-academic programs. A small difference was found for test anxiety at the end of preschool, with children attending the academically oriented schools showing greater test anxiety than children attending the developmentally appropriate programs. Correlational analyses revealed that mothers who scored high on an adult-instruction scale tended to have children rated as lower in creativity. (Mothers' belief in adult-instruction positively predicted children's performance on an academic skills test in preschool, but this relation disappeared in kindergarten.) Also, the greater the emphasis a preschool placed on adult-directed practices, the higher children's school-related anxiety. Finally, children who attended the developmentally appropriate schools were more likely to have a positive attitude toward school than were children who attended the high-academic programs.

Hyson and her associates cautioned that most of these effects, although statistically significant, were small in magnitude. Nevertheless, in general, there were no long-term benefits of an academically oriented preschool program and some evidence that such programs might have negative consequences (e.g., greater stress). Based on these findings, Hyson and colleagues concluded that there seems to be no defensible reason for encouraging formal academic instruction during the preschool years. Rather, for most children from middle-class homes, cognitive development and creativity can best be fostered in a developmentally appropriate preschool program that considers children's limitations as well as their abilities. In the opinion of these researchers, "it may be developmentally prudent to let children explore the world at their own pace rather than to impose our adult timetables and anxieties on them" (Hyson et al., 1990, p. 421).

Although the results of this and related (Weikart & Schweinhart, 1991) studies must be viewed cautiously, they are consistent with the perspective that cognitive development during the early years of life is best accomplished outside a formal, teacher-directed environment. Young children's cognitive system is immature, and because of this, learning and development might take place best in unstructured settings. The skills of young children are different from the skills of older children and might be ideally suited for the learning they need to do at this time in their lives (Bjorklund, 1997b; Bjorklund & Green, 1992). These findings argue that children ought not be hurried through a childhood that has purposes in and of itself.

PREPARED TO THINK

The philosopher John Locke's influential idea that the mind of a human infant is a tabula rasa, or blank slate, has long been rejected by academic psychologists. However, it is only recently that the implications of this "non-blank-slate" position have been recognized. If human infants are not empty vessels, waiting to be filled by the experiences provided by their culture, they must be born with some "knowledge." Few psychologists today would propose that infants enter the world with "innate ideas," at least not as conventionally understood. What infants come into the world with are processing biases and constraints—products of natural selection—that serve as the foundation for developing a human mind. Experience, of course, is necessary before the skeletal outlines for understanding objects, quantitative relations, space, or language, can be built to resemble those of mature members of the community. But infants have a head start and are born knowing, at an implicit level, something about the world they enter. Although the degree to which such knowledge is instantiated (e.g., Are constraints best thought of as representational or architectural in nature?) is much debated, there is little debate that human infants are prepared to learn some things more easily than others. Such preparations have been studied for cognitive domains that humans share with other animals (e.g., knowledge of physical objects and spatial cognition), for domains that are unique to humans (language), and for those that fall somewhere in between (mathematics).

In this chapter we presented only a sampling of the research performed within the domains reviewed and, for reasons of space, omitted other relevant domains (e.g., children's intuitive biology; Wellman & Gelman, 1998). One area we only indirectly alluded to, for which we believe there has been substantial preparation, is the social domain. Humans are social creatures, and it would be surprising if our cognitions were not shaped by natural selection to deal with important aspects of our social world beginning very early in life. We now turn to this topic.

7

SOCIAL COGNITION

We believe, as do many others (Alexander, 1989; Byrne & Whiten, 1988; Geary & Flinn, 2001; Humphrey, 1976; Jolly, 1966), that the evolution of the human species' unique intelligence was motivated by the need to deal with other members of our social group. To successfully maneuver the often stormy waters within small groups of long-lived conspecifics, adult humans must be able to represent the knowledge, desires, and intentions of others if they are to succeed. They must learn how to cooperate, how to compete, and which general social strategy is in their best interest. This requires what has been termed *social cognition*—cognition about social relationships and social phenomena. Other selection pressures, such as those related to changes in diet, technology, and responses to modifications in climate (Kaplan et al., 2000; Potts, 1998), also surely played interacting and contributory roles in shaping the modern human mind but, we argue, it was individuals who could both cooperate and compete successfully with one another, who were better adapted than their less socially facile compatriots, who became our great grandfathers and grandmothers.

It is simple enough to say that humans "evolved a keen social intelligence to deal with conspecifics," but it is another thing to specify what that intelligence is, how it differs from other forms of intelligence, how it is similar and different from the intelligence observed in other species, and how it develops. As we explained in our discussions of the nature of human cognition in chapter 5, we believe that there is not just one "social intelligence," but a set of hierarchically arranged, relatively specific abilities that evolved to deal with the variety of social problems faced by our ancestors. This domain-specific view is illustrated in Geary's (1998, 2001) categorization of individual and group modules presented in chapter 5 (see also Bugental, 2000). Specific social–individual modules in Geary's classification include processing information related to nonverbal behavior, language, facial processing, and theory of mind; specific social–group modules include processing information related to kin, in-group membership, out-group membership, and social ideologies.

In this chapter, we discuss what we believe to be the foundational abilities for human social cognition. First, social learning, both in humans and in nonhuman primates, is reviewed. Second, we examine theory of mind, looking at its development in human children, some of the factors that support its development, and its possible existence in the great apes. Finally, we look at a more sophisticated type of social cognition that social learning and theory of mind make possible—social reasoning, investigating the possibility that human children and adults are specially prepared to reason about problems involving social exchanges or violation of social rules.

SOCIAL LEARNING

Social learning refers to the acquisition of social information and behavior. It is ostensibly no different from other forms of learning, although historically it has been differentiated from classical and operant conditioning (Bandura & Walters, 1963; Miller & Dollard, 1941). A narrower definition refers to "situations in which one individual comes to behave similarly to others" (Boesch & Tomasello, 1998, p. 598).

Modern theories of children's social learning strongly emphasize cognitive processes. Albert Bandura (1986, 1989), the person most closely associated with the study of social learning, reformulated his theory to reflect cognitive, as opposed to associative-learning, mechanisms underlying children's acquisition of social behavior; and the most popular alternative account of children's social learning (and social cognition in general) is the social information-processing perspective (Crick & Dodge, 1994; Dodge, 1986; Rubin & Krasnor, 1986), which proposes that children engage in a series of cognitive processes (encoding, interpretation, searching for and evaluating response alternatives) when engaging in social interaction.

Neither of these approaches has focused on evolutionary issues, although much of the research generated from these theories has important implications for evolutionary developmental perspectives. In this section, we examine various ways in which juveniles acquire social information and behavior from observing the behavior of others. We take a comparative perspective, focusing on mechanisms of social learning used not just by humans but by other social primates, as well as aspects of social learning that seem to be unique to *Homo sapiens* and that may serve as the foundation for more complicated forms of social cognition.

Forms of Social Learning

Local Enhancement

One of the simplest forms of social learning is *local enhancement* (Thorpe, 1956). Animals (including humans) may observe other individuals engaging in some behavior with a particular object or at a particular location. Animals may then interact with the same objects or move to the particular location and, as a result, learn some specific behaviors via mechanisms of operant or classical conditioning. Learning has thus occurred mediated by observation, but nothing specific has been learned from observing the behavior of others. By attending to the actions of conspecifics, animals simply put themselves in a particular place or with particular objects and learn something from the experience.

Mimicry

If *imitation* is defined as the replication of an observed behavior, then mimicry qualifies as imitation. However, *mimicry* refers to the duplication of a behavior without any understanding of the goal of that behavior. Mimicry is observed in the behavior of some birds who are able to "parrot" the vocal responses of humans. It is also demonstrated by human children. For example, every time a barely verbal 2-year-old boy enters the grocery store, he steps on the 7-foot scale, looks up at its face, and waits for the needle to settle. His behavior is nearly identical to that of his father, but this child cannot understand the goal of this behavior (to obtain a measure of one's weight), for he has neither the concept of "one's weight" nor any idea of what the numbers on the face mean. He is accurately reproducing an observed behavior but does not comprehend the goal of the model.

We do not want to minimize the cognitive abilities that underlie mimicry. An individual must observe a model, store the representation of the behavior in memory, retrieve that representation when the proper environmental cues are present, and then reproduce the behavior. Individuals who mimic another can acquire complicated behaviors that may produce some positive consequences and thus become part of their behavioral repertoire, all without the need of understanding the goal of the model.

Emulation

In *emulation*, one individual observes another interacting with an object to achieve a specific goal. That individual then interacts with the object attempting to attain the same end but does not duplicate the same behavior as the model to achieve that goal (Boesch & Tomasello, 1998; Tomasello, 1996, 2000). For example, one chimpanzee may observe another getting ants from under a log by rolling the log. The chimp may then approach the same or a different log with the goal of obtaining ants. However, the second chimp (observer) will not duplicate the specific behaviors that it observed but will emit a variety of behaviors on the log, until the log is moved and ants are discovered. If successful, the ape will have learned, via trial and error, a valuable behavior, but not by imitation as conventionally understood. It has been suggested that local enhancement and emulation learning are the mechanisms underlying the "cultural transmission" of nut cracking observed in a group of West African chimpanzees (Boesch & Tomasello, 1998). The same argument could be made for the cultural transmission of termite fishing in other chimpanzee populations (Goodall, 1986).

Imitation

According to some theorists, true imitation requires that an observer be able to take the perspective of the model (Boesch & Tomasello, 1998; Tomasello, 1996, 2000; Tomasello, Kruger, & Ratner, 1993). It is not enough simply to observe and repeat the target behavior as in mimicry; for true imitation to occur, the imitator must understand the goal that the model had in mind (as in emulation) and reproduce important aspects of that behavior. According to comparative developmental psychologist Michael Tomasello and his colleagues (Tomasello, 1999; Tomasello et al., 1993), this ability is unique to humans. Although social learning is observed in many other primate species, only humans possess *cultural learning*—the transmission of acquired information and behavior within and across generations with a high degree of fidelity. "In cultural learning, learners do not just direct their attention to the location of another individual's activity; rather, they actually attempt to see a situation the way the other sees it—from inside the other's perspective, as it were" (Tomasello et al., 1993, p. 496). (Tomasello's claim that cultural learning is limited to humans may have to be rethought, given evidence of cultural transmission in wild chimpanzees, some of which we discuss briefly below; see also Whiten et al., 1999.)

Tomasello et al. proposed three stagelike levels of cultural learning in humans—imitative, instructed, and collaborative—with each stage involving a more advanced form of perspective-taking than the former. (Note that one stage does not replace earlier ones but affords individuals new means of social learning to accompany the earlier mechanisms.)

Imitative Learning. In *imitative learning* the learner internalizes something of the model's behavioral strategies and intentions for executing the observed behavior. The learner must understand the model's goal, which requires the ability to take his or her perspective. According to Tomasello and his colleagues, imitative learning is first seen during the latter part of the first year, when infants begin to imitate language sounds and actions with objects (Abravanel & Gingold, 1985).

Preverbal children engage in imitative learning, as is illustrated by experiments in which toddlers imitate not the precise actions of a model but their intended actions. In a study by developmental psychologist Andrew Meltzoff (1995), 18-month-old infants observed adult models successfully and unsuccessfully execute behaviors on objects. For one task, a model picked up a dumbbell-shaped object and made definite movements on the wooden-cube ends of the dumbbell, removing one of the cubes (successful condition). In the unsuccessful condition, infants watched a model pull on the ends of the dumbbell, but her hand slipped off the cubes, and the dumbbell did not separate. When later given the dumbbell, infants who had seen the successful demonstrations and infants who had seen the unsuccessful demonstrations removed the cube end significantly more often than did infants in control conditions. They seemed to realize what the model in the unsuccessful condition intended to do and imitated her behavior to achieve an inferred (but not witnessed) goal. In a second experiment, 14- and 18-month-old infants watched as either a person or a mechanical device acted on an object (e.g., a person or a vise removed the cube end of a dumbbell). Infants were twice as likely to imitate the actions when they had witnessed a person versus a machine perform the action, illustrating that by 14 months of age, infants understand that people (but not inanimate objects) have intentions (goals) that are sometimes worthy of imitating. In other research, 14- to 18-month-old infants observed adults engage in complex behavior sequences, some that appeared "intentional" as reflected by the model's vocal behavior and others that, based on what the models said, were "accidental." When later given the opportunity to imitate the model, the infants reproduced twice as many intentional as accidental behaviors (Carpenter, Akhtar, & Tomasello, 1998), again suggesting that infants have some understanding of the intentions of adults and will imitate intentional (goal-directed) actions but not accidental ones.

Imitation has been extensively studied in the great apes (see Custance, Whiten, & Bard, 1995; Galef, 1988; Parker & McKinney, 1999; Whiten, 1996). Although some convincing evidence exists for immediate imitation of action in sign-language-trained and wild chimpanzees (e.g., signs and facial expressions), less-convincing evidence exists of immediate imitation of actions on objects (Tomasello, Davis-Dasilva, Camak, & Bard, 1987; Whiten, Custance, Gómez, Teixidor, & Bard, 1996). For example, in one study, chimpanzees who were skilled at using a tool to acquire honey (honey fishing) were paired with naïve chimpanzees (Hirata & Morimura, 2000). Three of six naïve chimpanzees spontaneously observed their more experienced partner obtain honey either before making any attempt themselves or after a failed attempt to obtain honey. These chimpanzees rarely observed the more experienced partners following a successful attempt at honey fishing. These findings suggest that chimpanzees understand, at some level, that observing a more skilled conspecific is a necessary step in social learning. However, only two of five chimpanzees succeeded in their first attempt following an observation of their successful partner, suggesting that observation, although necessary, is not sufficient for imitation to take place. Moreover, having the chance to observe the problem being solved did not accelerate the learning process: Chimpanzees paired with an experienced partner took as long to solve the honey-fishing problem as chimpanzees who solved the problem alone (i.e., without the benefit of observing how it is done; but see Bard, Fragaszy, & Visalberghi, 1995, for a possible exception). Thus, although tool use in chimpanzees clearly seems to involve social learning (Matsuzawa, 1999), there is little indication that it involves true imitation.

Yet, it is incontrovertible that chimpanzees are marvelous social learners. Research with both free-living and laboratory animals has consistently shown that chimpanzees acquire complex behaviors, often involving the use of tools, in social contexts (e.g., Boesch, 1991; Call & Tomasello, 1994; Goodall, 1986; McGrew, 1992; Visalberghi, Fragaszy, & Savage-Rumbaugh, 1995). Recent claims have even been made for chimpanzee culture, with 39 distinct behaviors being identified as culturally transmitted, including fishing for ants and termites, cracking nuts, and grooming (Whiten et al., 1999). Results of several recent studies suggest that some aspects of imitative learning may indeed be observed in some chimpanzees to varying degrees.

As we noted in chapter 5, convincing evidence of deferred imitation of a model's actions, not attributable to emulation or local enhancement, has been observed only in enculturated apes (Bering et al., 2000; Miles et al., 1996; Tomasello, Savage-Rumbaugh, & Kruger, 1993). Although we believe that these studies provide solid evidence of deferred imitation in these animals, it could be argued that the apes were merely mimicking the model's behavior and did not actually understand the model's intention. Several studies involving chimpanzees with significant contact with humans have suggested that this was not the case. In one study, four chimpanzees, three with "language-training" experience, saw a human model perform a series of actions to open a box and remove food from it (Whiten, 1998b). Actions were performed in specified sequences (e.g., open bolt 1, open bolt 2, rotate pin, turn handle). The apes witnessed three demonstrations before they were first permitted access to the box themselves. After this first trial, the chimpanzees witnessed another demonstration, followed immediately by their second attempt to open the box, followed by another demonstration and a third and final attempt. Although two of the chimpanzees opened the box on their first attempt and three on the second attempt, there was no evidence that they matched the action sequence of the model on the first two trials. That is, although they opened the box, they did not organize their actions in the same sequence as did the model. However, the action sequences of the apes did match those of the model to a statistically significant degree for the third trial (in which all four apes opened the box), although they did not copy with great fidelity the particular behaviors within those sequences. (It is interesting to note that the single non-language-trained chimpanzee performed the poorest on this task.) Apparently, the repeated demonstrations, along with their previous efforts to open the box, resulted in an increase in their imitation of the actions of the model and greater success in retrieving the food. According to Whiten (1998b), "Certainly the imitators in this experiment seemed capable of extracting from what they saw the basic plan of the action sequences. . . ." (p. 280).

In a related study, juvenile chimpanzees were given a tube into which food was placed and dowels to retrieve the food (Bard et al., 1995). For one group of apes, a model demonstrated how to use the dowels to get the food before permitting them to attempt to retrieve the food themselves, whereas a second group was given the task without modeling. The 3- and 4-year-old chimpanzees, but not the 2-year-olds, in the modeling condition were more successful in solving the food-retrieval problem and in generalizing their behavior to a more difficult task. One 4-year-old solved the problem on his first try after observing a fellow chimpanzee, rather than a human, model the solution. The absence of a baseline period, during which the apes in the modeling condition attempted to solve the problem prior to any demonstration, precludes a definitive statement about whether imitative learning, rather than other social-learning mechanisms, was involved, but the results do suggest that imitation may be involved in the acquisition of complex behaviors in some chimpanzees.

In the study that we believe provides the most convincing evidence to date of imitative learning in great apes, enculturated juvenile chimpanzees were shown actions on one set of objects (e.g., banging cymbals together to make noise), and after a brief delay given a different set of objects (e.g., a pair of trowels). If the chimpanzees understood the intention of the model (to make noise by banging the objects together), then they should generalize that behavior to similar but different objects (Bjorklund et al., 2001). This is exactly what the animals did, strongly suggesting that the deferred imitation observed in this and likely earlier studies reflects true imitative learning rather than mimicry or emulation.

The results of this research with enculturated chimpanzees suggests that the representational abilities underlying imitative learning may be realized only when apes are reared in a humanlike environment, in which imitation is encouraged and in which they are engaged in triadic interactions with themselves, a caretaker, and an external referent. Ontogenies including shared-joint attention, referential communication, and, perhaps, language are often considered crucial in instantiating normal human patterns of social cognition (see discussion below).

Although imitation in human infants and toddlers has been much investigated (see Bauer, 1997; Meltzoff & Moore, 1997), most of this research has ignored the distinctions among emulation, mimicry, and imitative learning and has focused simply on children's reproduction of observed behavior. For example, research that evaluated preschool children's imitation of tool use reports that children will imitate the target behaviors but continue to do so even when those behaviors do not produce functional outcomes (Nagell, Olguin, & Tomasello, 1993; Whiten et al., 1996). This suggests that the children may not understand the intention of the adults using these tools, but merely were mimicking the modeled behavior. A recent experiment by Want and Harris (2001) suggests that this interpretation may be true for the youngest children but likely not for older preschoolers. Want and Harris demonstrated that 3-year-old, but not 2-year-old, children understand the goals of a model in a tool-using task. In their study, children watched as an adult attempted to remove a treat from a tube using a stick to push the treat out the end of the tube. Success on this task depended on which side of the tube the stick was inserted: If inserted from the wrong side, the treat fell through a hole and was not obtained. When children were shown both a correct and incorrect solution to the problems, 3-year-olds were later able to imitate the actions to retrieve successfully the treat. Two-year-old children, in contrast, performed essentially at chance on this task, suggesting that they did not fully understand the relationship between the model's actions with the tool and the outcome.

Instructed Learning. Imitative learning is followed in Tomasello, Kruger, and Ratner's (1993) classification by *instructed learning,* in which a more accomplished individual instructs a less accomplished one. Note that all cases of instruction do not qualify for instructed learning. Instructed learning requires that "children learn about the adult, specifically, about the adult's understanding of the task and how that compares with their own understanding" (p. 499). What distinguishes instructed learning from other aspects of "learning from instruction" is that in the former, children will reproduce the instructed behavior in the appropriate context to regulate their own behavior. That is, as in imitative learning, children must understand the purpose of the behavior—the adult's purpose when he or she initially taught the behavior. Children must internalize the adult's instruction, not just repeat a behavior on demand. If the 2-year-old boy in the scale example above were taught this behavior, it would reflect simply learning a "trick" and not be an example of instructed learning as Tomasello and his colleagues defined it.

The cognitive abilities underlying instructed learning may be the same underlying theory of mind (Whiten, 1998a; see also below). *Theory of mind,* as typically defined, requires that individuals understand that other people have beliefs and desires and that these may conflict with their own (Wellman, 1990). Similarly, instructed learning requires that the learner appreciate the perspective of the teacher. According to Tomasello and his associates (1993),

> To learn from an instructor culturally—to understand the instruction from something resembling the instructor's point of view—requires that children be able to understand a mental perspective that differs from

their own, and then to relate that point of view to their own in an explicit fashion. (p. 500)

There is limited evidence of teaching (instructed learning) in free-living chimpanzees (Boesch, 1991, 1993; Greenfield, Maynard, Boehm, & Schmidtling, 2000). For example, primatologist Christopher Boesch observed female chimpanzees in the Taï forest of the Ivory Coast "instructing" their infants in the skill of nut cracking. Nut cracking is limited to only a few populations of chimpanzees and qualifies, by some definitions, as a culturally transmitted behavior (Whiten et al., 1999). The action involves placing a nut on one rock, which serves as an anvil, and striking it with another, which serves as a hammer. This skill requires many years to master and is performed mostly by adult females (McGrew, 1992). On several occasions, Boesch observed mother chimpanzees positioning the nut, anvil, and hammer in such a way that all the infant had to do was strike the nut to break it and extract the meat. This is something adult chimpanzees without infants have never been observed to do. On other occasions, mother chimpanzees go through the motions of nut cracking more slowly when their infant is watching than when it is not. As impressive as these and related observations are, they are rare, and have been observed in only a handful out of thousands (perhaps millions) of interactions that primatologists have recorded over the past century. Thus, one must be cautious in interpreting these as "proof" that chimpanzees engage in instructed learning (Bering, 2001).

Collaborative Learning. The third stage of cultural learning in Tomasello and colleagues' (1993) model is *collaborative learning,* in which the less accomplished individual does not simply learn from the more accomplished one as in imitative and instructed learning, but learning occurs when two people work together to solve a common problem (Rogoff, 1998). Tomasello and his colleagues argued that collaborative learning is a process of cultural creation rather than cultural transmission. Collaborative learning requires that a child be able to represent his or her peer as a reflective agent, capable of reflecting on his or her own thoughts and the thoughts of fellow interactants. An example of reflective, or recursive, representation is when a child understands not only that another person can have a different point of view from the child's, but also that the other person can think about other people's perspectives. For example, "Peter thinks that I think that he likes Kelly" is a form of recursive thinking. This is necessary for collaborative learning, which often occurs in the middle of disagreement and conflict. For example, children must be able to evaluate and comment on another child's criticism of their previous suggestion if collaborative learning and the creation of a new joint perspective is to take place. According to Tomasello and his colleagues, this is first achieved during the early school years, between 6 and 7 years of age.

Collaborative learning follows the percepts of Soviet psychologist Lev Vygotsky (1978), who argued that children learn through participation in culturally relevant activities, particularly through their interaction with more competent adults and peers (see Rogoff, 1990, 1998). Educators have reported that the achievement of people working cooperatively is generally greater than for those working alone (Azmitia, 1992; Johnson & Johnson, 1989; Teasley, 1995) and that the children who gain the most from collaboration are those who were initially less competent than their partner (Manion & Alexander, 1997; Tudge, 1992).

One interesting finding about collaborative learning is that young children frequently make *source-monitoring errors*, remembering the behaviors of others as their own. For example, in research with preschoolers, children took turns with an adult putting pieces on a collage (Foley & Ratner, 1998; Foley, Ratner, & Passalacqua, 1993). After completing the collage, children were unexpectedly asked who had put each item on it, themselves or the adult. Four-year-olds were more likely to take credit for putting a piece on the collage that the adult had actually put on than vice versa (an "I-did-it error"), re-coding the actions of the adults as their own. Foley and Ratner suggested that this bias might lead to better learning of the actions of others, partly because misattributing the actions of others to oneself could cause children to link the actions to a common source (themselves) and thus produce a more integrated and easily retrievable memory. Consistent with this interpretation, 5-year-old children who performed a collaborative task with adults (placing furniture in a doll house) later made many I-did-it errors, but they also displayed greater memory for the location of the furniture in each room than did children in a noncollaboration group (Ratner, Foley, & Gimbert, in press). Thus, collaboration led to greater learning, but not in a way that might have been expected. Rather, young children's immature cognition resulted in many source-monitoring errors, which actually produced better learning. This is consistent with the argument that immature aspects of young children's cognition are actually well-adapted for their stage in development and may actually foster not hinder learning in some contexts (Bjorklund, 1997b; see also chapter 6).

Adaptive Nature of Poor Meta-Imitation

Consistent with Ratner and Foley's findings that young children's poor cognition produces some benefits on collaborative learning tasks, evidence exists that children's limited understanding of their imitative abilities (meta-imitation) may have some hidden benefits. For example, in a naturalistic study in which preschool children were asked to evaluate their own imitative skills, they overestimated how well they thought they would be able to imitate a task (prediction) on 56.9% of all attempts. Underestimation was infrequent (5.1%). Children were a bit better at evaluating their own behavior following an attempt at imitating a model (postdiction) but still overestimated on 39.6% of the attempts. Again, underestimations were rare (2.5%; Bjorklund et al., 1993, Study 1). Further, young children's knowledge of their imitative abilities was related to IQ, but the nature of the relation varied with age (Bjorklund et al., 1993, Study 3). Three-, 4-, and 5-year-old children observed a model performing a task (juggling one, two, or three balls) and were asked to judge how well they thought they would be able to imitate the behavior. They were then asked to perform the task and to estimate how well they thought they had done. Children's meta-imitative accuracy (how much they overestimated their performance—very few underestimated) was related to a measure of verbal IQ and is presented in Figure 7.1. (Because almost all children overestimated their abilities, lower scores represent less overestimation, and thus greater accuracy, and higher scores represent more overestimation, and thus greater inaccuracy.) Five-year-old children who were more accurate in predicting and postdicting their imitative abilities (who overestimated less) had higher IQs than did less-accurate children, a pattern found in older children on metacognition tasks (Schneider, Körkel, & Weinert, 1987). In contrast, the 3-year-old and especially 4-year-old children with higher IQs most overestimated their imitative abilities. These findings suggest that poor meta-imitation may actually be beneficial for young children in some contexts.

Young children are notorious overestimators of their cognitive and physical abilities (Plumert, 1995; Schneider, 1998; Stipek, 1984). Their immature metacognitive knowledge permits them to imitate a broad range of behaviors without the knowledge that their attempts are inadequate. As a result, bright young children continue to experiment with new behaviors and practice old ones, improving their skills at a time when trial-and-error learning is so important. As children's motor skills improve, so do their metacognitive abilities, which later in development are associated with more advanced cognition. As we have argued previously (see chapters 2 and 6; also Bjorklund, 1997b), aspects of children's immature cognitions may be well-suited to their developmental niche and should not necessarily be seen as deficiencies that must be overcome.

THEORY OF MIND

Perhaps the single most basic ability underlying human social interaction is the understanding that other people have knowledge and desires that may be different from one's own. *Theory of mind* has been used to reflect this knowledge and has been perhaps the single most investigated topic in

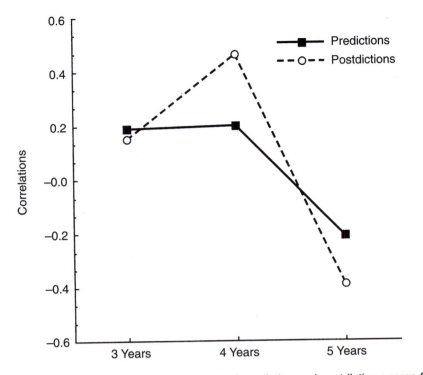

Figure 7.1. Correlation between IQ scores and prediction and postdiction scores for 3-, 4-, and 5-year old children (note that negative correlations imply that children with higher IQs overestimated less, that is, were more accurate, than did children with lower IQs).

Note. Figure from "The Role of Immaturity in Human Development," by D. F. Bjorklund, 1997b, *Psychological Bulletin, 122,* pp. 153–169. Copyright 1997 by American Psychological Association. Data from *Emerging Themes in Cognitive Development: Vol. II. Competencies,* (pp. 79–102), by D. F. Bjorklund, J. F. Gaultney, and B. L. Green, 1993, New York: Springer-Verlag. Copyright 1993 by Springer-Verlag. Reprinted with permission.

developmental psychology during the 1990s. A theory of mind refers to the tendency to construe people and their behaviors in terms of mind-related constructs, such as beliefs, desires, and intentions (Lillard, in press). Our interactions with others are based on what Wellman (1990) has termed *belief-desire reasoning*: We understand that our own behavior is based on what we believe (know, expect) and what we desire (want, wish)—and we assume that others' behavior is based on their beliefs and desires—and that our beliefs and desires can be different from those of others.

Children are not born with a belief-desire theory of mind, but most children develop an adultlike understanding of mind by 4 years of age (Perner et al., 1987; Wellman, 1990). Theory of mind is best reflected by performance on *false-belief tasks*. We described several versions of false-belief tasks in chapter 5. For example, in a widely used false-belief task (Wimmer & Perner, 1983), children watch as a treat is hidden in a specific location (e.g., in a box). Another person (Maxi) is present when the treat is hidden but then leaves the room, at which time the treat is moved to a new location. Children are then asked where Maxi will look for the treat when he returns. Most 4-year-old children can solve the problem, stating that Maxi will look where the treat was originally hidden, whereas most younger children state that Maxi will look for the treat in the new hiding place, apparently not realizing that Maxi's knowledge is different from their own.

Theory of mind is necessary for everyday social exchanges. For example, in one study, children played a game with "Mean Monkey" (Peskin, 1992). They were shown a series of stickers, some more attractive than others. "Mean Monkey," a hand puppet controlled by the experimenter, asked children which of the stickers they really wanted and which stickers they did not want. Mean Monkey then selected the children's favorite sticker, leaving them with the least desirable ones. Children 4 years of age quickly caught on to the game and told Mean Monkey the opposite of their true desires. Three-year-old children, however, never seemed to understand the social dynamics of the task, always pointing to their favorite sticker and consistently getting the booby prize. They behaved as if they never understood that Mean Monkey had a different goal in mind than they did.

Modules of Theory of Mind

Consistent with the domain-specificity perspective of evolutionary psychology, several researchers have proposed that theory of mind consists of a series of highly specialized modules that develop over the preschool years (Baron-Cohen, 1995; Leslie, 1994). For example, British psychologist Simon Baron-Cohen (1995) has proposed four separate, interacting modules involved in theory of mind, some of which are found in other animals and others of which are unique (apparently) to humans. For instance, the *intentionality detector* (ID) module permits one to infer that a moving object may have some intent toward one (it may bite me or groom me). The *eye-direction detector* (EDD) module interprets eye gazes (if an organism's eyes are looking at something, that organism then *sees* that thing). These modules develop in infancy (by age 9 months). The *shared-attention mechanisms* (SAM) module involves three-way interactions among the child, another person, and an object, so that if Person A and Person B are both looking at Object C, they both see the same thing. This develops by about age 18 months. The *theory-of-mind module* (TOMM) is roughly equivalent to belief–desire reasoning described earlier and develops around age 4 years.

Support for the modular perspective for theory of mind comes from a group of people who often display relatively normal intellectual functioning but have particular deficits in theory of mind. Baron-Cohen proposed that the more advanced forms of mindreading (SAM and TOMM) are typically absent in people with the childhood psychiatric disorder autism. Autistic children (and later adults) often seem to be in a world of their own and have a difficult time in most forms of social interaction. Baron-Cohen proposed that the primary characteristic of these children is an inability to read minds, or what he called *mindblindness*. Evidence for this comes from studies in which high-functioning autistic children are presented with false-belief and other theory-of-mind tasks and consistently fail them, despite performing well on other, nonsocial tasks (Baron-Cohen, 1989; Baron-Cohen, Leslie, & Frith, 1985; 1986; Perner, Frith, Leslie, & Leekam, 1989). In contrast, children with mental retardation, such as Down syndrome, perform the theory-of-mind tasks easily, despite often doing poorly on other tasks that assess more general intelligence (Baron-Cohen et al., 1985). Most autistic children perform well on the simpler tasks requiring the ID or EDD modules but fail tasks involving the SAM and especially the TOMM modules. According to Baron-Cohen, autistic children are unable to understand other people's different beliefs, and as a result, the world of other humans must be confusing and frightening, even for those who are functioning at a relatively high intellectual level. Baron-Cohen and his colleagues (1999) have recently demonstrated deficits in theory of mind in three highly successful autistic (Asperger's syndrome) adults, despite above-average performance on tests related to IQ and executive functioning and reasoning, further bolstering the independence of theory of mind and general intelligence. Recent research has shown that neurophysiological deficits in people with autism are located in the same brain region (left frontal lobe) as normal adults' processing on theory of mind tasks (Sabbagh & Taylor, 2000), further suggesting the domain specificity of theory of mind abilities.

Additional support for the modular view for theory of mind is found in behavioral genetic research. Hughes and Cutting (1999) reported substantial genetic influence on theory of mind tasks among 199 three-year-old same-sex twins. Importantly for the domain-specificity perspective, most of this genetic effect was independent of a general measure of verbal ability, consistent with the idea that theory of mind is not simply a function of general intellectual ability.

FACTORS INFLUENCING DEVELOPMENT OF THEORY OF MIND: THE SOCIAL ENVIRONMENT

Although theory of mind develops at about the same time and in the same sequence in most children around the world (Avis & Harris, 1991; Tardif & Wellman, 2000), the rate of its development is related to aspects of children's social environment. For example, level of children's teacher-rated social skills and performance on theory-of-mind tasks are correlated among 3- to 6-year-old children (Watson, Nixon, Wilson, & Capage, 1999), and the number of adults and the number of older peers that a preschool child interacts with daily are positively related to their scores on false-belief tasks (Lewis, Freeman, Kyriakidou, Maridaki-Kassotaki, & Berridge, 1996). Also, performance on theory-of-mind tasks is related to family size (Jenkins & Astington, 1996; Perner, Ruffman, & Leekam, 1994, but see Cutting & Dunn, 1999, for an exception), particularly the number of older siblings a child has. The more older siblings (or peers) a child has, the earlier he or she is apt to pass false-belief tasks (Ruffman, Perner, Naito, Parkin, & Clements, 1998). Possible reasons for the positive effect of older siblings and peers on theory-of-mind development include the greater opportunities for discussions of mental states, managing social conflict, pretend play, and reasoning about social issues, among others (Lewis et al., 1996; Ruffman et al., 1998; P. K. Smith, 1998). For example, Ruffman et al. argued that having older siblings stimulates fantasy play, which helps children represent "counterfactual states of affairs," a skill necessary for solving false-belief tasks. Leslie (1987) has argued that pretend play is an indicator of metarepresentational abilities as early as age 18 months and that such abilities are important for understanding that someone else may represent things differently (have different knowledge, or beliefs) from oneself. In support of this, researchers have found a relationship between the amount of cooperative social play preschoolers engage in (often with a sibling or parent) and later understanding of other people's feelings and beliefs (Astington & Jenkins, 1995; Youngblood & Dunn, 1995). (Although others looking at the same findings conclude that the evidence is not strong [Jarrold, Carruthers, Smith, & Boucher, 1994; Lillard 1993a], and it is, of course, possible that children with advanced theory-of-mind skills more readily engage in cooperative social play than their less advanced agemates.) From a different perspective, Cummins (1998a) suggested an explanation based on dominance theory. Siblings are always competing for resources, with older ones typically having the advantage because of their greater size and mental abilities. Younger children would be motivated to develop whatever latent talents they have to aid them in their social competition with their older siblings, and understanding the mind of your chief competitor sooner rather than later would certainly be to the younger siblings' advantage. A similar argument can be made for interacting with older peers.

Theory-of-Mind–Acquisition–Support System

Although children are clearly prepared to develop a theory of mind, they require a supportive social environment for these abilities to develop.

British developmental psychologist Peter Smith (1998) proposed a "theory-of-mind–acquisition–support system" analogous to what Bruner (1983) proposed for language acquisition. At this point in our evolutionary history, it seems that any "normal" human social environment will suffice. However, individual differences in children's social experiences, particularly during infancy and the early childhood years, may lead not only to differential rates of theory of mind development, but perhaps to different types of theory of mind, conducive to the type of environment (supportive, nonsupportive) in which children develop.

Do Great Apes Have a Theory of Mind?

We have suggested earlier in this chapter and in others (chapter 4) that one important impetus for the evolution of human intelligence may have been the need to cooperate and compete with other members of our species. That is, although finding food and avoiding the fangs of predators surely played a significant role in hominid cognitive evolution (Kaplan et al., 2000), our unique intelligence evolved primarily to deal with conspecifics (Alexander, 1989; Bjorklund & Harnishfeger, 1995; Byrne & Whiten, 1988; Humphrey, 1976). This hypothesis is made more credible by evidence of complex social organization and political life in great apes, particularly chimpanzees (*Pan troglodytes*; de Waal, 1982; Goodall, 1986). In fact, many primatologists now believe that chimpanzees possess *culture*, defined as local customs that are passed on via mechanisms of social learning from one generation to another (Boesch, 1993; Boesch & Tomasello, 1998; Whiten et al., 1999). It makes sense, then, to look for the roots of human social–cognitive competence in the great apes, and theory-of-mind research has been a focus of cognitive primatology over the past decade (see Byrne, 1995; Russon, Bard, & Parker, 1996; Tomasello & Call, 1997). In fact, the term "theory of mind" was first coined by primatologists David Premack and Guy Woodruff (1978).

Deception and Theory of Mind

Despite more than 20 years of research, comparative psychologists have not reached a consensus on whether species other than humans possess the cognitive abilities necessary for theory of mind. Some researchers, interpreting data collected mainly in naturalistic or seminaturalistic settings, are convinced that great apes (mostly chimpanzees) understand that others have intentions and knowledge that often differ from their own and monitor and adjust their own behavior accordingly (Goodall, 1986; Whiten & Byrne, 1988). For example, deception is an important social skill, and there is substantial observational evidence that great apes are able to deceive another individual when it is to their benefit. Whiten and Byrne (1988) sent questionnaires to 115 primatologists asking for any evidence of deception in their subjects. The primatologists responded with 79 examples, most involving concealment or distraction. Below we provide a few examples from this literature (from Bjorklund & Harnishfeger, 1995):

> Goodall (1986) reports that male chimpanzees would sometimes inhibit their distinctive cry during orgasm when copulating with a favorite female. In this way, they avoid having to share the female with other males.
>
> Premack (1988) and Savage-Rumbaugh and McDonald (1988) each report evidence of deception in language-trained chimpanzees. For example, Austin and Sherman of the Yerkes laboratory have perfected ways of escaping from their cages, but have never allowed humans to knowingly watch them while doing it (Savage-Rumbaugh & McDonald, 1988). Premack (1988) reports that chimps quickly learned to distinguish between a trainer who provided them with food rewards and one who did not, and suppressed responding in the presence of the hostile trainer (or, in one case, actively misled the trainer), but not in the presence of the benign one.
>
> In research by Menzel (1974), one chimp, Belle, was shown the location of hidden food and would then lead the small group of chimps to it. She ceased leading the group directly to the food, however, when Rock (the dominant male) was present, because Rock would kick or bite her and take the food. When Rock was present Belle would wait until Rock left before uncovering food. On a few occasions, Belle actually led the troop in the opposite direction from food, then, while Rock was searching, she doubled back to get the food. On other trials, the experimenter hid extra pieces of food in a second place. On those trials, Belle would lead Rock to this second, smaller source, and then go to the main cache. A similar observation of not retrieving or looking at a banana until after a more dominant chimp had left the area, was reported for the juvenile chimp Figan by Goodall (1971). (pp. 165–166)

Evidence of such deception in great apes reveals a complex social intelligence and suggests that such intelligence has deep evolutionary roots, likely extending to the common ancestors of modern apes and humans and perhaps beyond. However, such deception, although consistent with a theory of mind, does not necessarily require that one animal know what another animal is thinking. Rather, the apes could have learned a particular response in a specific situation (e.g., when I go directly to the cache of food when Rock is around, I lose it). In these situations, it may not be the case that an animal is reading the mind of another but only that a particular behavior has been unsuccessful in a particular situation, and inhibiting that behavior is in its best interest (Bjorklund & Kipp, 2001).

In fact, in more controlled laboratory situations, unambiguous evidence of theory of mind in apes is rare. For example, in a nonverbal version of a false-belief task, Call and Tomasello (1999) hid food treats behind a barrier, outside of the apes' view, but within the view of a human "communicator." When the barrier was removed, the communicator assisted the ape in locating the treats by placing a marker on the correct container. After the apes learned to track the movement of the treat from one container to another, they learned to ignore the communicator's marker when she placed it on an incorrect container (on a container in which the ape knew the treat was not hidden). Once apes were able to perform these tasks, the false-belief task was administered. The communicator watched as the treat was hidden in one of the containers. The communicator then left the room, after which a second person switched the location of the treat while the ape watched. The communicator then returned and placed a marker where she had seen it hidden previously. If the ape possessed a theory of mind comparable to that of a 5-year-old child, it would understand that the communicator had a false belief of the location of the treat. To obtain the treat, the ape must understand that the communicator has different (and false) knowledge about the location of the treat and select the container not marked by the communicator.

This version of the false-belief task is a difficult one; most 4-year-old children did not pass it, although 5-year-olds did. However, none of the two orangutans or five chimpanzees passed the task. Although aspects of this complicated procedure may have made it difficult for the apes to fully understand the task demands, Call and Tomasello (1999) believed otherwise, suggesting that these results are consistent with the interpretation that great apes do not possess a theory of mind similar to that of 5-year-old human children but rather solve complex social–relational problems via associative learning.

Do Chimpanzees Understand that "Seeing Is Knowing"?

Other primate research has assessed seemingly less sophisticated aspects of theory of mind, for instance, what great apes know about visual attention (Call & Tomasello, 1994, Experiment 3; Povinelli & Eddy, 1996; Tomasello, Call, & Hare, 1998). For example, solid experimental evidence exists that apes and monkeys reliably follow the gaze of conspecifics, suggesting that they understand that the direction of another animals' gaze implies that that animal is seeing something (Tomasello et al., 1998). However, such knowledge can be the result of a learning history that associates another animals' gaze with some important or interesting outcome. It does not necessarily imply that one animal understands that "looking" implies "seeing," which corresponds to the EDD module in Baron-Cohen's theory.

In fact, laboratory research has suggested that chimpanzees may not truly understand that the eyes are the source of knowledge. In a study by Povinelli and Eddy (1996), nursery-raised chimpanzees were taught to reach to a human to receive a food treat. In a series of conditions, two familiar caretakers were in front of the chimp, one who could see the ape making the reaching gesture and who could (and did) respond by giving the food to the ape, and another who could not see the ape. A reaching response made to this latter person would not be seen and a treat would not be received. The apes rarely made a reaching response to a person who had her back to the animal (i.e., was facing a wall away from the ape). In such cases, the animal consistently made a reaching response to the sighted person (i.e., the one watching the ape). However, the apes were less able to discriminate between a sighted person and one who had a bucket over her head, wore a blindfold, or had her eyes closed. Most animals for most of these conditions performed at chance levels, responding to the person who was unable to see as frequently as to the sighted person. Also, the chimpanzees responded randomly in a condition in which both experimenters had their backs to them, but one was looking over her shoulder at the ape (see also Reaux, Theall, & Povinelli, 1999). These findings suggest that apes understand only that a person with her back toward them cannot see them (which is often, but not always, correct). However, they apparently fail to understand that the eyes are responsible for acquiring useful knowledge about the immediate environment.

Yet, laboratory research that involved a more naturalistic setting and the evaluation of what a conspecific, and not a human caretaker, can and cannot see, suggests that chimpanzees, in some situations, do indeed possess knowledge that "looking is seeing." In a series of experiments by Hare, Call, Tomasello, and their colleagues (Hare, Call, Agentta, & Tomasello, 2000; Hare, Call, & Tomasello, 2001), a dominant and a subordinate chimpanzee were housed in connected cages, with food placed at various positions in the cages. When food was placed so that it was visible to both the dominant and the subordinate chimp, the dominant chimp obtained the food nearly every time. Subordinates successfully obtained food, however, in situations in which they could see where some food was located but the dominant chimp could not. Hare and his colleagues varied the experiments in several ways to eliminate alternative interpretations (e.g., the subordinate chimp was only monitoring the dominant chimp's behavior and retrieved the food only when the dominant chimp was not moving in that direction already) and concluded that, at least under these conditions, chimpanzees know what other chimpanzees can and cannot see and make decisions based on that knowledge.

The research by Hare and his colleagues fits with observations of and our intuitions about the social life of chimpanzees in the wild or in groups in captivity. They follow the gaze of others, and many observations of deception, some of which we described above, seem to require an understanding that "seeing is knowing." Yet we do not doubt the validity of the research by Povinelli and his colleagues. His chimpanzees clearly did not understand that the eyes gather knowledge. The seeming contradiction between the results of Hare's and Povinelli's research groups suggests that the EDD module postulated by Baron-Cohen is more complicated and more domain specific than he believed, at least in chimpanzees. Although we cannot specify exactly the conditions in the Povinelli and Hare studies that are responsible for the difference, the more naturalistic setting (competing with a conspecific over food) in research by Hare and his colleagues is likely a key factor. We find it a bit surprising that laboratory chimpanzees who have interacted with humans since early infancy fail to understand that a human caretakers' eyes are the source of seeing, whereas they can make that inference about conspecifics. This pattern suggests that understanding seeing is tied not simply to experience (which it must be) but also to some critical contexts (obtaining or competing for food) and, perhaps, to a recognition of conspecifics.

The Enculturation Hypothesis

One topic that has recently played a central role in the debate about great apes' possession of a theory of mind is that of enculturation. Call and Tomasello (1996) defined *enculturation* in great apes as a rearing environment that includes "something close to daily contact with humans and their artifacts in meaningful interactions" (p. 372). These environments alter species-typical ontogenetic trajectories, and there is some evidence that such exposure results in cognitive abilities more similar to those of human children than is found in mother-reared apes (Call & Tomasello, 1996; Miles et al., 1996). For example, the most successful of the language-trained apes have all experienced long periods of human-like rearing conditions (e.g., the chimpanzee Washoe, Gardner & Gardner, 1969; the bonobo Kanzi, Savage-Rumbaugh et al., 1993; the orangutan Chantek, Miles, 1994; and the gorilla Koko, Patterson, 1978). Evidence exists that immediate and deferred imitation are observed only in chimpanzees and orangutans that have had significant human contact (Bering et al., 2000; Bjorklund et al., 2000; Miles et al., 1996; Russon, 1996; Tomasello, Savage-Rumbaugh, & Kruger, 1993; see also discussion earlier in this chapter and in chapter 5). There is also tentative evidence that intentional communication and the understanding of intentions, which have never been observed in laboratory or mother-reared apes under controlled conditions (see Call & Tomasello, 1996), are displayed by enculturated apes (Call & Tomasello, 1994; Savage-Rumbaugh et al., 1986; Gómez, 1990), and we discuss this latter issue in more detail below.

Great apes use gestural communication with conspecifics in the wild for close-range social interactions (Goodall, 1986; McGrew & Tutin, 1978). However, there are no unambiguous observations of wild apes using gestures such as pointing to indicate distal objects (Goodall, 1986; Menzel, 1974). Furthermore, the gestural communication observed in most instances between apes in the wild has been attributed not to observational (social) learning but rather to the mechanism of ontogenetic ritualization (Boesch & Tomasello, 1998; Tomasello & Call, 1997). In *ontogenetic ritualization*, communicatory signals are learned as a result of the repeated interaction of two individuals. For example, Chimp A may instigate a play bout with Chimp B by slapping him on the head. Chimp B may notice that A always raises her hand before slapping him, so that, eventually, A must only raise her hand to initiate play with B. In another example, an infant may initiate nursing by heading directly for his mother's nipple, moving his mother's arm as he does. After many such episodes, the mother may anticipate her infant's intentions so that she is prepared to nurse at her infant's first touch. As a result, the infant learns to abbreviate his actions, so that a simple touch or gesture is sufficient to communicate his intention. In such cases, the two individuals shape each other's behavior, resulting in what appears to be complicated gestural communication but that can be explained by relatively simple mechanisms of associative learning.

In contrast to observations about wild and nursery-raised apes, enculturated members of each species of great apes have been observed to use pointing in interaction with humans (orangutans, Miles, 1990; common chimpanzees, Savage-Rumbaugh, 1986; bonobos, Savage-Rumbaugh, McDonald, Serak, Hopkins, & Rubert, 1986; gorillas, Patterson, 1978). (We should note, however, that rigorous controls are often lacking in these studies; see Povinelli, Bering, & Giambrone, 2001.) Part of humanlike rearing of apes involves pointing out distal objects (shared-joint attention), with social reinforcement when the apes look in the direction where the human pointed and when they point out objects to humans. Such routines of shared-joint attention are typical of human infant–adult interactions and may be necessary for children to understand referential pointing (Tomasello, Kruger, & Ratner, 1993). The central issue for enculturated apes concerns the mechanisms underlying referential pointing. Do they point at distal objects because they have learned that in doing so they get rewards (social or tangible ones, e.g., toys or food treats), or do they realize that, by pointing, they are directing the attention of another individual to an unobserved or unknown object (from the perspective of the other individual)? The latter interpretation is consistent, we argue, with possessing the rudiments of a theory of mind. The apes would understand that they possess knowledge that the other individual does not and that they can direct the attention of another by pointing.

Although nursery-raised apes can be trained to point, they apparently do so only because they have learned that such pointing is associated with rewards and not because they understand it as a means of conveying information to another individual (Call & Tomasello, 1994). For example, Call and Tomasello (1994, Experiment 1) trained a nursery-reared orangutan to point to the source of hidden food, so that a human caretaker, who did not know the location of the food, could retrieve the food for her. But when a tool was needed to retrieve the food and that tool was hidden by a first experimenter, the orangutan did not reliably point to the location of the hidden tool to a second experimenter. Moreover, nursery-raised great apes seem not to comprehend the use of pointing by humans (Call & Tomasello, 1994; Povinelli, Reaux, Bierschwale, Allain, & Simon, 1997). For example, Povinelli and his colleagues (1997) demonstrated that nursery-raised chimpanzees used the proximity of a human's hand to a desired reward rather than the pointing gesture as a communicative device symbolizing the location of the reward. Only the latter, of course, would reflect an understanding of the referential nature of the gesture.

In contrast, enculturated orangutans and chimpanzees in controlled studies have been shown to understand referential pointing as a means of directing another's attention to an unobserved object (Call & Tomasello, 1994; Povinelli, Nelson, & Boysen, 1992). For example, in the pointing experiment by Call and Tomasello (1994) described briefly above in which a nursery-raised orangutan failed to use pointing to direct a human's attention to a hidden tool, Chantek, an enculturated orangutan, performed near ceiling level on the second day of testing. In a second study with an enculturated and nursery-raised orangutan (Call & Tomasello, 1994, Experiment 2), an experimenter hid food in one of several containers and then pointed to the target container before leaving the room. A second experimenter then entered the room and stood in front of the containers. Only the enculturated orangutan pointed at the target container significantly more often than expected by chance. Similar levels of performance have been reported for enculturated chimpanzees in a related study (Povinelli et al., 1992).

Clearly the question of whether great apes have a theory of mind is not a simple one. Each species has evolved a cognition suited to its ecological niche, although the phylogenetic relationship among humans and the great apes makes it plausible that our common ancestor may have possessed the rudimentary abilities for theory of mind. And if great apes are found not to have a theory of mind, from what did such an ability evolve in humans? We think that the research with enculturated great apes may be revealing. In chapter 4, we suggested that animals with large brains, living in socially complex groups, and having extended juvenile periods may develop species-atypical cognitions and behaviors in response to species-atypical environments. If these novel environments remain stable (and thus lose their novelty), cognitive and behavioral patterns can also become stable, producing new phenotypes and setting the stage for evolutionary change. Presumably, orangutans and chimpanzees have the requisite cognitive flexibility, when raised in humanlike environments, to develop some critical humanlike cognitive abilities associated with theory of mind. Such cognitive abilities and their expression in species-atypical environments may reflect a *preadaptation* for social learning (Greenfield et al., 2000). If the common ancestor of contemporary great apes and humans also displayed this high degree of cognitive plasticity in response to rearing environments, it suggests one of several mechanisms by which cognitive evolution that resulted in *Homo sapiens* may have occurred.

DEVELOPMENT OF SOCIAL REASONING

We view theory of mind as a set of cognitive abilities that are necessary for sophisticated social interaction in human groups. It is difficult to imagine a successful adult (or 10-year-old child) in any human culture who did not possess at least the rudiments of belief–desire reasoning. But theory of mind is simply the foundation for more advanced forms of social cognition that are performed in everyday life in every village, town, and city of the world. One area in which theory of mind is a necessary but not sufficient condition for effective social interaction involves social exchanges—making deals, if you will—and the ability to detect people who may be breaking the rules.

Evolutionary psychologists Leda Cosmides and John Tooby (1992) proposed that several specific cognitive abilities are needed to successfully form social contracts and avoid cheaters. These include the abilities to recognize many different people, to remember one's past interactions with people, to communicate one's beliefs and desires to others, to understand the beliefs and desires of others, and to represent the costs and benefits of items or services that are being exchanged. The final test of whether someone can detect social cheaters, however, boils down to figuring out the logic of an exchange, and formal logic is something that humans are not particularly good at, unless the logic is in the context of social exchange.

Cosmides and Tooby (1992) summarized a series of experiments they performed contrasting the logic people use to solve abstract problems with the same logic they use to solve social-contract problems. First, the abstract problem: Cosmides and Tooby used variants of the Wason (1966) task. Adult participants were shown four cards on a table, such as the ones displayed below:

A   G   2   7

Participants were then given the following rule: "If a card has a vowel on one side, then it must have an even number on the other side." They were told that they must determine if the set of cards in front of them conformed to the rule or not and should turn over the fewest cards possible to determine the truth of the rule. Most people turned over the card with an "A" or the cards with an "A" and a "2." The "A" card is clearly correct. According to the rule, a card with a vowel on one side must have an even number on the other side, and the only way to test this is to turn over the "A" card. But turning over the "2" card is useless. The rule does not state that all cards with an even number on one side must have a vowel on the other, so knowing what is on the other side of the "2" cannot prove or disprove the rule. The critical card here was the "7." If there is a vowel on the other side of the "7," the rule is violated. Thus, to solve this problem "logically" in the fewest moves, one must turn over the "A" and the "7" cards and no others.

This is a difficult problem, and at least one of us is hesitant to present this to a classroom of students for fear of getting it wrong. But look what happened when Cosmides and Tooby asked participants to apply the same logic to a social-contract problem. For example, adults were given the following cards:

Beer   Coke   16 years old   25 years old

They were then asked to test the following rule: "If a person is drinking alcohol, then he or she must be at least 21 years old." As in the previous problem, they were told that they should turn over the fewest cards possible to determine the truth of the rule. Most adults solved the problem easily, turning over the "Beer" and "16 years old" cards. They recognized immediately that what is on the other side of the "Coke" card is irrelevant, as is what is on the other side of the "25 years old" card. Cosmides and Tooby demonstrated that adults in different cultures around the world could solve easily social-contract problems like the "Beer/16 years old" problem but usually failed to use the identical logic to solve more abstract or nonsocial-contract problems. Cosmides and Tooby (1992) proposed that the reason for the discrepancy in performance between the "abstract" and "social-contract" versions of the same problem is that people do not use a general problem-solving ability to solve all logical problems but rather have evolved domain-specific "cheater detectors" that are limited to social contracts. From a similar perspective, the social-contract problems reflect *deontic reasoning*, which is reasoning about what one may, should, or ought to do, whereas the abstract problems reflect *descriptive*, or *indicative reasoning*, which implies only a description of "facts" and no violation of social rules.

A variant of this argument has been advanced by Bruner (1986), who suggested that "narrative thought" was concerned with personal and deontic matters, whereas "paradigmatic thought" was concerned with more logical mathematical forms of reasoning. From this view, individuals have access to both forms of thought. Use is determined by immediate factors in the environment. At the proximal level, narrative thought can be "triggered" by breaches of conventionality and by personal struggles marked by deontic modal auxiliaries to verbs, addressing issues of personal value and obligation, such as "must" or "should." In short, humans seem to have two clearly distinctive forms of logic at their disposal, one more formal and abstract and one addressing matters specific to social interaction.

Deontic reasoning is found very early in childhood. Several experiments have presented 3- and 4-year-old children with simplified versions of the four-choice task just described and reported above-chance levels of performance when the problems are presented as reflecting potential social violations versus simply descriptions of a situation (Cummins, 1996b; Harris & Núñez, 1996). For example, Harris and Núñez (1996, Experiment 4) told short stories to children, some of which involved breaking a prescriptive rule and others that had the same content but without breaking any rule. For instance, in the prescriptive (or deontic) condition, children were told that "One day Carol wants to do some painting. Her Mum says if she does some painting she should put her apron on." Children in the descriptive condition were told, "One day Carol wants to do some painting. Carol says that if she does some painting she always puts her apron on." Children were then shown four drawings, with the experimenter describing each of the drawings, for example, Carol painting with her apron, Carol painting without her apron, Carol not painting with her apron, and Carol not painting without her apron. Children in the deontic conditions were then told "Show me the picture where Carol is doing something naughty and not doing what her Mum said." Children in the descriptive condition were told "Show me the picture where Carol is doing something different and not doing what she said." Although both 3- and 4-year-old children selected the correct picture of the four most frequently in both the deontic and descriptive conditions, both groups of children were correct more often in the former condition (72% and 83% for the 3- and 4-year-olds, respectively) than the latter (40% for both the 3- and 4-year-olds). Similar to adults, young children were able to reason correctly about a problem in which a social contract was being violated but not about a problem in which no such social obligation was mentioned.

Cummins (1996a, 1996b, 1998a, 1998b) has argued that children's deontic reasoning ability is innate and that it evolved in the context of dominance hierarchies within primate groups. To survive within a dominance hierarchy, individuals must know what someone of one's rank is permitted to do and not to do and to recognize when others are following or breaking the rules, which could have consequences for one's standing in the hierarchy. This alone does not require a theory of mind or social reasoning but, Cummins argued, social reasoning evolved out of the combined forces of a complex, hierarchically organized primate social system and a large brain. We concur but add that an extended childhood was also likely necessary for deontic reasoning to evolve. In this light, we argue not that such reasoning is "innate" but that children are predisposed to attend to and are sensitive to feedback related to social contracts or exchanges, which facilitates the development of deontic reasoning.

SOCIAL INTELLIGENCE AND THE CREATION OF CULTURE

When one thinks of the great intellectual accomplishments of the human species, he or she tends to focus on the invention of some new form of technology, the discovery of a medical procedure or a cure for a debilitating disease, or on abstract or mathematical discoveries such as Einstein's theory of relativity. Yet, our species' most remarkable form of intelligence, at least in the big picture, may be reflected in our day-to-day interactions with other people. Culture as we understand it could have evolved only within a social species that was able to infer the intentions of fellow members. Without social learning at least of the complexity of imitative learning as described by Tomasello and his colleagues, rituals, technologies, and important ecological knowledge could not have been transmitted efficiently from one generation to another. Without theory of mind, social relations would remain on the level displayed by 3-year-old children or perhaps members of a chimpanzee troop. Such relations can be quite complicated, involving dominance hierarchies and memories for many individuals and histories of past interactions, but any culture emerging from groups of such "mindblind" individuals would be greatly limited in its complexity and cohesiveness.

Children are not born with the ability to infer the intentions of others, but these abilities develop over early childhood, so that by 4 or 5 years of age, human children possess a form of social cognition that is seemingly never achieved by great apes. The exception to this may be for apes reared under humanlike conditions. For some tasks, enculturated chimpanzees and orangutans appear to be able to infer the intentions of their human caretakers. Although the verdict is still out on this "enculturation hypothesis," these findings suggest that the roots of *Homo sapiens'* social intelligence likely date back at least to the common ancestor we last shared with orangutans, perhaps 15 million years ago. Changes in social complexity and extension of the juvenile period may have promoted modifications of rearing conditions, which in turn led to the ability to understand the intentions of others and eventually to the creation of culture.

8

ALL IN THE FAMILY:
PARENTS AND OTHER RELATIONS

Few things are as crucial to a child's well-being as parents. A child's very survival is dependent on the nurturing received from his or her parents, extending well past infancy into the juvenile period and perhaps beyond. But the benefits of parenting go both ways. There can be few things so crucial to an adult's fitness as the survival and reproductive success of his or her child. From an evolutionary perspective, children are the most direct means of continuing one's genetic line, the true test of natural selection. Children and their parents have an enormous amount invested in the parent–child relationship, and developmental psychologists (Bornstein, 1995; Collins, Maccoby, Steinberg, Hetherington, & Bornstein, 2000) and evolutionary psychologists (Geary, 2000; Keller, 2000) have recognized this.

According to evolutionary developmental theory, infants and children have evolved psychological (and physical) mechanisms to enhance the nurturing they get from their parents and to adjust aspects of their ontogeny to best match the conditions of their local ecology. Parents, in turn, have evolved mechanisms for evaluating the worth of infants and determining how much and for how long they should invest in an offspring, dependent on local conditions. On the one hand, parents and children have the same self-interests—seeing that the child, who possesses copies of the parents' genes—reaches reproductive age. There would seem to be no other relationship, except perhaps that between identical twins, that should promote greater cooperation among interactants than that of parent and child. On the other hand, the interests of children and their parents are not identical. For the child, his or her personal survival is all-important. From an inclusive fitness perspective, a child can "benefit" by the survival and reproductive success of a close relative but, in general, nothing spells success like growing up to be a reproductive member of one's species. For a parent, however, any particular child is only one of potentially many offspring who may carry his or her genes into the next generation. Although we think it is fair to say that every parent wishes that all of his or her children will be successful, the reality is that some offspring will have a better chance of success than others. All children, then, are essentially competing with their siblings, and even potential siblings, for their parents' attention and investment of resources. Correspondingly, parents must guard against "overinvesting" in specific children. That is, parents must balance costs associated with care of a specific child against resources that can be used for other children and for the parents themselves. Parents have a bigger picture in mind (although not likely in the conscious part of their minds) than does any individual child. As shown below, the result is competition, both between parents and children and between children and their siblings (Trivers, 1974).

The evolutionary theory that best accounts for many aspects of parenting in a wide range of species, including humans, is evolutionary biologist Robert Trivers's (1972) *parental investment theory*, which has been discussed briefly in previous chapters. This theory accounts for the amount of investment males and females put into parenting (all actions related to raising an offspring to reproductive age) versus mating (including the seeking, attaining, and maintaining of a mate). Differences in parenting and mating investment vary among species, between males and females, and as a function of environmental conditions. In the first major section of this chapter, we discuss parental investment theory as it relates to human child-rearing. We then take a closer look at some of the factors influencing the decisions parents and other people make for investing in children. In a brief section, we ask "How important are parents for healthy psychological development?" In the final section, we look at a different type of family relationship—that of siblings—and examine how evolution may have prepared children to deal with brothers and sisters.

PARENTAL INVESTMENT THEORY

There is more to life than mating and parenting but, from an evolutionary perspective, few things are more important. Mating and parenting are different but related aspects of reproductive effort. Both are concerned with fitness, or the replication of one's genes. Although a person can serve his or her genetic fitness by promoting the survival of kin such as siblings or cousins, the most direct way of ensuring one's genetic immortality is to have sex and produce offspring. However, having babies is only part of the evolutionary story. For species whose young are dependent on their parents for succor, parental care is critical in growing up to sexual maturity to continue the life-sustaining process. Individuals must make "decisions" (although not necessarily conscious ones) about how much of their efforts and resources they should put into seeking and maintaining mates and how much they should put into parenting. (Of course, animals must also devote effort to avoiding predators, seeking shelter, and obtaining food or otherwise acquiring resources. However, these resources are often used to gain or keep a mate or to provide sustenance to an offspring.)

This is the central idea behind parental investment theory. But from such simple ideas, great explanatory power can come. Parental investment theory has been applied frequently and successfully to the study of sex differences (see Buss & Schmidt, 1993; Geary, 1998; Trivers, 1985). The investment of male and female animals is not equal; for most mammals, for example, investment in an offspring is far greater for females than for males (see Clutton-Brock & Vincent, 1991). Even for very simple animals, differences in the size of the gametes reflect greater female investment. Eggs are larger than sperm and thus require more cytoplasm (and thus more energy) to produce. The sex differential in investment is substantial in mammals, with conception and gestation occurring inside the female, and the sole source of early postnatal nutrition being provided exclusively by the lactating mother. Male investment can theoretically end following copulation. As such, males have higher potential reproductive rates, in that, following insemination of a female, they can seek additional mating opportunities; in contrast, once conception has occurred, females' mating opportunities end (at least temporarily) and their parenting efforts begin. The end result is that males typically invest more in mating than parenting, whereas the reverse pattern is found for most females.

Because of differences in ecology and the maturational course and needs of the offspring, the amount of postcopulatory investment males make in their offspring varies with species. Males of some species provide no obvious support to their offspring or its mother after sex, whereas the males of other species provision their mates and offspring and may even spend considerable time in "child care" activities, such as carrying the offspring on their backs. However, in greater than 95% of mammals, males provide little or no postnatal investment to their offspring (Clutton-Brock, 1991).

In contrast to most mammals, men frequently interact with and provide resources for their children (Geary, 2000). However, women in all cultures support and interact with their children more frequently than men (Eibl-Eibesfeldt, 1989; Whiting & Whiting, 1975). For example, in a study of six cultures (Kenya, India, Mexico, the Philippines, Japan, and the United States), children were in the presence of their mothers between 3 and 12 times more frequently than in the presence of their fathers (Whiting & Whiting, 1975). This pattern persists in Western societies in which women work outside the home (Hetherington, Henderson, & Reiss, 1999) and has even been found for a group of fathers in Sweden who requested paternal leave from their jobs to be the primary caretaker for their newborns (Lamb, Frodi, Hwang, & Frodi, 1982). Because of changes in values and, especially, the increased involvement of women in the workforce, the amount of time many fathers in contemporary Western cultures spend with their children is approaching that of mothers; however, at the same time, the number of children living in homes headed by females has increased fourfold since 1960 (see Cabrera, Tamis-LeMonda, Bradley, Hofferth, & Lamb, 2000). Thus, social forces in today's world influence degree of paternal investment, but the overall pattern is still women devoting more of their time in child care than men, even in the most liberated of families and countries.

The consequences of such differential investment in offspring are reflected in sex differences in behavior. For human females, sexual intercourse brings with it possible conception and 9 months of pregnancy; in traditional societies today (and for our ancestors) it also means several years as the sole source of food (breast milk) for the infant. Thus, parental investment is obligatory and substantial for females; with the sci-fi exception of test-tube babies, infants cannot survive both pre- and postnatally without investment from their mothers. The potential investment for men in our environment of evolutionary adaptedness was (and is today) substantially less. This differential causes women to be more cautious than men in assenting to sex (Oliver & Hyde, 1993); a woman must evaluate not only the physical qualities of the potential father of her children (Does he look healthy, strong, fertile?) but also his access to resources (Is he wealthy, of high status, or otherwise capable of supporting a family?) and the likelihood of his investing those resources in her and her offspring. Men, in contrast, are less concerned with the resources a future mate may posses or her likelihood of sharing them with him; he is more concerned with her genetic fitness (Is she healthy?) and her ability to conceive, give birth, and care for a child. These are not necessarily conscious concerns of either sex, for they are reflected in the behavior of nonhuman animals as well (Clutton-Brock, 1991; Trivers, 1972, 1985).

The second sex difference in behavior is that members of the less investing sex (usually males) compete with one another for access to the more investing sex (usually females), and the greater the differential in investment between the sexes, the more intense the competition will be. The result of such competition in many animals where size and strength are used to compete is a physically larger male. Large size and the accompanying strength afford males a competitive advantage over other males and are associated, in many species, with higher social status and greater sexual access to females (see Geary, 1998, 2000).[1] This does not mean that females do not compete with one another over males; they do (see chapter 9). But

---

[1] High-status or otherwise successful males do not simply "take" females as mates; rather, by competing successfully with other males, they possess traits that females, over evolutionary time, have come to prefer. To a large extent females select successful males, rather than successful, stronger males forcing smaller females into submission. This is the process of *sexual selection* (Darwin, 1871), in which female choice of male characteristics leads to heritable reproductive advantages.

the competition among males is usually more physical and fiercer, often resulting in injury or death for some. Other males are prevented access to reproductive females and thus are "shut out" of the Darwinian game altogether.

Third, whereas maternity is always certain, paternity never is. It is within the woman's body that the child is conceived and carried to term, making maternity a sure thing. Males, in contrast, could unwittingly spend their resources investing in another man's biological child (cuckoldry), which is not adaptive from an evolutionary perspective.[2] This sex difference is somewhat exaggerated in *Homo sapiens*. Unlike most other female mammals, women are potentially sexually receptive throughout their menstrual cycle. Thus, willingness to copulate is not necessarily a sign of ability to conceive. Similarly, whereas females of many other primate species display physical signs of sexual readiness and fertility (e.g., swellings in the genital area), human females give no such signals. Not only can men not tell when a woman is ovulating and thus most likely to conceive, but neither can most women.

Parental investment theory has been applied to humans to help explain and discover sex differences in a range of behaviors (Bjorklund & Shackelford, 1999; Buss & Schmidt, 1993; Keller, 2000). Although birth control, equal rights, and an information-age economy may have reduced the magnitude of some of the sex differences in parental investment (and even reversed it for others), contemporary men and women are still sensitive to many of the same cues as were their ancestors in the environment of evolutionary adaptedness. Male and female psychologies evolved in these ancient environments, and although our behavior today (as in the past) is flexible and responsive to the specific environments in which we develop, we still possess sex-specific biases that underlie our behavior. For example, men around the world want as long-term partners women who are attractive, intelligent, and kind. Women also desire attractive, intelligent, and kind men as hus-

---

[2]In modern human societies, presumably to reduce a father's fears of cuckoldry and to increase the likelihood of paternal investment, mothers and maternal kin are apt to comment on how much a new baby resembles the father. In the first study to assess this, of all remarks made about the appearance of the baby, 80% concerned the resemblance to the father, whereas only 20% concerned the resemblance to the mother (Daly & Wilson, 1982). This pattern has been found repeatedly and in different cultures (Brédart & French, 1999; Christenfeld & Hill, 1995; McLain, Setters, Moulton, & Pratt, 2000). Although one study reported that an independent group of people did in fact rate babies as looking more like fathers than mothers (Christenfeld & Hill, 1995), this finding has not been replicated (Brédart & French, 1999). However, the bottom line seems to be that people, particularly the mother in the presence of the domestic father (McLain et al., 2000), emphasize the resemblance of the baby to the father, seemingly to alleviate any concern of cuckoldry the purported father may have. In line with this, research has shown that among a group of men participating in a program for domestic violence, their ratings of the quality of the relationship with their children was positively associated with the degree to which the children resembled the father and negatively associated with the severity of the injuries suffered by their mates (Burch & Gallup, 2000). This suggests that men use paternal resemblance as an indication of paternity.

bands but rate financial resources as more important in a prospective mate than do men (Buss, 1989; Feingold, 1992). As another example, both men and women get jealous, apparently with equal ardor. However, research has shown that men show greater distress when they envision a mate having "meaningless" sex with another man than when they envision a mate becoming emotionally (but not sexually) involved with someone else. Women display the opposite pattern (Buss, Larsen, Westen, & Semmelroth, 1992). The seeming reason for this difference is that men can never be certain of paternity, which makes their partner's possible sex with another male (even meaningless sex) a great threat. Although women also do not like the idea of their partner having meaningless sex with another woman, of greater concern to them is losing the support of their mate, which is more likely to happen if he becomes emotionally involved with another woman.

Some of the elevated aggression and violence displayed by young men of reproductive age can be attributed to their competition over females, either directly or indirectly (see chapter 9). Other, more subtle, behavioral sex differences have also been attributed to differential parental investment, specifically the ability to inhibit unwanted behavior (Bjorklund & Kipp, 1996). Because of women's greater potential investment in sex, it might be in their reproductive interests to have greater control over their sexual arousal to evaluate more closely the value of a man before assenting to sex. Ancestral women also may have needed greater political skill to keep sexual interests in other men hidden from a mate. Male response to suspected female infidelity can be violent, and even when infidelity does not lead to aggression, it often leads to divorce which, historically and in contemporary societies, is more detrimental to a woman and her offspring than to a man (H. E. Fisher, 1992). In support of this, limited evidence exists that women are better able to inhibit sexual arousal than are men (Cerny, 1978; Rosen, 1973) and strong evidence that women are better able to control the expression of their emotions than are men. This latter result is found despite the fact that females are more emotionally expressive than males. For example, when people are asked to display a positive emotion after a negative experience (e.g., pretending that a foul-tasting drink tastes good) or vice versa, females from the age of 4 years are better able to control their emotional expressions (i.e., fool a judge watching their reactions) than are males (Cole, 1986; Saarni, 1984).

Females likely require greater inhibitory skills to deal with the difficulties presented by infants (Bjorklund & Kipp, 1996, 2001; Stevenson & Williams, 2000). Caring for infants and young children often requires the suppression of aggressive and other emotional responses and the delaying of one's own gratification, and child care has always been the purview of women. Similar to the findings of sex differences in social situations (e.g., control of facial expressions), evidence exists that females perform somewhat better than males on tasks that involve resisting temptation (Kochanska, Murray, Jacques, Koenig, & Vandegeest, 1996; Slaby & Park, 1971) and delaying gratification (Kochanska et al., 1996; Logue & Chavarro, 1992), exactly the pattern that one would predict if pressures associated with taking care of young children were greater on hominid females than males (see Bjorklund & Kipp, 1996).

The basic contention of evolutionary psychologists is that the sex differences predicted by parental investment theory are based on evolved, genetic differences between males and females that cause them to process, interpret, and value information related to mating and parenting differently. We concur. However, such "innate" biases do not appear de novo in adolescence or adulthood but are sensitive to environmental factors and develop as a function of the unique experiences an individual has over childhood. That is, sex differences in parental investment develop, as do all aspects of human (and nonhuman) behavior and cognition. However, the range of expected phenotypes is not infinite. Given the importance of mating and parenting decisions for both sexes, ontogeny will progress in one of several directions that will optimize, under prescribed local conditions, an individual's inclusive fitness.

All behavior is multidetermined (Wachs, 2000), and behavior related to parental investment is no exception. It is not enough to state that "males and females have different evolved psychologies that result in different patterns of behaviors." Rather, one must ask what are the developmental mechanisms that result in different patterns of adaptive (and sometimes maladaptive) behavior. We attempt to keep this in mind throughout this chapter.

INVESTING IN CHILDREN

Although grandparents, siblings, aunts and uncles, and often unrelated group members will sometimes care for orphaned children, no one has a greater interest in children's survival than their parents (other than the children themselves). In traditional societies (and in historical times in Western culture), children without the support of fathers are more likely to die before reaching adulthood than are father-present children. The incidence of child mortality is even higher for children without mothers (see Geary, 1998). As we made clear earlier, rearing children requires substantial investment from parents, particularly mothers. Parents provide physical, social, and psychological care to infants, allocating effort and resources to their offspring that could otherwise be devoted to mating effort or their own sustenance and acquisition of resources. However, allocating resources to an infant not only limits one's own ontogeny and mating efforts but also compromises opportunities to invest in other offspring, both those born and unborn (Bjorklund, Yunger, & Pellegrini, 2002; Keller, 2000). Although it may seem obvious that parents, especially mothers, always want the best for every one of their children, the health of a child, the conditions of the local economy and ecology, the presence of additional children, the age and reproductive status of the parents (particularly the mother), and the amount of social support available to help raise a child, among other factors, influence the amount of investment parents are willing and able to devote to any given offspring.

In this section, we look at how some of these factors influence decisions women and men make in information-age and traditional cultures (and in the recent and ancient past) about investing in their offspring. Genetic parents, of course, are not the only people concerned with rearing children. We also look at the investment that grandparents make in their grandchildren, and, importantly for contemporary societies with high divorce rates, the role of stepparents. Throughout this section, we rely on evolutionary theory, particularly parental investment theory, to provide a framework for understanding the complex patterns that so significantly affect the survival and success of parents and their children. But first we describe briefly what it takes (and what it took in our evolutionary past) to rear a child, specifically the role of alloparents.

Alloparents and Child-rearing in the Ancient Past

Former First Lady Hillary Clinton (1996) borrowed the African proverb "It takes a village to raise a child" to argue that it is the responsibility of the entire community, and not just the parents, to support children to become successful members of society. In all societies today women depend on the assistance of others to help rear their children. *Alloparenting* refers to the caring for children by individuals other than the genetic mother and has likely always played a significant role in human child-rearing (Hrdy, 1999). Human mothers in traditional societies are the sole source of nutrition for their infants during the first few years of life. However, the period of offspring dependency extends well past the typical age of weaning (in traditional cultures, usually between ages 3 and 4 years) into the childhood period. Because of this extended period of childhood dependency, women have relied on the assistance of others to help rear their offspring. Although professional day care serves this role for many women in modern societies, in the past it typically fell to familiar members of one's immediate social group, typically female kin.

Alloparenting is not unique to humans (see Hrdy, 1999; Packer, Lewis, & Pusey, 1992). For example, communal suckling is often found in animals that live in matrilineal groups, such as elephants, lions, Cebus monkeys, and bats (Packer et al., 1992). In such groups, the allomother who permits the offspring of other females to suckle is providing resources for nieces and nephews, thus enhancing her inclusive fitness.

Although wet nurses have been used throughout human history (see Hrdy, 1999), it is likely that most of the assistance women in traditional cultures receive (and received in the environment of evolutionary adaptedness) in caring for their children is indirect. For example, among the Ache, a hunter–gatherer group of people from the forests of eastern Paraguay, mothers with young children forage less than women without children, but the deficit is made up through the efforts of other women, most (but not all) of them blood relatives (Hill & Hurtado, 1996). In many societies, "babysitting" is done by pre-adolescent girls (often older siblings) and grandmothers (see Hrdy, 1999). (Grandparental investment is discussed in a section below.)

The point is that in traditional societies, and almost surely in our prehistoric past, child-rearing was performed by an interconnected group of mainly female relatives who had a vested interest in the survival of the children they helped care for. Fathers played a role (and are playing an increasingly greater role in information-age societies, see Cabrera et al., 2000; Collins et al., 2000), but most of the responsibility of caring for a child in ancient times fell to the mother and her female relatives, and this pattern continues today (Hetherington et al., 1999). We are not suggesting that traditional child-rearing was an equally shared communal effort; mothers have always had the primary responsibility for the care and provision of their offspring. But the nature of human development has made it necessary that a mother receive assistance from others, typically beyond whatever support is provided by the father, in rearing a child to adulthood (or at least to the juvenile period). Humans are social creatures, and this is perhaps most apparent (and critical) when it comes to rearing children (Bjorklund et al., 2002; Geary & Flinn, 2001).

Calculus of Maternal Investment

In the majority of contemporary, industrialized societies, people expect a mother to love and care for all of her children equally. But research from both modern and traditional societies has demonstrated that such societal expectations are frequently not met. Many factors involving both the child and the parent, as well as the local ecology, influence the amount of investment a mother puts into her various children. This differential care is predictable from an evolutionary viewpoint: Mothers will invest most in children who have the greatest chance of reaching reproductive age and thus carrying forth the mother's genes. Mothers who are adept at identifying cues to a child's future reproductive success are more likely to invest the most time, energy, and resources in those children. Henceforth, children who are highly invested in have an advantage over children who are not so highly invested in and are thus the most likely to reach reproductive age and produce future generations. Mothers who are not proficient at identifying such cues are likely to squander scarce resources on a child who may not make it to adulthood, no matter the degree of investment that is made. Indeed, natural selection has selected mothers who are skillful at identifying which children, as well as which circumstances, are best suited to raising a child to reproductive years (see Hrdy, 1999).

Reduced maternal investment can, and has, taken many forms. Children may be neglected, receiving less attention, medical care, and food than they need. Abuse may accompany the neglect. More extreme forms of investment reduction include fostering out the child with relatives or even strangers, wet-nursing, or oblation (leaving the child in the custody of some religious institution).[3] Infants and children in some cultures have been sold into slavery or, at the extreme, put to death (Hrdy, 1999).

Child's Health

The first, and perhaps the most salient cue to a child's fitness, is the health of the child. In some societies, mothers who invest heavily in unhealthy infants may be seen as wasting their efforts on a child who will not reach reproductive age. Our society, of course, puts a high value on life and expects parents to care for even the most sickly of offspring. In situations where this becomes an insurmountable hardship, the state steps in to shoulder the burden. However, not all societies share this view of the sanctity of life and, historically, few governments have seen it as their responsibility to care for children whom parents cannot or will not care for themselves. Moreover, even given the "official" moral and legal statutes concerning caring for sickly infants, attitudes and behavior of individuals in contemporary cultures continue to be influenced by the harsher dictates of millions of years of natural selection. Although our discussion below of the options of mothers toward infants with low-reproductive value is intended to be seen from a historical and cross-cultural perspective (i.e., it is not meant to describe what is "typical" for women in information-age societies), modern women nonetheless possess the same evolved psychology as their ancestral

---

[3] Oblation was an alternative to infanticide or infant abandonment for many European women between the 15th and 18th centuries. Institutions were usually operated by the church and were established to reduce the number of abandoned infants. Women could leave their babies anonymously at the home, thus assuaging their guilt and believing that their infants would be well cared for. However, the death rate at these institutions in times before baby formula—when wet nurses had to be found or when infants were fed an inadequate diet of gruel—often exceeded 60% (Hrdy, 1999). Although economic conditions in Europe today are far better, Hamburg, Germany, opened its first foundling home since the 1700s in March 2000 (C. J. Williams, 2000).

grandmothers and contemporary women in traditional societies. Thus, although some of the options of minimizing investment in unhealthy children are not legally available to mothers in modern cultures, these women are not immune to the thoughts and feelings that such children may invoke in them and, particularly under economically and socially trying circumstances, they may behave in socially unacceptable ways.

Let us begin our look at the calculus of maternal investment with cases of neglect or abuse in modern societies. An abundance of evidence exists that children with mental retardation or other congenital defects, such as Down syndrome, spina bifida, cystic fibrosis, or cleft palate, are abused at rates 2 to 10 times higher than are nonafflicted children (see Daly & Wilson, 1981, for review). Furthermore, when these children are placed in institutions, parental interest rapidly decreases, and many are not ever visited by family (Daly & Wilson, 1988b). Although such descriptions seem to be about cruel and uncaring parents, they reflect the reality that mothers do differentiate among offspring and divest resources from children they deem undeserving of parental investment.

In other situations, differential investment in unhealthy children may be less drastic. In a study involving seven pairs of premature and extremely low-birthweight twins in the United States, Mann (1992) tested the idea that mothers would demonstrate more positive maternal behaviors, such as talking to, playing with, gazing at, kissing, holding, and soothing, toward the healthier of the two infants. Observations were made of the mothers interacting with both of the infants at age 4 months and again at 8 months of age. Results at 4 months showed a slight preference for the healthier infant. However, striking results were found when mother–child interactions were observed four months later: Every mother demonstrated more positive behavior toward the healthier of the two infants. This effect held even though the sicker of the twins was often more vocal and responsive to the mother. Mann thus found support for her "healthy-baby hypothesis"— maternal preferences were clearly linked to the baby's health status, mediated by the differential behavior and appearance of the two siblings (Sameroff & Suomi, 1996).

The most extreme form of reduced parental investment is infanticide. However, it is useful for our analysis because it is also the least ambiguous and, whereas child abuse and neglect frequently go unrecognized or unreported, child homicide rarely does. Although looking at such extreme behavior may not accurately describe what is "normal," it can serve as a guide to the underlying psychological mechanisms that influence less extreme behavior in people.

Infanticide is sanctioned in many traditional societies and, in fact, is expected by most members of some societies, including the parents, under certain circumstances (Daly & Wilson, 1984). Data provided by the Human

Relations Area Files (HRAF), a compendium of anthropological data, reveal that of the 60 major cultures included in the data set, 35 described instances of infanticide (Daly & Wilson, 1988b). The intentional killing or abandonment of children who were deformed or seriously ill was noted in 21 of the 35 societies. In many of these societies, deformed children are viewed as ghosts or demons, and thus it is the parent's right, in fact obligation, to kill them.

Child's Age

Another cue available to mothers concerning the amount of investment to devote to a particular child is the child's age. Mortality rates of infants are higher than those of older children, due to the prevalence of childhood diseases and mishaps. This was especially true in ancient times, before the marvels of modern medicine. Thus, the reproductive value of a child increases with age, with the implication that mothers should be more likely to invest in older versus younger children (Daly & Wilson, 1988b). Looking at the cross-cultural evidence compiled in the HRAF files, evolutionary psychologists Martin Daly and Margo Wilson (1988b) found that when a child was born into a situation of scarce resources, he or she was more likely to be killed at birth. When there was an older sibling, the younger of the two was always the one put to death. It seems that rather than risk the survival of the more reproductively valuable older child by sharing already scarce resources, infanticide of the younger child is committed.

Of course, infanticide is not sanctioned in developed cultures today, and the murder rate of infants and young children is low relative to the murder rate of teenagers and adults. But infanticide (murdering a child less than 1 year of age) does happen, and when it does, a natural parent is seven times more likely than a nonrelative to be the perpetrator, and the mother is more likely to commit the homicide than the father (Daly & Wilson, 1988b). The probability of a child being killed by a parent decreases sharply after the first year and continues to fall until young adulthood, when it is effectively zero. This is in contrast to the pattern of homicide by a nonrelative, which increases with the age of a child (see Figure 8.1). It is thus not the case that children are simply better able to defend themselves as they get older. Rather, infanticide occurs most frequently when the reproductive value of the child and the amount of previous investment in the child is the least.

All in all, parents have evolved psychological mechanisms that allow them to pick up on cues about a child's ability to transform parental resources into future reproductive success. Children who are more likely to succeed in this arena should be invested in, by mothers in particular, at much higher rates than children who are less likely to do so.

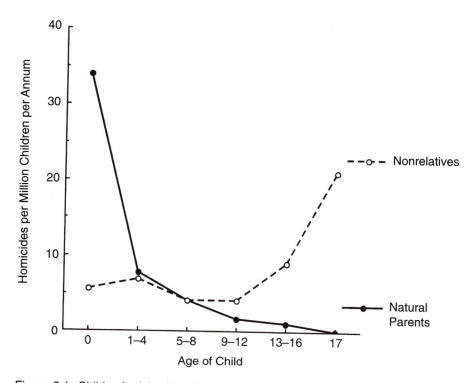

Figure 8.1. Children's risk of homicide by a natural parent in relation to the child's age, Canada 1974–1983.

Note. Adapted from *Homicide* by M. Daly and M. Wilson, 1988b, New York: Aldine de Gruyter. Copyright 1999 by Aldine de Gruyter. Adapted with permission.

Mother's Reproductive Status

Several factors germane to mothers themselves, independent of infant status, may elicit differential maternal investment. One may be linked to the woman's reproductive status. A young, fertile woman can expect to have many more child-bearing years ahead. Thus, an infant who is unwanted may be invested in minimally without great cause for worry—when circumstances are better, there will be other opportunities for producing viable offspring. As a woman's reproductive years diminish, the cost of abandoning a child increases. Thus, selection pressures would be expected to produce younger mothers who are more stringent in which infants they choose to invest in, whereas mothers nearing the end of their reproductive years should invest immediately in their offspring and not be as discriminating.

One of the best predictors of child abuse in the United States is the age of the mother, with young mothers being more likely to neglect and abuse their children than older mothers (Lee & George, 1999). Moreover, maternal age is also a potent predictor of infanticide (Daly & Wilson, 1988b; Overpeck, Brenner, Trumble, Trifiletti, & Berendes, 1998). Data provided by Daly and Wilson based on homicide rates in Canada between 1974 and 1983 indicate that teenage mothers are more than four times as likely to kill their infants as are mothers in their 20s. Although the emotional immaturity and lack of social support of young mothers are certainly factors in this statistic, it is a pattern that is predicted by evolutionary theory. Data from traditional cultures support this pattern (Bugos & McCarthy, 1984): Among the Ayoreo, a society of nomadic foragers from the border regions of Bolivia and Paraguay, infanticide is a common occurrence when the child is judged to be defective or economic prospects are poor. Bugos and McCarthy found that given these circumstances, younger women are more likely to kill their infants. As in the Canadian data, the risk to infants is highest when the mother is a teenager.

Social Support

Another cue mothers use when deciding to invest in their children is the amount of social support that is available to them. Support from the community, particularly alloparents consisting mostly of maternal kin, is one factor to be considered. But perhaps more important is the amount of help a mother can expect from her mate. Lack of a marital partner may indicate restricted availability to resources and thus a limited chance for the child to survive and prosper. From a strictly inclusive-fitness perspective, mothers should only attempt to raise a child alone when conditions are favorable, such as having plentiful economic resources (Lancaster, 1989). But when a mother lacks appropriate support or resources, lower parental investment in children is predicted. To this end, a mother's marital status can be regarded as a cue to her availability of resources, and thus unmarried women would be expected, on average, to have less social support and limited access to resources.

If one looks at the extremes again, evidence from infanticide statistics can be seen to support this contention: Among infanticidal women are many more single mothers than would be expected by chance (Daly & Wilson, 1988b). Moreover, this effect is not diminished when the age of the mother is considered (single mothers are more likely to be younger than married mothers). Based on Canadian data for 1974–1983, the rate of infanticide decreased with maternal age and was greater for single than for married mothers at all ages between the teenage years and mid-30s. Although the magnitude was the greatest for teenage mothers, the difference did not disappear at later ages. Similar patterns are seen in the HRAF files for nonindustrialized societies. In several societies, infanticide is sanctioned when a women is unmarried or when she is in discord with her husband. Presumably, lack of paternal support is indicative of a mother's inability to raise her offspring to reproductive age. Given this situation, from a strictly evolutionary perspective, a mother's optimal strategy is to immediately curtail investment in her offspring by means of infanticide and to try again later when the situation is more favorable (Daly & Wilson, 1988b).

Parent–Child Conflict

For reasons stated above, mothers are the dominant caregivers of children, and any care provided by a mother to her offspring greatly benefits her direct fitness. Following this logic, a mother might be thought to be in perfect harmony with her child—any amount of investment that she imparts to the child also benefits her. However, mother–child relationships are rife with discord. But how can this be the case given that mothers greatly benefit when their offspring prosper?

The answer to this question is provided by Trivers's (1974) theory of parent–offspring conflict. Trivers proposed, on the basis of the cost–benefit ratio suggested by Hamilton (1964), that any offspring would attempt to maximize its own fitness by endeavoring to extract as many parental resources as possible. Parents, on the other hand, would not want to convey all of their resources to just one offspring; doing so might jeopardize the welfare of current or future offspring, or the parents themselves. In some cases, offspring may not be equally competent at using parental resources in a way that maximizes the parent's return, so parents may choose to invest in some offspring more than others (Daly, Salmon, & Wilson, 1997). In particular, Trivers (1974) proposed that parents and their offspring would have conflict over three major issues:

1. the period of time parents should continue to invest in a given offspring;
2. the amount of investment a parent should impart to an individual offspring; and
3. the amount of altruism or egoism that an offspring should demonstrate to other relatives, particularly siblings.

Accordingly, children have evolved mechanisms, both psychological (Lummaa, Vuorisalo, Barr, & Lehtonen, 1998) and physiological (Haig, 1993), to garner as much investment from parents, mothers in particular, as those parents are willing or able to give. Some researchers have suggested that these types of conflict have evolved as early as conception (Haig, 1993) and cell division (Trivers, 1974). But mothers have also evolved mechanisms to counter their offsprings' manipulations. Trivers (1974) likens the results to a "tug-of-war" in which both child and parent attempt to extract from each other, or curtail, their ideal amounts of parental investment.

In an interesting elaboration of Trivers's theory, Haig (1993) has detailed the mechanisms involved in human mother–child conflict during pregnancy. From the moment of conception, the mother's body may attempt to abort any fetus deemed unworth investing in, owing to several possible inadequacies (e.g., chromosomal abnormalities). The fetus in turn endeavors to remain in the mother's womb by producing the hormone human chorionic gonadotrophin. This hormone prevents the mother from menstruating and thus shedding her uterine lining along with the newly implanted embryo. The fetus also projects trophoblasts into the mother's endometrial arteries (the arteries that supply nutrients to the now developing fetus) that prevent the arteries from constricting and consequently reducing blood flow to the fetus. Once the fetus has control of the blood flow to the placenta, it releases human placental lactogen to decrease the effects of insulin in the mother's blood, thus keeping higher amounts of glucose (fuel) available to itself. In effect, the fetus has strong-armed the mother into providing adequate resources to ensure its healthy development, even though these fetal manipulations may be harmful, in some cases, to the mother.

As discussed above, extrinsic circumstances that may not be ideal for child-rearing sometimes lead to infanticide. Often, the competition between siblings leads to the scarcity of resources, thus endangering the well-being of both siblings. One such circumstance that provides grounds for this type of parent–offspring conflict is the occurrence of twins. In some traditional societies, when resources are scarce, only one of the twins is allowed to survive, permitting the parents to invest their already scarce resources into only one individual (Daly & Wilson, 1988b). The child who is killed will often be the second born, the weaker, or the female. Another circumstance identified by Daly and Wilson involves infanticide as a solution to the problem of scarce resources when the spacing between two children is too short. If a child is born while an older sibling is still nursing, it will not only take resources away from its sibling but also limit the amount of resources available to itself, thus endangering the survival of both offspring. It seems that parent–offspring conflict involves not only disagreement over the amount of resources any given offspring receives from his or her parent but also disagreement concerning any amount of investment that the parent should convey to his or her siblings.

We have used infanticide data frequently in this section to illustrate the extremes that mothers sometimes go to when making decisions about parental investment. When viewed from the broader animal literature, humans are far less likely to participate in infanticide than many other animals (more likely than some others), although the circumstances under which humans practice infanticide are similar to those for many other species (Hrdy, 1999). Primatologist Sarah Hrdy (1999) suggested that mothers may kill their own infants when other means of birth control are unavailable and they are unwilling or unable to commit themselves to further care of the infant. However, with notable exceptions, mothers rarely plan to kill their babies. To quote Hrdy (1999),

> Rather, abandonment is at one extreme of a continuum that ranges between termination of investment and the total commitment of a mother carrying her baby everywhere and nursing on demand. Abandonment is, you might say, the default mode for a mother terminating investment. Infanticide occurs when circumstances (including fear of discovery) prevent a mother from abandoning it. Although legally and morally there is a difference, biologically the two phenomena are inseparable. (p. 297)

Men's Contributions to Their Offspring

As we have mentioned, for the vast majority of mammals, males provide little or no parental care (Clutton-Brock, 1991). Humans are an exception to this trend, although not the only one (see discussion of paternal care and longevity below). Although in contemporary culture it may be appropriate to ask why men do not spend more time with their children than they do, from a broader, species perspective, the better question is, why do men contribute to the care of their children at all? (Clutton-Brock, 1991; Geary, 2000).

Looking across species, males are most likely to provision or otherwise care for their infants when paternity certainty is high, additional mating opportunities are low, and survival of their offspring is fostered by such investment (Clutton-Brock, 1991; see Exhibit 8.1). There could be little adaptive advantage to a male of devoting substantial resources to an offspring who is genetically unrelated to him (unless it affords him increased mating opportunities, see discussion below) or whose success is unrelated to his parenting efforts. In such cases, a male would reap the greatest benefits by mating with many different females and leave the job of parenting to them. However, if a male's inclusive fitness is dependent on the resources he supplies his offspring, multiple matings will not afford any reproductive benefits unless, through his efforts, he can increase the chances of his offspring reaching adulthood and becoming reproductive members of the species.

Aspects of *Homo sapiens*' life histories are conducive to moderate levels of paternal investment. The period of dependency for humans is vastly extended relative to other mammals. In addition to the 2- to 4-year period in which mothers (in the days before baby formula) provide infants their only source of calories via nursing, dependency is extended for another several years, when children are unable to eat an adult diet but must have food specially prepared for them (Bogin, 1999). Under these conditions,

EXHIBIT 8.1
Factors Associated With the Evolution of Paternal Investment in Species With Internal Fertilization

Offspring Survival

1. If paternal investment is necessary for offspring survival, then it is obligatory, that is, selection will favor males who invest in offspring.
2. If paternal investment has little or no affect on offspring survival rate or quality, then selection will favor male abandonment, if additional mates can be found.
3. If paternal investment results in a relative but not an absolute improvement in offspring survival rate or quality, then selection will favor males who show a mixed reproductive strategy. Here, within-species variation is expected, with individual males varying their degree of emphasis on mating effort and parental effort, contingent on social (e.g., male status, availability of mates) and ecological (e.g., food availability, predator risks) conditions.

Mating Opportunities

1. If paternal investment is not obligatory and mates are available, then selection will favor the following:
   A. male abandonment, if paternal investment has little affect on offspring survival rate and quality.
   B. a mixed male reproductive strategy, if paternal investment improves offspring survival rate and quality, that is, variation in degree of emphasis on mating effort and parental effort contingent on social and ecological conditions.
2. Social and ecological factors that reduce the mating opportunities of males, such as dispersed females or concealed (or synchronized) ovulation, will reduce the opportunity cost of paternal investment. Under these conditions, selection will favor paternal investment, if this investment improves offspring survival rate or quality, or does not otherwise induce heavy costs on the male.

Paternity Certainty

1. If the certainty of paternity is low, then selection will favor male abandonment. Given that any level of parental investment is likely to be costly (e.g., in terms of reduced foraging time), indiscriminant paternal investment is not likely to evolve.
2. If the certainty of paternity is high, then selection will favor paternal investment, if
   A. such investment improves offspring survival or quality, and
   B. the opportunity costs of investment (i.e., reduced mating opportunities) are lower than the benefits associated with investment.
3. If the certainty of paternity is high and the opportunity costs, in terms of lost mating opportunities, are high, then selection will favor males with a mixed reproductive strategy, that is, the facultative expression of paternal investment, contingent on social and ecological conditions.

Note. Adapted from "Evolution and Proximate Expression of Human Paternal Investment," by D. C. Geary, 2000, *Psychological Bulletin, 126,* pp. 55–77. Copyright 2000 by American Psychological Association. Adapted with permission.

paternal provisioning increases the chance of survival for their offspring, and, in most contemporary hunter–gatherer societies, fathers provide the majority of calories consumed by their offspring and their mates (see Kaplan et al., 2000). In fact, the extended period of immaturity for humans and paternal investment likely coevolved (Geary, 1998; S. J. Gould, 1977; Lancaster & Lancaster, 1987). A slow-growing, dependent, large-brained creature, who requires a long time to master the demands of a complex social community, could not likely have evolved without paternal support. Conversely, females would not have selected high-investing males unless it afforded greater survival value for their long-dependent offspring. Although many synergistically interacting factors have contributed to the evolution of modern *Homo sapiens* and our marginally monogamous/marginally polygamous mating patterns, increased paternal investment, relative to our presumed ape ancestors, was likely one critical component without which modern human beings would not exist. Increased paternal investment permitted human females to rear multiple dependent offspring and to cut the childhood mortality rate in half in comparison to other primates and group-hunting carnivores (Lancaster & Lancaster, 1987).

Humans' mating strategies have also led to a relatively low level of cuckoldry, at least in comparison to our closest genetic relatives, the promiscuous chimpanzees (*Pan troglodytes*) and bonobos (*Pan paniscus*). Although findings are not definitive, estimates of paternity discrepancy (in which the domestic father is not the genetic father) have been found to be between 7% and 15% in several studies from a wide range of countries (see Bellis & Baker, 1990; Lerner & von Eye, 1992). Thus, although women clearly engage in extramate copulations (surely enough for men to have evolved mechanisms to guard against cuckoldry), they apparently do not frequently make cuckolds of their mates. This relatively low rate of cuckoldry is a seemingly necessary condition for high paternal investment to evolve.

Geary (1998, 2000) has examined evidence for the role that a father's investment has had and still has today on the survival and well-being of children. Most of the evidence, of course, is correlational, making statements of causality difficult. But in general, data from traditional hunter–gatherer societies (Hill & Hurtado, 1996), from the historical records of Western Europe (Reid, 1997), and from contemporary developing countries (United Nations, 1985), provide a consistent pattern: Children's mortality rates are higher and their social status is lower when fathers are absent. Moreover, in contemporary America, research has shown that quality of a father's active and supportive involvement in his children's lives is positively associated with academic achievement, emotional regulation, and social competence (see Cabrera et al., 2000; Lamb, 1997; Pleck, 1997).

A father's social status not only provides physical resources for his offspring but also is associated with the "quality" of the mother, at least in modern hunter-gatherer societies. High-status and successful men are apt to mate with high-status and successful women, who may be healthier and better providers (e.g., better foragers; Blurton Jones, Hawkes, & O'Connell, 1997). Thus, a mother's efforts are sometimes confounded with a father's direct provisioning, clouding the independent contribution of a father. Similarly, the higher death rates and lower social status of the children of unmarried mothers may be attributed in part to the possibility that women who cannot maintain a mate may be less fit than women who can. As such, factors other than the absence of the father likely also influence child mortality and success. However, historical data, data from traditional cultures, and data from contemporary societies, all indicate a significant role of paternal investment in the survival of children (see Geary, 1998, 2000).

Paternal investment would seem to be most important under conditions of limited resources, high stress, and intense competition. By investing in his offspring rather than seeking other mating opportunities, fathers increase the chance of their children's survival and social competitiveness. Following evolutionary psychological theory, men should have evolved a flexible reproductive strategy that leads them to invest in their offspring, at least in times of high stress, thus enhancing the opportunities and social status of their children (Geary, 2000; Lancaster & Lancaster, 1987; MacDonald, 1997). According to Geary (2000), "in environments with intense competition over scarce resources and with fluctuating and therefore unpredictable mortality risks, paternal investment in children's social competitiveness is, in a sense, insurance against unforeseen future risks" (p. 65).

Interesting evidence exists that male caregiving in a species is related to male longevity. As we have noted, human males devote more care to their offspring than most other mammals, and the amount of paternal investment is related to offspring success. Humans are not the champions of male caregivers among the primates, however. Males of several other species spend comparable time interacting with their offspring, and in two species of New World monkeys (owl monkeys and titi monkeys), males carry the offspring from birth, thus actually spending more time in contact with their offspring than the females do (Allman, Rosin, Kumar, & Hasenstaub, 1998).

Biologist John Allman and his colleagues (1998) compared the female–male survival ratio for a sample of primates as a function of how much care males of a species typically provided; these data are presented in Table 8.1. Larger numbers reflect greater female longevity, a ratio of 1.0 reflects comparable survival rates for males and females, and numbers less than 1.0 indicate greater male longevity. As can be seen, the survival rates of males relative to females increase as a function of the degree of paternal care. The slight advantage in survival rates for women relative to men is well-known in contemporary industrialized societies but is also found in historical records from Sweden dating back to 1780 and in traditional societies (Hill & Hurtado, 1996).

Allman (1999; Allman et al., 1998; Allman & Hasenstaub, 1999) speculated that one difference between male caregivers and noncaregivers is an aversion to risk. Males who are caregivers may be risk aversive in species in which they provide the bulk of the care. When survival of the

TABLE 8.1
Female and Male Survival Rates as a Function of Degree of Paternal Care for Selected Primates

Primate	Female/Male	
	Survival Ratio	Male Care
Chimpanzee	1.42	Rare
Spider monkey	1.27	Rare
Orangutan	1.20	Rare
Gibbon	1.20	Pair living, but little direct role
Gorilla	1.13	Protects, plays with offspring
Human	1.07	Supports economically, some care
Goeldi's monkey	0.97	Both parents carry infant
Siamang	0.92	Carries infant in second year
Owl monkey	0.87	Carries infant from birth
Titi monkey	0.83	Carries infant from birth

Note. Adapted from *Evolving Brains* (p. 184), by J. M. Allman, 1999, New York: Scientific American Library. Copyright 1999 by Henry Holt & Co. Adapted with permission.

offspring is highly dependent on the survival of the caregiver, natural selection may favor males who avoid risk. Males who are noncaregivers, in contrast, can take increased risks which, in many species, is associated with heightened social status and mating opportunities (Daly & Wilson, 1988b). It has similarly been argued that females, in general, have evolved a risk-aversive strategy, placing a high value on protecting their own lives, because their offsprings' survival is highly dependent on their maternal care (Campbell, 1999).

The human data from Allman et al. (1998) indicate two age-related peaks in the female survival advantage over males. The first peak occurs around age 25, and the second, smaller peak appears during the postreproductive years. This pattern has been found in contemporary industrialized countries and in historical data from Sweden. Such peaks are also apparent in the life cycles of chimpanzees and gorillas (see Allman & Hasenstaub, 1999). Allman has argued that the peak in early adulthood corresponds to the period of greatest child-rearing responsibility for women and the greatest risk seeking by men. This increased risk taking, Allman proposed, is what is primarily responsible for the differential death rate at this time. The second peak, occurring in early old age, corresponds to the time when many women become grandparents and is related to the greater risk of stroke, heart disease, and cancer among men. This peak in differential female survival may be attributed to the long-term effects of hypertension and vascular disease brought on by greater stress in males. Related to this is the idea that exposure to testosterone, which promotes the development of secondary sexual characteristics and plays a role in male status competition and aggressiveness, can suppress the effectiveness of the immune system (see Folstad & Karter, 1992). Thus, characteristics associated with success in young adulthood may contribute to poor health in later life.

Grandparental Investment

Other than the father, the individuals most likely to be willing and able to assist a mother and her children are the children's grandparents.

The "Grandmother Hypothesis"

Several theorists have proposed that grandparents' investment (particularly grandmothers') in their grandchildren contributed to our species' extended life span (Alexander, 1974; Gaulin, 1980; Geary & Flinn, 2001; W. D. Hamilton, 1966; Hawkes, O'Connell, & Blurton Jones, 1997; O'Connell, Hawkes, & Blurton Jones, 1999; Sherman, 1998). That is, perhaps long-lived people, although no longer of reproductive age themselves, may foster their grandchildren's survival. This possibility is consistent with Hamilton's (1964) concept of inclusive fitness, which posits that individuals can serve their reproductive ends not only by producing offspring but also by increasing the likelihood that their more distant relatives (e.g., grandchildren) survive. This argument is particularly germane for women, whose fertility declines at a much faster rate than their general physiologic functioning. Most women, even in traditional societies, can expect to live more than 20 years beyond the age at which they last give birth.

Support for the "grandmother hypothesis" comes from evidence of reproductive risk factors associated with maternal age. Compared with younger women, older women who become pregnant are more likely to spontaneously abort a fetus, have children with low birthweight, have children who are stillborn, and are more likely to die themselves (see Kline, Stein, & Susser, 1989). Furthermore, given women's longevity, in general, a woman who continues to remain fertile after her mate dies invites the presence of a new mate, which in turn may endanger the welfare of her existing offspring, as well as the benefits she may be receiving from her deceased mate's family (Turke, 1997). In addition, menopausal women in traditional societies can invest more time and resources in their older children and in their grandchildren without having to distribute scarce resources to young infants and jeopardize the survival of older offspring and grandoffspring (Geary & Flinn, 2001; see section on the calculus of maternal investment). Each of these characteristics may contribute to older women realizing greater reproductive benefits by helping to rear their grandchildren than by having additional children themselves.

Some evidence exists that vervet monkeys (Fairbanks, 1988), baboons, and lions (Packer, Tatar, & Collins, 1998) benefit by the presence of a grandmother, and research with traditional human groups also provides some support for the grandmother hypothesis. For example, evidence that postmenopausal women contribute to the success of their grandchildren comes from observations of the Hadza, a small group of foragers living in Africa's Rift Valley (Hawkes et al., 1997; O'Connell et al., 1999). Hawkes and his colleagues noted that older women's foraging was particularly important for the nutrition of young children who had been weaned but who were not yet prepared to eat adult food (see Bogin, 1999; also chapter 4). Hawkes et al. (1997) reported that in Hadza families in which the mother was not nursing, children's nutritional status was related to their mothers' foraging efforts: The better their mothers were at obtaining food, the healthier children were. However, in families in which mothers were nursing, the nutritional status of weaned children was related to the foraging efforts of their grandmothers. If such a pattern reflects ancestral hominid populations, the result would have been to increase fertility by permitting mothers to wean a child earlier and become pregnant again. Without the grandmothers' support for weaned children, nursing would likely continue for several more years, reducing the number of children a hominid female could expect to have. Factors associated with increased life span would also be selected, at least in females. As we mentioned in chapter 2, normally, traits and behaviors characterizing postreproductive individuals are neither selected for nor against, because these individuals have already reproduced (Baltes, 1997). The exception would be when behaviors of older individuals increase the success of their offspring or grandoffspring. In such cases, longevity would be selected for because it increases the survival of progeny.[4]

Grandparental Investment and Paternity Certainty

A grandparent's provisioning of children will benefit his or her inclusive fitness only to the extent that such care promotes the survival of genetic kin. Thus, grandparents must evaluate the likely genetic relatedness of an

---

[4] Although it seems clear that the care provided by older women can benefit the survival of their grandchildren, it is less clear that such investment yields greater inclusive fitness than they would experience by having additional children (i.e., by not becoming menopausal). For example, in research by Packer and his colleagues (1998), grandmother lions often nursed their grandoffspring, but only when they had a litter themselves. There was no evidence of grandmotherly care and greater grandoffspring survival for "menopausal" grandmothers. Using data from the Ache people, Hill and Hurtado (1991, 1996) found no evidence that older women were better off, from an inclusive fitness perspective, investing in their grandchildren than in having more children themselves. They suggested that the benefits to menopausal women of investing in their adult children and grandchildren might be larger during periods of lower population growth. (Hill and Hurtado examined a time of relatively rapid population growth.) All researchers seem to agree that grandmothers do contribute to the success of their grandchildren and that this may have contributed to longevity (Hawkes et al., 1997; Hill & Hurtado, 1999; Packer et al., 1998). The debate centers on the function on menopause and whether it has any adaptive function or is merely a by-product of something else.

offspring to themselves. A paternal grandparent can be no more certain of paternity than the father can, whereas the maternal grandparents have no reason to doubt the genetic relationship of their grandchildren to them. On the basis of this reasoning, maternal grandparents, on average, should be more willing to provide resources and devote care to their grandchildren than paternal grandparents.

Evidence from a variety of Western societies supports this contention. For example, Euler and Weitzel (1996) asked 1,857 German adults (ages 16 to 80 years) to rate the amount of care they had received from each of their grandparents up to age 7. (Data from 603 of these participants, who had all four grandparents living during this time, were used in most analyses.) Consistent with predictions based on paternity certainty, maternal grandmothers were rated as devoting the most care to participants, followed by maternal grandfathers. This was true despite the fact that the maternal and paternal grandparents lived, on average, equally close to their grandchildren. The greater solicitude shown by the maternal grandfather than the paternal grandmother is especially compelling, given the greater role that women play in child care relative to men in all cultures (see Geary, 1998). These findings are consistent with other retrospective reports of adults who stated that they felt closer to their maternal than their paternal grandparents (Hoffman, 1978/1979; Rossi & Rossi, 1990; Russell & Wells, 1987) or actually had greater contact with their maternal than paternal grandparents (Eisenberg, 1988; Salmon, 1999; M. S. Smith, 1988). A similar pattern has been observed for investment in offspring by maternal versus paternal aunts and uncles, with the former pair seen by college student participants as expressing more concern for them than the latter pair (Gaulin, McBurney, & Brakeman-Wartell, 1997).

The only published exception to the general trend of greater care from and closeness to maternal than paternal grandparents that we are aware of was reported by Pashos (2000). In his study, adults from Germany and Greece were asked to evaluate how much each of their grandparents had cared for them. Results from participants from urban areas were similar to those reported by Euler and Weitzel (1996) and others: Maternal grandparents provided more care than paternal grandparents. However, exactly the opposite pattern was observed for rural Greeks, with paternal grandparents providing more care than maternal grandparents. Pashos explained the results, in part, as due to the traditional family customs and practices of rural Greece, in which children usually live closer to their paternal grandparents, often in the same house. Paternal grandparents also have the social obligation of caring for their grandchildren, particularly for their grandsons, who are their primary heirs. The increased physical closeness of the paternal family may result in greater paternity certainty than is the case in urban settings. When the movements of women (daughters-in-law) are known and controlled in part by the husband's family, there is little uncertainty about paternity. As a result, investment of the paternal family in the grandchildren can be expected to be enhanced, given the patrilineal traditions of the society.

Pashos's findings, we believe, do not contradict the predictions of parental investment theory but rather demonstrate that grandparent investment is influenced by paternity certainty, even when other cultural traditions also play a role. When social customs reduce paternity certainty, as in contemporary urban environments, paternal grandparental investment is reduced; when social customs increase paternity certainty by controlling the actions of women, as in rural Greece and likely in other patrilineal societies, paternal grandparents can invest in their grandchildren with relative certainty and follow comfortably the dictates of the local community.

Cinderella or Marsha Brady? Stepparent Investment

Stepfamilies seem to be becoming increasingly prevalent in information-age societies due to high rates of divorce and remarriage. But stepparental relationships are probably not novel to modern environments—death rates of parents were exceedingly high in ancient times, owing mainly to disease and warfare. On the death of a partner, the surviving parent is thought to have formed new relationships with other potential mates, thus providing a mother or father substitute for his or her offspring. This situation was common enough in the environment of evolutionary adaptedness to suggest that modern peoples may have mechanisms specifically adapted toward stepparenting.

According to parental investment theory, stepparents should show little interest in the welfare of children who are clearly not their biological offspring—any resources allocated to offspring who are not one's own seemingly provide no immediate benefit to one's reproductive fitness. Instead, resources spent on a stepchild help to ensure the well-being of another's offspring. The tendency to squander resources on another's offspring would surely be selected against. However, in most contemporary stepfamilies, the stepparent indeed contributes significantly to his or her stepchild's care. As demonstrated below, however, the amount of care provided often differs significantly from that given to one's biological children.

Although there are similar numbers of stepmothers and stepfathers, most analyses focus on the relationships between stepfathers and their stepchildren. This is mainly due to the fact that young children are most likely to remain with their mother on separation of the natural parents, and thus, situations in which the child resides with the mother and stepfather are more common and more accessible to researchers. Although most examples given below examine the dynamics between stepfathers and stepchildren, similar processes are expected to be operating for stepmothers.

Investing in Children or in Mating Opportunities?

Most evolutionary accounts of stepparent investment attribute the amount of care given to a stepchild to mating effort on the part of the stepparent (Anderson, Kaplan, & Lancaster, 1999a; Rowher, Herron, & Daly, 1999). It would be more advantageous, the argument goes, for a woman to form a relationship with a man who is interested in helping her raise her existing offspring than to pair with a man who has no vested interest in her offspring. Hence, the best strategy for a man when looking for a new mate may be to act solicitously toward the potential mate's children. In this way, the man is able to convince the woman that he will be a valuable commodity: not only a willing and able investor in her and her children but a willing and able investor in any future children they may have together.

Although it may seem somewhat harsh to accuse stepparents of having an ulterior motive to their stepparenting efforts, the dynamics may not be so different from what is happening in biological father–child relationships. The parenting efforts of natural fathers have been questioned concerning the actual motivation behind their behaviors. Some researchers have suggested that natural fathers contribute to their offspring's fitness as forms of parenting effort and mating effort (Marlowe, 1999), whereas others insist that fathers invest in children solely as a form of mating effort (Hawkes, Rogers, & Charnov, 1995; van Schaik & Paul, 1996). Although this controversy remains a point of contention, one thing should be relatively clear from a direct fitness perspective: Stepparents should not be expected to contribute more to a child's fitness than the natural parents would be expected to contribute. Thus it seems likely that the investments made by stepparents are more aimed at mating effort than actual parenting effort.

This is not to say that stepparents are unable to form strong emotional bonds with their stepchildren—many stepparents love their stepchildren deeply. But in the majority of cases, stepparents find it difficult to form an emotional attachment to their stepchildren that parallels feelings for their biological children. In one study of middle-class stepfamilies in the United States, only 53% of stepfathers and 25% of stepmothers claimed to have any "parental feelings" whatsoever for their stepchildren (Duberman, 1975). The fact that many stepparents do have close, enduring bonds with their stepchildren may be primarily due to motivation for such a bond on the part of the stepparent.

Stepparental Investment in Their Stepchildren

Empirical evidence from a wide range of cultures supports the idea that stepparents invest less in their stepchildren than in their natural children. Research in the United States by Anderson and colleagues (1999a) reported that stepchildren do in fact receive considerably less monetary help from their families for their college education than children who have two biological parents. In a related study, Xhosa high school students in South Africa were examined, and similar results were found—stepfathers spent significantly more money on their natural children than on their stepchildren (Anderson et al., 1999b). Another study involving data collected from the U.S. National Education Longitudinal Survey of 1988, a survey that contained a sample of eighth graders representative of the population of the United States, found similar patterns (Zvoch, 1999). Stepfamilies saved less money for their children's education, started savings accounts for children later, and expected to spend less money for their child's education in the future. Yet, stepchildren had a considerable amount of resources allocated to them by their stepfamilies, although this amount was much reduced over the amount of resources allocated to children of two biological parents.

Observations from the studies done by Anderson and colleagues also demonstrated that stepfathers spend significantly less time with their stepchildren than with their natural children. Data involving the U.S. sample revealed that fathers spend approximately 3 hours less per week with their stepchildren than with their natural children (Anderson et al, 1999a); data from the South African sample showed that stepfathers have approximately 84 more interactions a year with their biological children than their stepchildren (roughly 1.62 fewer interactions with stepchildren each week; Anderson et al., 1999b). Fathers in the South African sample were also less likely to help stepchildren with their homework or in practice speaking English than with a genetic child. Similarly, using data from the United States and South Africa, recent evidence indicates that less money is spent on food when a child is reared by an adoptive, foster, or stepmother than a biological mother (Case, Lin, & McLanahan, in press). Note that whereas most research has focused on the reduction of support to nonbiological offspring made by stepfathers, this study documents that a similar trend is observed for stepmothers.

In an observational study of the Hadza, Marlowe (1999) demonstrated that stepchildren receive considerably less care from stepfathers than biological children. Marlowe found that fathers spent more time in close proximity with, communicated more with, and nurtured more (held, fed, pacified, cleaned) their natural children than their stepchildren. The most striking difference between treatment of natural children versus stepchildren was found in the amount of time spent in play with the father: In not one instance did a Hadza stepfather engage in play with a stepchild. These results are particularly intriguing in that when interviewed, most group members reported that stepfathers care for and feel the same about both stepchildren and natural children. Interestingly, when asked directly, about half of the stepfathers admitted that their feelings for their stepchildren were weaker than those held for their natural children. Clearly there is some difference between the Hadza's ideas of what should be and what actually is. Marlowe interpreted this discrepancy as a type of deception that may be involved in promoting the ideology that stepfathers should be good fathers. Several studies from rural Caribbean kin-based communities have produced similar results (Flinn, 1988; Flinn, Leone, & Quinlan, 1999). Cross-culturally, the results seem markedly similar: Stepchildren are invested in less than biological children.

Perhaps these findings lend some credence to the tales of "wicked" stepparents in folklore and mythology. Fables based on evil stepparents are abundant worldwide. One extensive study of such fables revealed two major classifications of stepfathers, "lustful" and "cruel," and found that stepmothers are depicted as harsh taskmasters and murderesses (Thompson, 1955; cited in Daly & Wilson, 1988b, p. 85). It is perhaps telling that such similar tales have flourished across so many diverse cultures; from the Aleuts to the Indonesians, every culture appears to have its own version of Cinderella. But is the myth of the cruel stepparent just a myth? Empirical evidence suggests perhaps not.

When Children Are Abused

Although research has shown that child abuse is more common in stepparent homes than in natural parent homes, it is important to recognize that the vast majority of stepparents are not abusive; many, if not most, truly love and care for their new "offspring." Yet, it should be recognized that stepparents may have to exert effort in attempting to love and nurture their stepchildren, whereas strong emotional bonds tend to happen "naturally" and without substantial conscious effort between genetic parents and their children. This point fits well with parental investment theory: Parents are expected to care more about their genetically related offspring than about others' offspring. Although stepparents may have the best intentions toward their stepchildren, in fact they are less invested in those children, and the potential for abuse and neglect is higher.

Daly and Wilson (1988b, 1996) have extensively researched child abuse and child homicide as a function of living with a stepparent. In one of the most extensive studies of its kind, they interviewed 841 households containing children younger than age 17 and 99 additional abused children known to children's aid societies and reported to a provincial registry in Hamilton, Ontario, Canada (Daly & Wilson, 1985). The results were startling: Children were 40 times more likely to be abused if they lived with a stepparent versus two birth parents. This enormous difference remained even when possible influencing factors that may be associated with stepfamilies, such as poverty, the mother's age, and family size, were statistically controlled. Given these and similar findings (Wilson, Daly, & Weghorst,

1980), Daly and Wilson (1988b) concluded that "stepparenthood *per se* remains the single most powerful risk factor for child abuse that has yet been identified" (pp. 87–88).

But these results are still subject to criticism. Perhaps suspected cases of abuse are more likely to be reported when occurring in a stepparent home than in a home with two genetic parents. To overcome such biases, Daly and Wilson (1988b) examined cases of unmistakable abuse that were assumed to be less subject to reporting bias—child homicide. They thought that if there truly were a reporting bias of stepparent abuse, that bias would be eradicated when examining child murders by parents. Such cases of lethal abuse are more difficult to hide and even more difficult to interpret as anything other than homicide. Perhaps not surprisingly, they reported that the relationship between stepparents and abuse (in this case, homicide) did not diminish; it increased. In one study that examined a 10-year period in which 408 Canadian children had been identified as victims of homicide at the hands of either genetic parents or stepparents, the risk of being killed by a stepparent was far greater than that of a genetic parent (Daly & Wilson, 1988a). This difference was particularly high for children younger than age 2: The likelihood of being killed by a stepparent was 70 times greater than the likelihood of being killed by a natural parent. A similar study using a U.S. sample reported even stronger results: The youngest children, younger than 2, were 100 times more likely to be killed by a stepparent than by a birth parent (Daly & Wilson, 1988b).

When stepparents (particularly stepfathers) do invest heavily in their stepchildren, such investment can be viewed as a form of mating effort in that the stepfather hopes to gain mating opportunities with the child's mother by way of aiding her in the care of her children. Although the majority of stepparents in industrialized societies make fervent attempts at loving and caring for stepchildren, the reality is that these relationships are often not as reciprocally rewarding as those involving a natural parent and child. The result is that, on average, stepparents spend less time with, invest fewer resources in, and care less about their stepchildren than they would their own genetic children. In the most extreme cases, stepchildren are invested in so little as to be abused or even killed.

HOW IMPORTANT ARE PARENTS FOR HEALTHY PSYCHOLOGICAL DEVELOPMENT?

It seems incontrovertible that parents play a significant role in children's psychological development. However, an alternative, evolutionarily based theory suggests that normal individual differences in parenting style contribute little to important individual differences in children's personality and cognitive development. Developmental psychologist and behavioral geneticist Sandra Scarr (1992, 1993), in an extension of her genotype → environment theory (Scarr & McCartney, 1983), proposed that "ordinary differences between families have little effect on children's development, unless the family is outside of a normal, developmental range. Good enough, ordinary parents probably have the same effects on their children's development as culturally defined super-parents" (Scarr, 1992, p. 15). Scarr's claim is based on the strength of active genotype → environment effects, in which genetically based dispositions cause children to seek environments consistent with their genotypes (see earlier discussion in chapter 3). Although it is experiences in these environments that significantly shape children's personalities and intellects, these experiences are driven by their genes. Such genotype → environment effects increase over the course of ontogeny, as parents' direct (nongenetic) influences wane. Scarr's claim is that children around the world grow up to be productive (and reproductive) members of their societies despite considerable differences in child-rearing practices. A species that required high-quality parenting for survival would likely be extinct before too long. Thus, the species evolved so that children could tolerate great flexibility in child-rearing practices and still grow up to be "normal."

Although Scarr is surely correct at one level (humans are amazingly flexible to the vagaries of parenting behavior), this is not the level with which most psychologists are concerned, and Scarr's theory has been severely criticized (Baumrind, 1993; Jackson, 1993). Children in a wide range of environments will grow up to be reproductive members of their species, but their mating strategies are influenced by a history of parental interaction. Moreover, even if one gives the behavioral geneticists their due when they propose that most individual differences in personality and cognitive abilities can be attributed to direct genetic effects and nonshared family environments, which can be thought of as indirect genetic effects (see chapter 3), children must have the opportunities to seek their niche. If children's opportunities are restricted (e.g., a child with a disposition toward high literacy is reared in an environment without books and that does not reward academic achievement), they will not be able to "reach their full potential."

Scarr (1993) concurred with her critics that in some environments good-enough parents may not be sufficient. Children who lack opportunities and experiences associated with the dominant culture will show intellectual and social detriments relative to children in the majority culture. Scarr noted that these effects can often be ameliorated by education. The difference between Scarr and her critics seems primarily to be one of focus: Scarr is looking at how individual children become functioning members of their species; her critics have focused on individual differences among children within a culture.

In support of both Scarr and her critics is research on "resilient" children—children who develop social and intellectual competence despite growing up in impoverished and "high-risk" environments (Masten & Coatsworth, 1998). The single most important factor contributing to the success of resilient children is competent parenting, starting early in life. This fact is consistent with Scarr's critics (e.g., Baumrind, 1993; Jackson, 1993), who have pointed out that individual differences in how parents interact with their children are essential to their eventual success. However, Masten and Coatsworth pointed out that, although many parents do fail to provide adequate warmth and support for their children, it does not take extraordinary efforts to raise a competent child, even in unfavorable environments. "Through the process of evolution, parenting has been shaped to protect development; nature has created in ordinary parents a powerful protective system for child development" (Masten & Coatsworth, 1998, p. 213).

The bottom line is that parents do play an important role in fostering the psychological health of their children. However, human children have evolved the ability to tolerate a wide range of parental behaviors and still grow up to be functioning members of their species. This does not mean that all adults function equally well, particularly when adaptation to modern economic conditions, and not just procreation, is considered. It is at this level of individual differences that patterns of parenting contribute importantly to psychological development. Moreover, children's psychological development is influenced by factors beyond their family as well (Harris, 1995; see also chapter 9), so that predicting adult adjustment solely as a function of parenting style is difficult if not impossible. Parents can take neither all the credit nor all the blame for their children's lot in life.

LIFE WITH SIBLINGS

Siblings share the same proportion of genes with one another as they do with their parents. Thus, from a "selfish-gene" perspective, relationships among siblings should have many of the same qualities that are seen in parent–child relationships. Like parents and children, siblings' genetic similarity should result in their promoting the interests of one another. Yet, because they share, on average, only 50% of their genes, they do not have identical self-interests, and conflicts will arise. There are, of course, important differences between parent–child and sibling relationships. The most obvious is that parents must devote considerable time, energy, and resources to their young offspring to ensure their survival. Although older juveniles may be given child care responsibilities for younger brothers and sisters, they rarely play the role of "major provider" for their siblings. In addition to the conflict children have with their parents over investment, they are also in conflict with their siblings for the distribution of those resources. The lament of many an adult to his or her sibling, "Mom always did like you best," is spoken only half in jest.

Although there are certain things that children can "expect" (e.g., having a lactating mother and receiving social support from a father or kin), there is much variability in human families, and siblings are part of that variability. Will there be older or younger siblings? What will be the spacing between siblings? What are the ecological conditions for supporting a multiple-child family? How will parents and others respond to different children in a family? What responsibilities, if any, will one sibling have toward another? How much competition for resources is there outside of the family in comparison to within the family? Although having siblings was a fact of life for most children throughout our species' history, one could not easily predict what the nature of sibling relationships would be. Some conflict was unavoidable, if for no other reason than that siblings lived in close quarters and competed for their parents' attention. (Anyone who has grown up with identical twins realizes that having identical genes does not eliminate conflict.) Further, the "closed-field" nature of sibling relationships (i.e., they are not free to physically leave the family until a certain point in their development) may exacerbate conflicts. But this same physical closeness and genetic relatedness should also promote companionship, psychological closeness, and cooperation, at least under some circumstances. For these reasons, it is likely that children have evolved some mechanisms for dealing with the "problems" that siblings bring. However, such mechanisms should be relatively general in nature and flexible. We do not believe that psychological mechanisms have evolved for being a first-born or a later-born child. Rather, natural selection should have operated on mechanisms that promote kin recognition and for dealing with competitors early in life that vary in status (e.g., younger or older siblings), particularly as they are concerned with acquiring resources from one's parents. In the remainder of this section, we examine aspects of conflict and cooperation among siblings, possible birth-order effects, and incest avoidance, all from an evolutionary developmental perspective.

Conflict and Cooperation

We do not need scientific research to know that sibling relationships are complex and multifaceted. At times, siblings are our closest and dearest companions, our protectors from bullies and other ecological hazards, and our allies against parental control; at other times, they are our most bitter enemies, competing with us for scarce resources, being given control over us (or put in our unwilling charge), and being a focus of jealousy. Empirical research has confirmed what most of us growing up in multichild families know from experience. Conflict between siblings is common, but so is warmth and companionship, for the same children at the same time (Furman & Buhrmester, 1985; Garcia, Shaw, Winslow, & Yaggi, 2000).

Although evolutionary theory predicts that siblings should display conflict and cooperation (as do most other theories), conflict, or rivalry, should be most apparent when a first-born child suddenly has his or her status changed from an "only child" to an "older sibling." Every first-born child is an only child for a time, and the birth of a sibling is likely to be a momentous event is his or her life. The phenomenon of *regression*—displaying behavior characteristics of a younger child, such as having toileting accidents, whining, or wanting to be clothed or fed by a parent—is familiar to many parents who bring a new baby home to a family that already has one child in it. Regression can be seen as a tactic on the part of the older child to get more attention and resources from the parent. The child's position of sole object of parental attention has been usurped, and acting in a more dependent way serves to increase the amount of attention he or she receives (and often to increase the ire of the parents).

The introduction of a second child to a household clearly changes the relation between the parents and the firstborn. One effect of having a new baby in the house is to decrease the security of attachment between the firstborn and his or her mother. Teti and his colleagues (1996) assessed security of attachment between first-born preschool children and their mothers, once late in the mother's pregnancy with her second child and again 4–8 weeks after the baby was born. Security of attachment declined for the preschoolers between the two testings, with the decline being greatest for children 2 years of age and older. The authors speculated that young children did not have the cognitive ability to feel threatened by the introduction of a younger sibling, at least not to the extent that the older children did, accounting for the age difference in response to the new sibling. It should not be surprising that 2-year-old children display substantial distress to the birth of a new sibling. In traditional cultures, and surely for our ancient ancestors, such children would still be nursing, inhibiting ovulation in their mothers, and ensuring their "youngest child status" for another year or two. Other research has shown that first-born preschool girls who had a positive relation with their mothers before a second child was born reacted particularly negatively to the birth of a new sibling (Dunn & Kendrick, 1981). Fourteen months after the birth of their sibling, these girls were hostile and negative to their new baby brothers and sisters, and the secondborns similarly had developed a negative attitude toward their older sibling.

Resentment, of course, is not the only emotion that older siblings feel toward the new arrival to the family. Many firstborns become caregivers and attachment figures for their younger siblings. For example, in one study, mothers, their preschool children, and their infants participated in a modified

Ainsworth Strange Situation (Stewart & Marvin, 1984). For 4 minutes of a 54-minute session, the preschool child was alone with his or her younger sibling, and for another 4 minutes the child and infant were alone with an unfamiliar adult. Half of the preschool children (51%) displayed some form of caregiving behavior toward their younger sibling within 10 seconds of the mother departing. The children most likely to provide caregiving were those who had passed a test of perspective-taking (who were also likely to be somewhat older), reflecting the role of social–cognitive factors in the caregiving responses of young children. Also, girls were somewhat more likely to provide care to their distressed younger sibling (57%) than were boys (43%).

That girls were more likely than boys to provide caregiving to younger siblings is consistent with cross-cultural work indicating that girls from a relatively early age in traditional cultures are frequently given (or sometimes take) the responsibility of caring for younger siblings (Eibl-Eibesfeldt, 1989; Hrdy, 1999). This is also found in contemporary societies. For example, McHale and Gramble (1989) reported that girls spent nearly twice as much time per day helping to care for their younger siblings as did boys. (Girls and boys spent comparable amounts of time, however, in caregiving for a disabled younger sibling.) Whether girls are more oriented toward childcare than boys or they are merely complying with pressure to fulfill a socially stereotypic role is difficult to determine (see further discussion in chapter 9). However, the greater amount of time that girls spend caring for their younger siblings helps prepare them for the adult role of child caregiver which, in all cultures, is the primary responsibility of women.

It is well known that siblings, throughout life, cooperate with one another, form alliances, and provide each other social and economic support. Most cultures strongly encourage this sort of affiliation among siblings, making it difficult to claim that the basis for such cooperation is found in evolved mechanisms rather than cultural traditions. Although research from the animal literature makes it clear that kin selection plays a critical role in social behaviors in general, and altruism specifically (Dawkins, 1976, see chapter 9), it is difficult to provide incontrovertible evidence that positive relations among siblings are attributed to evolutionarily based mechanisms that foster one's inclusive fitness rather than to adherence to social conventions. In most of the world, the two are perfectly confounded: Societies encourage "family values," both legally and morally.

A recent study by sociologists William Jankowiak and Monique Diderich (2000) went a long way toward teasing apart effects attributable to cultural conventions from inclusive fitness. They conducted extensive interviews of families from a polygamous Mormon community in the southwestern United States. In this community, the father is the head of the family, and the official dogma of family life is that all of a man's children are equal.

The unity of the family, with the father as head, is consistently stressed in church services, Sunday school, and local schools. Children (officially) get their family identification through the father, not the mother, and cooperation within the polygamous family is paramount. The social stereotype, then, is for a high degree of within-family cooperation, with no distinction between full and half-siblings (same father, different mother).

Jankowiak and Diderich interviewed 70 individuals from 32 polygamous families who had both full and half-siblings about issues related to solidarity. For example, adult interviewees were asked about lending money to or baby sitting for a sibling (functional solidarity); they rated their closeness to their various siblings and were asked to nominate their favorite baby in the paternal household (affectional solidarity); and they were asked a series of questions concerning how frequently they interacted with one another, such as attending birthday parties and weddings. Figure 8.2 presents the percentage of full and half-siblings who were listed for each dependent measure. All differences were statistically significant, with full siblings being mentioned more often than half-siblings for each of the measures. These results are striking, not only for their consistency, but for the fact that preference for full siblings is counter to the community's values, which stress that all offspring of the head of the family should be regarded equally.

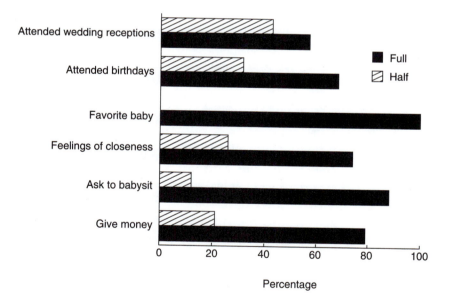

Figure 8.2. Percentage of full and half-siblings nominated for different measures of solidarity.
Note. Data from "Sibling Solidarity in a Polygamous Community in the USA: Unpacking Inclusive Fitness," by W. Jankowiak and M. Diderich, 2000, Evolution and Human Behavior, 21, pp. 125–139. Copyright 2000 by Elsevier Science. Adapted with permission.

These data argue strongly that feelings and actions of sibling solidarity are not based simply on growing up in the same household together or to social sanctions encouraging sibling cooperation. Rather, consistent with inclusive fitness theory, genetic similarity appears to play a crucial role in sibling solidarity. This does not mean, however, that children "instinctively" know who their siblings are and develop feelings of solidarity toward them. We are certain that there are proximal effects that bond full siblings together more so than half-siblings. For example, Jankowiak and Diderich commented that mothers often show preferential treatment to their own offspring, something that children in the family readily identify. For instance, mothers in many families would take their children into their bedroom with them to watch television or read. Although other children were not excluded from such gatherings, they were more apt to congregate with their own mothers and full siblings. Thus, it appears that behavior of mothers in polygamous families promotes greater solidarity among full siblings.

From a father's perspective, all of his offspring possess half of his genes and are thus equally related to him. Cooperation among any and all of his children will be to the father's inclusive fitness. Mothers in polygamous families have a different perspective. Despite social and religious values to the contrary, a mother's best inclusive interests are served by encouraging solidarity among her offspring, full siblings in a polygamous family.

Does Birth Order Really Matter?

If children are sensitive to the investment their parents make in them relative to their siblings, a child's birth position in a family should be an important factor in influencing behavior. We saw this earlier in the results of studies looking at the reactions of firstborns to the birth of a new sibling (Teti et al., 1996). The role of birth order in personality development has a long history within psychology, producing a data set that is full of contradictions (Adler, 1927; Ernst & Angst, 1983; Sulloway, 1996). It is likely true that any particular individual's birth rank in a family is a salient factor in shaping that person's development; however, research findings have questioned the extent to which being a firstborn, lastborn, or middle child, for example, has universal, as opposed to idiosyncratic, effects on development (Ernst & Angst, 1983; J. R. Harris, 2000).

Historian of science Frank Sulloway (1996) has interpreted the effect of birth order on personality development from an evolutionary perspective, arguing that natural selection has made children sensitive to resource-related factors associated with their relative position in the family. Specifically, children compete with one another for niches within the family, which in turn shapes their personalities. Sulloway proposed that firstborns obtain parental attention and resources via traditional societal routes and as a result tend to be high in conscientiousness but low in openness to experience. However, they must also defend their high-status position from threats from younger siblings, which makes them low in agreeableness. Laterborns, in contrast, must seek alternative routes for success and parental attention and resist the oppression often imposed on them by their higher-status, older siblings. As a result, they tend to develop greater empathy, a more individualistic style and, according to Sulloway, are more likely to be "rebels." Sulloway proposed that birth order is only one of several interacting child and family characteristics that influence personality development, although it is an important one. In addition to birth order, amount of parent–child conflict, number of siblings, gender, age, age at paternal loss, social class, and temperament influence personality in an interactive (although predictive) manner.

Support for Sulloway's model has been mixed. For example, in his own research, Sulloway (1996) provided evidence that leaders of scientific and political "revolutions" are more likely to be laterborns, whereas supporters of the status quo are more apt to be firstborns (but see Townsend, in press, for criticism of this work). Research examining personality factors predicted by Sulloway to be characteristic of firstborns (high conscientiousness, low openness to experience) and laterborns (high agreeableness, rebelliousness) has sometimes supported his hypotheses (Paulhus, Trapnell, & Chen, 1999; Salmon & Daly, 1998) and sometimes not (Freese, Powell, & Steelman, 1999; Michalski & Shackelford, 1999; Parker, 1998). Behavioral geneticist Judith Harris (2000) has argued, consistent with Sulloway, that birth-order effects may play an important role in determining a child's behavior within a family; however, departing from Sulloway, she argued that children do not generalize these behaviors outside of the family. Rather, she proposed, learning is context dependent, such that behaviors and attitudes acquired within the family are useful only (or primarily) in that context and do not influence substantially behavior outside of the family. Delroy Paulhus and his colleagues' (1999) study, which assessed adults' perceptions of their and their siblings' personalities, has reported the largest birth-order effects consistent with Sulloway's hypothesis. Thus, when making comparisons within families, birth-order effects consistent with the predictions made by Sulloway are found; results are less robust when between-family contrasts are made, consistent with Harris's hypothesis.

From the perspective of evolutionary developmental psychology, it would be surprising if a child's birth position in his or her family did not have a significant impact on his or her development. Competition for resources among siblings is real, and parents do tend to favor, consciously or otherwise, one child over another. A firstborn has the advantage of being older and thus having already survived for several years by the time a second child is born. In environments where infant mortality is high, this gives the older sibling an immediate (if perhaps short-lived) advantage. However, although we acknowledge that birth order can affect a child's behavior and personality, we concur, in general, with Harris's (2000) contention that the effect of birth order will be strongest within a family. Unlike Harris, we do believe that experiences within the family, including interactions with siblings, can have significant and more global effects on personality and development, yet we recognize that much of what children learn is context dependent and that this may be particularly important when differentiating children's behaviors within and outside of the family (J. R. Harris, 1998).

Incest Avoidance

The fundamental purpose of sex is the generation of variability. By mixing genetic characteristics of two individuals, the resulting offspring will inherit not only favorable characteristics of each parent, but different offspring will inherit different combinations of characteristics. It is on this variability that natural selection works. One apparently important aspect of variability afforded by sexual reproduction is resistance to parasites. Rapidly reproducing microorganisms that are able to infect a parent organism may be less likely to infect an offspring who has acquired half its genes from a second parent. The mixing of genes associated with the immune system permits individuals to keep a step ahead in the war with parasites (Ridley, 1993). A second and more familiar reason for genetic variability is the avoidance of detrimental recessive genetic traits. Parents who share many of the same alleles likely possess several recessive copies of deadly genes, and inbreeding increases the chance that these genes will find their way into the phenotype of the next generation, resulting in increased rates of mortality or morbidity (the presence of illness or disease).

All species that are subject to this inbreeding depression have developed ways of minimizing its effects, mostly by reducing the likelihood that close relatives (siblings, parents, and offspring) will mate. In most social primate species, for example, one of the sexes typically leaves the home troop prior to adulthood and joins another. In most New and Old World monkeys, it is the males that emigrate. In baboons, chimpanzees, and prehistoric humans (see Owens & King, 1999; see also chapter 2), the females leave the home base in search of a mate. Humans, in addition, universally possess incest taboos—cultural proscriptions against mating within one's family. Freud (1952) proposed that such proscriptions were necessary because children, in particular, have a strong desire for sexual relations with their opposite-sex parents (the Oedipus and Electra complexes), and these desires extend to opposite-sex siblings. An evolutionary perspective, in contrast, proposes that these cultural taboos are built on evolved mechanisms. Rather than being biased toward incestuous relations, as Freud proposed, we, as other animals, have biases that inhibit the likelihood of inbreeding.

An evolutionary explanation of incest avoidance extends back to the late 19th century and the writing of Finnish anthropologist Edward Westermark (1891). In his work, *The History of Human Marriage*, Westermark noted that people who had extensive contact with one another as children rarely ever married, despite being available as mates. The reason is that such childhood familiarity breeds not contempt necessarily, but a lack of sexual attraction. The likelihood of being closely related to someone you have grown up with is high, so having an aversion to mating with such people reduces the chance of inbreeding depression. Thus, childhood familiarity becomes the proximal mechanism for incest avoidance.

Two sources of evidence are most frequently cited in support of the Westermark effect. The first is based on the work of anthropologist Arthur Wolf (1995), who studied, over nearly 40 years, the tradition of "minor marriages" in Taiwan. In the late 19th and early 20th centuries, marriages were arranged between families for their young children. In such cases, the girl would be adopted by the boy's family, and the future bride and groom would be reared together as brother and sister. (Note that such an arrangement serves as a form of mate guarding by the paternal family, increasing paternity certainty.) Wolf observed that the future wife frequently objected to the marriage with her adopted brother, particularly when she was adopted prior to 30 months of age. When such marriages did occur, the divorce rate was three times higher, they produced 40% fewer children, and the wives were more likely to admit to having extramarital affairs, all compared to "major marriages."

The second frequently cited source of evidence for the Westermark effect comes from studies of people reared communally in the Israeli kibbutzim (Shepher, 1983). Unrelated children spent many hours together throughout childhood and even engaged in frequent heterosexual play prior to reaching puberty. Yet, as adolescents and adults, sexual intercourse between members of the same kibbutzim was extremely rare, and of 2,769 couples from 211 kibbutzim there were no marriages between members of the same kibbutzim (Shepher, 1983).

These studies suggest that unrelated people who are raised together from early childhood avoid one another as sexual partners later in life, much as full siblings do. But what about siblings? The only systematic research examining incest avoidance in siblings (as opposed to unrelated people) was performed by Bevc and Silverman (1993, 2000). In their initial study, based on surveys completed by 500 university students, Bevc and Silverman reported that the incidence of "mature" post-childhood sexual activities, which included completed or attempted genital, anal, or oral intercourse, was more frequent among siblings who had been separated early in childhood than for nonseparated siblings. Separation did not predict the incidence of "immature" postchildhood sexual behavior, however, such as fondling, exhibitionism, and touching. But the level of "mature" sexual activities in the sample was absolutely low, which prompted their second study, in which they recruited volunteers by placing adds in major newspapers in Toronto to answer survey questions about sexual relations between brothers and sisters. This technique resulted in a smaller overall sample (178 individuals) but a higher number of people admitting to genital intercourse with a sibling (54 individuals). Of the 54 cases of intercourse, 35 reported vaginal intercourse with ejaculation, 10 vaginal intercourse without ejaculation, and 9 cases of attempted vaginal intercourse. Nonintercourse sexual activities (that were referred to as "immature" in the 1993 paper) were also reported. The results of this second study were similar to those reported earlier: Sibling pairs who were separated during early childhood were more likely to have engaged in genital intercourse than nonseparated pairs, although there was no effect of separation for "immature" sexual behaviors.

Bevc and Silverman (2000) proposed that their findings require a modification of the principal interpretation of the Westermark effect. Early cohabitation results in reduced incidence of reproductive sexual behavior (e.g., intercourse) but not necessarily in a reduction of nonintercourse sexual behaviors (e.g., fondling and exhibitionism). They further hypothesized, consistent with the proposal of Wolf (1995), that the sensitive period for the incest inhibition effect is before age 3.

Although nearly a third of the people answering Bevc and Silverman's (2000) survey reported having genital sex with a sibling, this high value does not, of course, represent the rate of incest in the general population. However, the greater-than-zero incidence of postchildhood sexual relations (both "mature" and "immature") between siblings indicates that the incest inhibition between cohabitant siblings is not total.

Noting that early cohabitation and "familiarity" are the proximal causes of the Westermark effect does not reveal the biological or psychological mechanisms for incest inhibition. What is it about cohabitation or familiarity that results in reduced sexual interest in a sibling? Current research suggests that olfactory cues carry the messages that result in aversions of sexual relations among early cohabitants (Schneider & Hendrix, 2000). For example, animal research indicates that differences in aspects of the immunological system (specifically the major histocompatibility complex; MHC) are perceived via olfactory cues. Animals tend to prefer as mates members of the opposite sex that differ from themselves with respect to the MHC genotype and detect these differences by smell (see Apanius, Penn, Slev, Ruff, & Potts, 1997; Brown & Eklund, 1994). (Recall that variability in the immunological system is associated with greater resistance to parasites.) Similarly, in humans, evidence exists that women prefer the smell of men who differ from them in terms of the MHC region of their genome (Wedekind & Füri, 1997; Wedekind, Seebeck, Bettens, & Paepke, 1995), and one study reported that of 411 Hutterite couples, people tended to marry those whose MHC types were different from their own (Ober et al., 1997). Schneider and Hendrix (2000) proposed that children develop sexual aversions to the odors (reflective of the MHC) of those with whom they cohabit, although there is a sensitive period for such aversions (or preferences for those with different odors) to develop. Although more research on a variety of levels needs to done to sort out the proximal causes of incest inhibition and the conditions under which it will and will not be expressed, it seems clear that evolved mechanisms interact with environmental factors and developmental constraints to produce what is generally adaptive behavior.

EVOLUTION AND FAMILY RELATIONSHIPS

American politicians in the latter portion of the 20th century touted the importance of "family values" for a sound society. Although the values and families these politicians had in mind may have been more a creation of Madison Avenue marketing than reality, the family is, in some form, vital to the survival and success of children (and thus society). Parents are indispensable to children, and children are parents' route to genetic immortality. Each has a vested interest in the other, and each has been prepared by hundreds of thousands of years of evolution to make the best of their relationship. But although there may be epigenetic programs shaped by natural selection for dealing with the problems that having children (and having parents and siblings) may bring, these mechanisms are sensitive to environmental conditions. No predetermined outcomes exist within families. In fact, it is within families that the interaction of "innate" and postnatal experiential factors in shaping development is perhaps most obvious. The result is an environment in which cooperation, conflict, and competition are all simultaneously present, fueled by evolutionarily based mechanisms geared to maximize each participant's fitness.

9

INTERACTIONS, RELATIONSHIPS, AND GROUPS

As we have commented repeatedly throughout this book, humans are a social species, and the need to cooperate, compete, and understand conspecifics has been responsible, in large part, for the evolution of the modern human mind. *Homo sapiens* have been prepared by evolution for life in a human group; but such preparation does not mean that children have ready-made solutions to recurrent social problems. Because of the great variability in human social environments, people must develop flexible ways of dealing with others with whom they interact frequently. The existence and use of alternative social strategies is an important exemplar of the way that we conceptualize the environment in which organisms develop. Environments do not simply "trigger" a behavior or set of behaviors in the Lorenzian sense. Instead, organisms have options, and they choose or shape an environment to maximize compatibility. Accordingly, individual development is conceptualized as the process of change that is a result of the transaction between individuals and their environments, which means that individuals solve developmental problems in a variety of ways. The systems theory idea of *equifinality* is an important implication of this position (von Bertalanffy, 1968). By this we simply mean that individuals can reach the same developmental hallmark through a variety of developmental routes: There is no one "royal route" through development. Given the diversity of niches into which humans are born as well as genetically based individual differences, it is imperative that they have a variety of strategies for approaching recurring developmental problems.

Social behavior can be conceptualized in terms of strategies used to gain and maintain access to resources. Resources, in turn, are relevant to reproduction and survival. From this view, social behaviors can be categorized along a variety of dimensions. For example, the valence of the behavior can be described; behaviors may be positive, reflecting friendship, altruism, and cooperation, or negative, such as aggression. Social behavior can also be described in terms of more molar categories. Ethologist Robert Hinde (1976, 1980, 1983), for example, proposed that social behaviors be described in terms of interactions, relationships, and structure, the latter referring to behavior in social groups. We follow Hinde's explicitly evolutionarily based model here, examining the development of positive (cooperation) and agonistic social behaviors as they are found in interactions, relationships, and groups, both for human children and for other social species. Before assessing the specifics of social behavior, however, we first discuss certain aspects of evolutionary theory particularly pertinent to social behavior, specifically ways in which social behavior can be evaluated in terms of its role in natural selection and fitness.

FITNESS AND THE DEVELOPMENT OF SOCIAL BEHAVIOR

Since its origin, evolutionary theory has been concerned with the ways in which individuals separately and in groups interact in the "struggle" to survive and reproduce. Indeed, the "function" of a behavior or strategy, in the evolutionary sense, refers to its reproductive value. In this regard, the value of a behavior is judged by its ability to maximize benefits in terms of reproduction and survival of offspring in ecologies with varying amounts of resources (mates and food) and competition for those resources. The ability to maximize benefits is considered in relation to costs, or risks, associated with the benefits reaped.

Within this system, it is quite easy to explain one set of social behaviors, aggression, as it is easily understood in terms of one individual gaining advantage over another, especially in cases involving low levels of risk. In such cases, natural selection would clearly favor those individuals who used aggression effectively to secure mates and other resources. In economic terms, natural selection favors those strategies where benefits outweigh costs. So, aggression would be naturally selected if associated costs, for example, risk of death and injury, were outweighed by benefits, for example, access to resources and mates. It would not favor cases where costs associated with aggression outweighed benefits; for example, incurring a debilitating injury for the sake of securing a mate.

Although explaining the fitness value of aggression is relatively straightforward, evolutionary theory had a difficult time, initially, reconciling the existence of cooperative and altruistic behaviors. How could natural selection favor a set of behaviors, such as altruism, where the costs (self-sacrifice) outweigh the benefits? Group selection theory (Wynne-Edwards, 1962) attempted to solve this problem. Briefly, *group selection theory* suggests that the unit of natural selection is the group, not the individual. Consequently, cooperation and altruism could be explained in terms of benefits to the whole group. Thus, altruism may be very costly to an individual but greatly benefit the group.

This theory was challenged by discoveries in molecular biology and the cracking of the DNA code (Dawkins, 1976; G. C. Williams, 1966; although see Lewontin, 1982, and D. Wilson, 1997, for alternative arguments). Briefly, these advances suggested that the unit of natural selection was the individual (or the gene, more exactly), not the group (see discussion in chapter 2). Thus, evolutionary theory was revisited by the conundrum of explaining altruism.

An initial explanation for altruism was provided by Hamilton's (1964) theory of *inclusive fitness*, later expanded by Trivers (1971), which we discussed in chapters 2 and 8. According to inclusive fitness theory, individuals' cooperative, altruistic, and antisocial behaviors will vary according to the degree of genetic relatedness among interactants. Individuals will be more cooperative with those closely related to themselves, compared with more distant kin and unrelated people ("non-kin"). So, for example, the theory predicts that parents and their offspring (who share 50% of their genes) will cooperate better than will grandparents and grandchildren (who share only 25% of their genes) or first cousins (who share only 12.5% of their genes). In this way, any cost associated with social behavior, although borne by the actor, will be balanced by associated benefits to kin, as well as the actor. Consistent with this argument, kin comprise a sizable portion of adults' social networks in contemporary Great Britain (Dunbar & Spoors, 1995). Dunbar and Spoors found that adults nominate a high proportion of kin, relative to non-kin, for help and support.

Getting from cooperative interactions with kin to interactions with non-kin involves making inferences about the environments in which these strategies evolved. In the environment of evolutionary adaptedness, individual social groups presumably contained a relatively high proportion of kin (Hinde, 1980). Consequently, the groups in which individuals interacted comprised relatives with whom they shared not only genes, but also a social history and a social future. They had knowledge of past interactions and that they would probably continue to interact in the future. Indeed, it has been suggested that familiarity is the mechanism that enables individuals to recognize kin (Cheyne & Seyfarth, 1990; Smuts, 1985). This familiarity and the consequences associated with future meetings should result in cooperation with conspecifics. In other words, familiarity and the promise (or threat) of future interaction may have become a proxy for kinship.

That social behavior tends to be cooperative among social actors who are familiar with each other and meet repeatedly, has been expressed in the theory of *reciprocal altruism* (Axelrod & Hamilton, 1981), that is, social actors will cooperate with those with whom they will interact in the future. Costs associated with cooperative and altruistic behaviors will be minimized, quid pro quo, by others reciprocating the good turn. Similarly, aggressive or deceptive acts will be reciprocated. In this way, costs associated with cheating outweigh benefits when dealing with individuals in a stable social group. Where actors are not related or familiar with one another, and where there is little chance of future meetings, individuals act in their own immediate self-interest. Such circumstances would reward deception and discount cooperation and altruism. In short, cooperating with kin and familiar conspecifics is favored by natural selection, because benefits outweigh costs. By extension, the ability to detect "deception" and "cheating" are important cognitive skills that probably evolved in response to such pressures (Cosmides & Tooby, 1992; Humphrey, 1976; see also chapter 6).

Although economic models of social behavior in which costs and benefits are evaluated are useful in making inferences about those strategies favored by natural selection, they do not explain the usefulness of specific strategies by individuals in specific circumstances. For example, the optimality of one strategy (cooperation or aggression) will vary as a function of the order of that behavior in the stream of behavior between interactants (Is it a first move or a response?), the nature of the other behavior (Is it cooperative or aggressive?), the value of the resource around which the interaction is occurring, and the frequencies of the behaviors in the population (Archer, 1988). *Game theory* calculates costs and benefits associated with behaviors, especially aggression in relation to other social behaviors (Archer, 1988). It is useful in explaining behavioral choices at this level but also helps explain the probability of it becoming an *evolutionarily stable strategy* (ESS). An ESS is one which, in a specific situation, cannot be replaced in a population (Maynard Smith, 1972). For example, in situations where two individuals meet in the presence of a resource valued by both, they have two choices: attack or retreat. The value of either choice is dependent on the frequency of each in the population and the strategy of an opponent. An ESS is one that cannot be bettered in those specific circumstances. "Attack first" when the response is "retreat" is an ESS. This example is expanded below. The aim of game theory is to determine an ESS in a specific circumstance. The value of a strategy is calculated in a payoff matrix where values for resources and estimated frequencies of strategies in a population are entered. The most basic game involves hawks and doves and tests the value of aggression and cooperation as an ESS.

As a simple illustration, we assume only two strategies (hawks or doves) and that they exist in equal numbers in the population (see Figure 9.1).

	H	D
H	-25	+50
D	0	+15

Figure 9.1. Payoff matrix.

Hawks attack rapidly and inflict damage, and doves "display," or threaten but retreat when attacked. The payoff matrix presented in Figure 9.1 (Archer, 1988, p. 161) represents two strategies, hawks (H) and doves (D), meeting either hawks or doves. In this scenario, hawks always beat doves (+50, the benefits of victory), doves always lose to hawks (0 payoff to doves), and doves fare better in encounters with other doves (+15) than hawks do in encounters with other hawks (−25). By extension, when the value of a resource is high, participants are more likely to use a costly, hawkish strategy rather than a dovish one. Similarly, when costs are low, a dovish strategy is more likely.

For example, given the above payoff matrix with a different proportion of hawks (10%) and doves (90%) in the population, there would be fewer high-cost encounters between hawks (−25) and a high proportion of victories of hawks over doves (+50). For the doves, only 10% of their encounters would be against hawks where they would always lose (0), and the other 90% would be against doves (+15). In this case, the proportion of hawks would increase in the population, increasing the likelihood of hawks encountering other hawks, thereby increasing costs to hawks. At some point neither hawks nor doves would have an advantage (Archer, 1988).

In general, behaviors and strategies will be selected if benefits outweigh costs. Both cost–benefit and game-theory models have individuals calculating (unconsciously) the value of a strategy in relation to several other factors, such as costs to themselves, the value of the resources, and the probability that they will encounter specific individuals again. Further, we stress that natural selection probably exerts pressures at different phases in individuals' developmental histories; thus behaviors and strategies should be evaluated in terms of their value for those periods in which they are observed.

PEERS AS SOCIALIZING AGENTS OVER AN EXTENDED CHILDHOOD

The complexity of these social strategies may be responsible, in part, for the extended childhood that characterizes humans and other primates (see chapter 4). During this time, children learn a variety of social skills by interacting with adults and peers. When human children and other primates (Harlow & Harlow, 1962; Harlow, Harlow, Dodsworth, & Arling, 1966; Harlow & Zimmerman, 1959) are deprived of these opportunities, their social development is greatly hampered. Further, the effects of early social deprivation can be remediated when juveniles are provided with opportunities to interact with younger peers (Suomi & Harlow, 1972). This sort of safe and nonthreatening context supports individuals' trying out, learning, and honing new social skills.

Although parents and other adults surely play a role in fostering children's social development, by the time children are 4 or 5 years of age, adults are replaced by other children as the primary agents of socialization, although juveniles' viability remains, of course, dependent on adult-provided protection and resources. One well-developed version of the argument that children's peer groups, rather than their families, are the primary socializing agents is J.R. Harris's (1995, 1998) *group socialization theory*. According to Harris, even influences of parents and teachers are filtered through children's peer groups. Children seek not to be like their parents so much as to be like their peers. The waning influence of the home environment on personality and intellectual development has been repeatedly demonstrated (McCartney et al., 1990; Scarr, 1992; see also chapter 3). Harris suggested that such a trend makes good evolutionary sense, in that, as adults, children will operate outside the home and compete and cooperate with agemates of their group. Becoming too well adapted to the home and too agreeable to the demands of one's parents is not (usually) conducive to one's inclusive fitness.

A complete description of group socialization theory is beyond the scope of this chapter. However, particularly pertinent to our discussions here is Harris's (1995) assertion that human group behavior is predicated on four evolutionary adaptations that humans share with other primates: (a) group affiliation and in-group favoritism, (b) fear of or hostility toward strangers, (c) within-group status seeking, and (d) the seeking and establishment of close dyadic relationships. These "evolutionary adaptations" are thought of as built-in predispositions, which are operating early in life but nonetheless develop over childhood.

CLASSIFYING SOCIAL BEHAVIOR

In this section we examine exemplars of social behavior, following Hinde's (1976) model of social relationships. This model, derived from human social sciences and rooted in evolutionary theory, was presented as a heuristic for the study of nonhuman primate behavior. Hinde insisted that the study of humans' social behavior, compared to that of other animals, should consider the role of higher cognitive and linguistic processes. We consider cooperative behavior and aggression from infancy through adolescence and evaluate each in terms of their phylogenetic histories.

Social behavior, according to Hinde, should be described along three dimensions: interactions, relationships, and structure. *Interactions* occur between two individuals and contain one or more types of behavior. Descriptions of interactions should include what the individuals are doing (content) and how they are doing it (quality). The interaction can be described in terms of physical movements and muscle contractions (e.g., slap) or consequences (seeking proximity to a peer) of both participants in the interaction. In this regard, who initiates and who responds is important. For example, the meaning of a child initiating a play bout or an aggressive act is different from one in which a child responds to a playful or aggressive bout. The quality of the interaction should also be considered. Was it rough or gentle? Facile or rudimentary? Again, differences in quality tell us something about the meaning of the interactions.

Relationships are interactions between two individuals over time. Different relationships reflect different interaction histories, where individuals have interacted with each other in the past and anticipate future interactions. Common relationships include attachment between mothers and their children, sibling relationships, friendships, and romantic relationships.

That relationships affect the ways in which individuals act toward each other has implications for evolutionary theory. Kin should be more cooperative than nonkin, as discussed in chapter 8. By extension, one specific relationship—friendships—should be more cooperative than relationships with nonfriends. These predictions, based on kin selection theory and the theory of reciprocal altruism, suggest that behavior in stable peer groups reflects the environment of evolutionary adaptedness where such groups comprised mostly kin and that familiarity was the basis of kin recognition and subsequent behaviors.

As in interactions, the content and quality of relationships are described across time. In addition, the patterning of interactions is important in relationships. Patterning is represented by the frequencies and sequences of different interactions and can be summarized in terms of the interactions characteristic of different attachment relationships. For example, a secure attachment relationship is represented by mothers' responding to children within a specific time span and by a pattern of reunion behavior after separation. Similarly, certain peer relationships, such as friendships, are represented by a high proportion of conflicts and resolutions (Hartup, 1996).

Because relationships, by definition, have a temporal dimension, they are dynamic. They change across time as individuals accumulate more shared history and have differing goals, and possibly as a function of developing cognitive and linguistic capabilities.

Structure is the description of social behavior at the group level. Here we also describe content, quality, and patterning of interactions, but at a level that generalizes across individuals and relationships. Some level of theory is needed to organize this level of generality. For instance, Dunbar (1988) suggested that group size of humans has been limited as a result of the capacity to remember information about each member (e.g., remembering who can and cannot be trusted) and their ability to speak and hear in more immediate groups (limiting immediate group size to four).

At a more abstract level, *dominance* is a construct that can be used to explain *social structure*, or the processes by which social groups stay together. Dominant individuals have greater access to limited resources (be it food, mates, or toys in the case of preschool children) and will use a variety of techniques to attain and maintain their preferred status (Hawley, 1999). Thus, in many ways, dominance is an ideal measure of social development because it explains the ways in which an individual's behavior is related to more general group structure, or cohesion. Dominance is a variable that reflects the ways in which specific individuals interact and the effects of those interactions on access to resources. We stress that dominance reflects interactions among specific rather than all members of a group, because not all members of a group, especially if it is large, interact with each other (Archer, 1992). From this individual-selection (rather than group selection) point of view, individuals must only know their status relative to those with whom they interact most frequently. Their status in the whole group is not relevant from this perspective (Archer, 1992).

Dominance status in children can be measured in ways similar to those used with nonhuman animals by constructing a dominance hierarchy from dyadic or triadic observations of individuals (McGrew, 1972; Strayer & Noel, 1986). In these matrices, interactions are scored in terms of winning and losing. Dominance hierarchies are formed based on the history of competitive interactions, with the frequent winners occupying the top of the hierarchy and the frequent losers occupying the bottom. Hierarchies are usually described in terms of the linearity of the transitive relations among all individuals in a group, where A is dominant over B and B over C; thus A should also dominate C. Dominance hierarchies reduce antagonism within the group, distribute scarce resources, and focus division of labor (Savin-Williams, 1979).[1]

Dominant, or alpha, individuals, and their kin have preferred access to resources, such as food and mates. Individuals establish and maintain dominance through a combination of aggressive and cooperative interactive strategies (Hawley, 1999). The choice of a strategy is influenced by the

---

[1] Archer (1992) has forcefully presented the limitations of this procedure for defining dominance hierarchies as it is generally used. Briefly, he suggested that measures of linearity are typically inflated or invalid for several reasons. First, not all members of a group, especially if the group is large, typically interact with each other. Relatedly, this whole-group orientation assumes that the group stability, which is an outcome of dominance, is beneficial to the group, not the individual, and in contrast to individual selection theory. At the opposite end of the group size issue, linear measures of small groups are typically inflated. Dominance hierarchies should be used and interpreted with these limitations in mind. For example, limitations associated with expressed linearity and transitivity should be recognized.

ecology of the group as well as associated costs and benefits. Further, kin status may be relevant such that the offspring of a dominant individual may have high status by virtue of his mother or father, even though the offspring lacks fighting and social skills.

That cooperative behaviors are included in a dominant individual's repertoire also departs from more traditional conceptualizations of dominance. As noted above, dominance has typically been defined in terms of agonistic encounters and in some cases by attention structures (Chance, 1967; Vaughn & Waters, 1981). Force alone, however, does not explain why groups of individuals stay together. For example, why would a stronger and more dominant individual soothe and comfort an individual he or she has just publicly defeated, as researchers sometimes observe? As deWaal (1982) has so convincingly illustrated, dominant animals have a dominance style, using varying degrees of force and reconciliation to maintain their affiliations. Cooperative and reconciliatory strategies are used in situations where the dominant individual needs his or her subordinates, and the subordinates are free to leave the group.

Similar findings have been reported with children. For example, in "open-field situations" (where children are free to leave the group) more conflict resolution and cooperation are observed relative to that in "closed-field situations" (where children cannot leave the group; Laursen & Hartup, 1989). Viewing dominance in terms of affiliation and agonistic behaviors is also consistent with data from the period of early childhood and early adolescence, where dominance is positively and significantly correlated with popularity (Pellegrini & Bartini, 2000; Rodkin, Farmer, Pearl, & Van Acker, 2000; Vaughn, 1999; Weisfeld & Billings, 1988).

As we noted in chapter 7, the presence of dominance hierarchies may have played a critical role in the evolution of social intelligence (Cummins, 1998b). Individuals must be aware of their rank in the social group relative to those of others, know what behaviors are permitted and which are not allowed, and be aware of when others are breaking the rules and perhaps when one can cheat without getting caught. From this view, individuals use aggression strategically. This "Machiavellian" use of aggression has been observed in children and adolescents who are "bullies" yet are quite capable of taking the perspective of others and appear to be quite capable in theory-of-mind tasks (Sutton, Smith, & Swettenham, 1999). It is under these conditions, we suggest, that social reasoning would evolve in a large-brained creature with an extended childhood (Bjorklund & Pellegrini, in press; Cummins, 1998b).

In the sections to follow, we examine children's interactions, relationships, and groups at different periods of development (i.e., infancy, childhood, and adolescence), interpreted in terms of the evolutionary developmental psychological principles and theories discussed above. One focus of our discussion is sex differences in social behaviors, particularly as they may relate to preparations for adult roles (see chapter 2).

INTERACTIONS

Infancy and Toddler Years

Until relatively recently, infants were characterized as egocentric beings, with little desire or ability to interact with agemates. However, they do exhibit rudimentary social interaction, such as mutual gaze at around age 2 months (Cairns, 1976). During the first year of life they also begin more sophisticated forms of social interaction that are the roots of cooperative behaviors, such as intentional smiles to and intent observations of peers as well as responses to peers (Rubin, Bukowski, & Parker, 1998). During the toddler period, children engage in "games," characterized by mutual gaze, patterned exchanges, turn-taking, and reciprocal imitation (Eckerman & Whitehead, 1999; Rubin et al., 1998).

One robust sex difference relevant to social interactions concerns females' greater orientation toward other people, and this difference can be seen during infancy. For example, female infants from shortly after birth orient themselves to faces and voices more frequently than male infants (Haviland & Malatesta, 1981). Infant and toddler girls have also been found to be more empathetic, responding to the distress of others more so than boys. For example, Zahn-Waxler and her colleagues (1992) examined 12- and 20-month-old children's response to the distress of other people. They reported that girls more often tried to comfort the distressed individual and sought information about the person's distress ("What's wrong?") than did boys. Girls also displayed facial expressions, such as sad looks, and made sympathetic statements or gestures to indicate their concern more than boys, who were more apt to be nonresponsive to the distress of others.

These basic elements of social interaction are also observed in many nonhuman primates. Smiles and play faces are used by monkeys and apes to communicate cooperative and playful overtures (Fagen, 1981). Similarly, patterned exchanges, in the form of play, are observed in the social interactions of many mammalian species. For example, the social play of rats and hamsters is typified by reciprocal turn-taking and stereotypical motor routines (Pellis & Pellis, 1998).

Conflictual behavior also appears during infancy. Indicators of anger (in the form of facial expressions) appear during infancy and exhibit cross-cultural uniformity, suggesting a biological basis. By 4 months of age, infants reliably express anger and use these expressions to communicate socially (see Coie & Dodge, 1998).

During the second year of life, anger and aggression are directed first toward caregivers and then toward peers. Conflicts with mothers, too, can be explained through evolutionary theory. From this view, conflict can be motivated by mothers' concerns with costs she incurs with prolonged care of the infants and infants' desire for continued protection (Trivers, 1974; see also chapter 8). Although the expression of anger is influenced by differences in individuals' temperament, in one study, toddlers expressed anger an average of 7% of the total time they were observed by mothers in their homes (Radke-Yarrow & Kochanska, 1990).

Peer-directed aggression is first observed at the end of the first year of life and typically occurs in the context of object disputes (Coie & Dodge, 1998). Indeed, a fair amount of the interaction between toddlers is conflictual (up to 50%), although not aggressive, per se. Also during this period, youngsters are beginning to use language in conflicts and prosocial behavior to resolve conflicts (see Coie & Dodge, 1998).

That conflicts and aggression first appear in the context of disputes over resources is certainly consistent with the phylogenetic record. Aggression can be a very effective strategy or it can be ineffective and costly, depending on several factors outlined above, such as cost, opponents' strategies, and the value of the resources. Also, toddlers' general acquisitiveness of the toys of other toddlers should not necessarily be viewed as maladaptive. Hawley (1999) suggested that, given toddlers' limited negotiation abilities, "taking" a toy is an effective means of acquiring resources and "may in fact indicate a healthy assertive approach to the world that may lead to material rewards that ultimately foster growth and survival" (p. 121).

Childhood

The study of preschool children's social interaction was pioneered by child psychologist Mildred Parten (1932) in her doctoral thesis at the University of Minnesota. She described children's social participation from solitary, parallel, associative, and cooperative interaction and hypothesized that this order represented a developmental continuum. Subsequent research (Bakeman & Brownlee, 1980; Barnes, 1971; Rubin, Watson, & Jambor, 1978) found, however, that these forms of social interaction did not represent an ontogenetic ordering, although cooperative interaction did increase across time. Parallel interaction seems, instead, to be a strategy used by children to enter social groups (Bakeman & Brownlee, 1980), and solitary behavior is not necessarily "immature" relative to cooperative behavior (Rubin, 1982).

Cross-species comparisons of social behavior indicate, too, that juveniles spend a considerable amount of time interacting with each other (Cheyne & Seyfarth, 1990; Goodall, 1986). As noted previously, some researchers have argued that an important function of the extended juvenile period typifying primates is to allow children to learn social skills and the cognitions associated with them (see chapter 4).

Physical aggression during early childhood, relative to toddlerhood, decreases while verbal aggression increases (Coie & Dodge, 1998). Like the earlier period, however, aggression for the preschooler is frequently a result of property disputes (P. K. Smith & Connolly, 1980). The relative decline of physical aggression during this period is probably a result of children's increased sophistication in the use of cognitive strategies (e.g., delaying gratification) and linguistic strategies (e.g., posing alternative strategies and compromise; Coie & Dodge, 1998). The cognitive sophistication of preschool-age children is also evidenced in their ability to use *relational aggression*. During this period, children, especially girls, begin to aggressively manipulate social relations by shunning and spreading rumors, among other strategies (see Crick & Grotpeter, 1995). As we discussed above, and see below, aggression and cooperative behavior come together in interesting ways to influence group structure in the form of dominance. We do not know, however, the place of relational aggression in dominance relationships, perhaps because females' dominance relationships have not been studied thoroughly.

Adolescence

A basic difference in the interactions between children and adolescents relates to differences in where and with whom each group spends its time (Rubin et al., 1998). Children, relative to adolescents, spend more time in close proximity to adults. Consequently, their behavior can be monitored, resulting in less boisterous, aggressive, and antisocial behavior (Fagot, 1974; Pellegrini & Smith, 1998). Adolescents, on the other hand, spend more time with peers and away from adult supervision. This lack of direct supervision may result in adolescents' adopting, at least temporarily, values of their peer groups, which can be polar opposites of adults. Abrupt discontinuity between children and parents during adolescence, however, is not common (Eccles et al., 1993). Where it does exist, increased rates of conflict generally center on issues related to autonomy and control (Collins, 1990).

Interactions with peers change from fantasy-based themes and rough-and-tumble (R&T) play bouts to physical games for boys (Pellegrini & Smith, 1998) and to sedentary social interaction for girls (Pellegrini, 1995b). Further, the level and frequency of boys' interactions with their male peers decrease. Also at this time, boys are spending more time with peer groups composed of both males and females (Pellegrini, 1995a). Declines in boys' physically vigorous behavior may be an accommodation to girls' more sedentary behavioral style. This cross-sex interaction corresponds to the time when each is maturing sexually and probably reflects interest in heterosexual relationships and ultimately mating.

The form of aggression that comes to the fore during late childhood and adolescence is bullying and victimization (Coie & Dodge, 1998; Olweus, 1993; Rubin et al., 1998). Bullying involves persistent and repeated cases of aggression committed by a more powerful individual over a less powerful one (Olweus, 1993). Bullies are more frequently boys than girls and represent about 10% of the elementary school population in most industrialized countries (Boulton & Underwood, 1992, in the United Kingdom; Olweus, 1993, in Scandinavia; and Pellegrini, Bartini, & Brooks, 1999; Pellegrini & Bartini, 2000, in the United States). Bullying increases in early adolescence as youngsters make the transition from primary to middle school and drops again as they progress through adolescence (Pellegrini & Bartini, 2000). It is probably the case that bullying tactics are used to establish dominance as youngsters make the transition into new social groupings in middle school. For example, it is probably the case that boys use physical aggression in bullying same-sex peers and girls use relational aggression with other girls, both in the service of acquiring resources. Rates of aggression decrease after dominance is established.

Bullies' victims tend to be physically frail children with few friends or affiliates (Hodges & Perry, 1999; Pellegrini et al., 1999; Pellegrini & Bartini, 2000). Indeed, this lack of social affiliation makes victims vulnerable to bullying. Vulnerability decreases, however, when victims are embedded in a social network or when they leave specific venues where bullies are present (Olweus, 1993). In fact, in different settings, children once considered victims appear to be quite "normal" (Olweus, 1993). This adaptability of victims demonstrates rather nicely the notion of alternative tactics. That is, resources can be accessed by members of a group, other than the most dominant, or bullies. For example, individuals of low rank can access resources by forming alliances with other low-ranking peers or with higher ranking peers who want to challenge the alpha individuals.

Males at all ages engage in more physical aggression than females, and the aggression that adolescent and young-adult males engage in is more likely to lead to serious injury and sometimes death than female aggression or aggression by older and younger males (Cairns & Cairns, 1994; Daly & Wilson, 1988b). This tendency of greater male violent behavior during adolescence and young adulthood is not limited to the United States or other developed countries but is universal (Daly & Wilson, 1988b) and has its origins in sex difference in parental investment (Trivers, 1972).

We discussed parental investment theory in some detail in chapter 8, and so we review its principle components only briefly here. Basically, the sex that invests the most in offspring is more selective in choosing a mate, and the sex that invests the least in offspring competes for access to the higher-investing sex. In humans, and mammals in general, females invest substantially more in an offspring, both before and after birth, and males actively compete for access to females. Of course, females compete with one another for males, but the intensity of that competition is not as fierce as it traditionally has been for males. This is particularly critical in a species, such as humans, that is marginally polygamous, with some males being able to monopolize more than one female and other males having access to no females, or only to less-desirable females (those with low reproductive value). Most mammalian females will find a mate, even if not a highly desirable one; in contrast, the fitness variance is larger for mammalian males, with many males being totally excluded from mating. As a result, there was selection pressure for a male psychology in which competitive risk-taking was favored (Daly & Wilson, 1988b; Wilson & Daly, 1985). Such risk-taking, and the violence that can accompany it, peaks when males are entering the reproductive market, which in humans is in adolescence (Byrnes, Miller, & Schafer, 1999).

Risk-taking and accidents are frequently the result of competitive or "show-off" behavior, with the purpose being to compete with other members of the same sex or to impress members of the opposite sex. For example, death rates due to automobile accidents rise rapidly for males during the late teens and continue to increase into the mid-20s before declining. Females show a similar pattern but at rates less than half that of males (National Center for Health Statistics, 1999). Although one hypothesis for this pattern relates to driving opportunities and driving experience (younger teenagers cannot drive, and older adults have more driving experience), the pattern for the females belies this. Death rates increase in the late teens but drop in the 20–24 age group; despite the comparable number of male and female drivers, the death rate is two to three times higher for males at the same age.

Another way of looking at age differences in risk-taking is to examine the types of injuries received by trauma patients. Figure 9.2 presents the percentage of admissions to a trauma unit that resulted from violence, relative to all admissions. The data were collected over a three-month period in a Los Angeles medical center (Cairns, Nemhauser, & Neiman, 1991). As can be seen, more than 60% of the teenagers seen incurred their injuries by way of violence (e.g., gun shot, knife wound); this frequency declined to a still substantial 42% for the next two age groups. And, not surprisingly, the ratio of male to female trauma patients was approximately 3 to 1. Although Los Angeles cannot be considered representative of the rest of the nation (or the world in general), these data reflect the extreme conse-

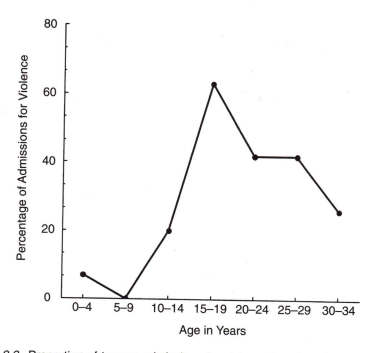

Figure 9.2. Proportion of trauma admissions for violence to a Los Angeles medical center for three months in 1990, by age.
Note. Adapted from "Aggressive and Violent Injuries Among Children and Young Adults," by R. B. Cairns, J. B. Nemhauser, and J. Neiman, 1991, *Annals of Emergency Medicine, 20,* p. 951. Copyright 1991 by Harcourt. Adapted with permission.

quences that adolescent and young-adult risk-taking can have in modern cities, where access to guns is easy.

Finally, adolescent and young adult males are more likely to be both the victims and perpetrators of homicide. This is a phenomenon that has been found in all cultures and time periods studied (Daly & Wilson, 1988b). Figure 9.3 presents the homicide victimization rate for males and females by age in the United States between 1995 and 1997 (National Center for Health Statistics, 1999). Data are presented for all males and females as well as for White and African American people. As can be seen, the likelihood of becoming a homicide victim is greater if one is male than female; the rates begin to climb in the late teens and early 20s and gradually decline thereafter. The likelihood of committing a homicide follows a nearly identical pattern (see Daly & Wilson, 1988b).

Again, this is the period when males are most actively competing for a mate. Despite the societal penalties and presumed stupidity of much of this behavior in contemporary culture, human males have inherited a

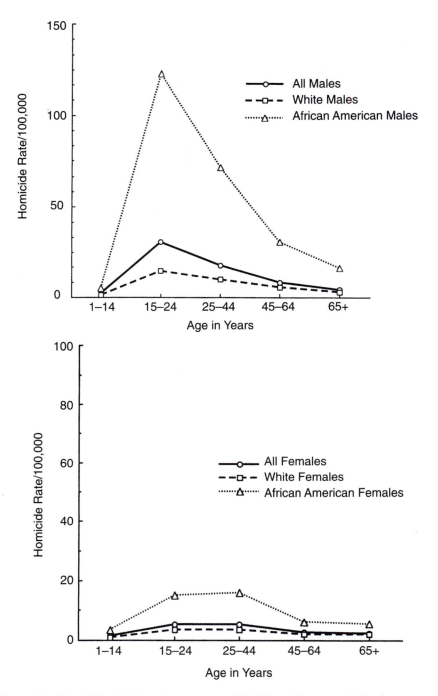

Figure 9.3. Homicide victimization rates per 100,000 resident population by age and sex in the United States, 1995–1997, presented separately for White and African American people.

Note. Data from *Health, United States* (pp. 189–191), by National Center for Health Statistics, 1999, Hyattsville, MD: U.S. Department of Health and Human Services. Copyright 1999 National Center for Health Statistics.

psychology that was adapted to different conditions in which risky competition, on the average, resulted in increased inclusive fitness. The benefits of aggression and risk-taking outweighed the costs (see Figure 9.1). Such behaviors are not, of course, "programmed" or "inevitable" but are shaped by experience and are more likely to be expressed in some environments than others. For example, when a young male has limited access to important cultural resources and when life expectancy is low, it makes more sense to compete vigorously for mates and what resources one can attain, rather than take a more cautious, long-term approach. Under such conditions, which typify impoverished communities in affluent nations, males can be expected to engage in elevated levels of risk-taking and violence against other males. This is exactly the pattern one sees in the United States for homicide rates of African American males (see Figure 9.3, Males). The age pattern is similar to that of White males, but the absolute rate is higher, associated with reduced access to educational and economic opportunities for many inner-city African Americans in comparison to White people.

RELATIONSHIPS

Infancy and Toddler Years

The first relationship for the infant is attachment with his or her primary caregivers. *Attachment* refers to the close, bidirectional emotional bond between an infant and his or her caretaker, usually the mother. Attachment has been studied in a wide range of species and is usually measured by some combination of three factors: proximity behaviors, distress on separation, and the extent to which the attachment figure can soothe a distressed infant.

The modern era of human attachment research can trace its roots to psychoanalyst and ethologist John Bowlby (1969), who incorporated psychoanalytic ideas about the nature of attachment with an evolutionary perspective. Bowlby saw a functional analogy between the behavior Lorenz (1943) observed in precocial birds (imprinting) and the human infant–mother relationship. Both resulted in close proximity of the infant to the mother during a period of high risk for the infant. Bowlby (1969) proposed that in *Homo sapiens'* environment of evolutionary adaptedness, behaviors that resulted in infants becoming attached to their mothers were positively selected, such that "attached" infants were more likely to survive than "unattached" infants. Enduring infant–mother bonds are not unique to humans, of course, and can be observed in other mammals, particularly primates (which became favorite laboratory subjects for assessing the effects of numerous variables on attachment, see Harlow et al., 1966). Bowlby's primary concerns were discerning the most adaptive style of attachment (a topic his colleague Mary Ainsworth made a career investigating; see discussion below) and the psychological mechanisms within both infants and mothers that promoted successful attachment.

In the sections below, we investigate some of Bowlby's claims about attachment, including the possible evolved mechanisms of infants and parents (particularly mothers) that serve to foster attachment. We evaluate further Bowlby's proposal that some forms of attachment are more adaptive than others and conclude by examining alternate attachment strategies, in which children take different developmental pathways as a function of the perceived availability of resources, including parental investment.

Infants' Appearance Promotes Caregiving

In chapter 6 we examined some perceptual biases and learning abilities of young infants that may orient them toward humans (specifically their mother), thus promoting attachment. However, some physical characteristics of infants and predispositions of adults may further bias the attachment process early in life.

Consistent with his ethological perspective, Bowlby proposed that physical characteristics and behaviors of infants activate caregiving behaviors in adults, particularly mothers. For example, consistent with the observations of Lorenz (1943) for a variety of species, Bowlby believed that caretaking behavior in humans is triggered by infants' immature features. Lorenz noted that infants of many species share certain characteristics, such as a head that is proportionally larger than the body, a forehead that is large in relation to the rest of the face, large eyes, round cheeks, a flat nose, and short limbs. As we commented in chapter 4, most adults find such characteristics endearing, which may facilitate attachment. Some support for this idea is found in studies of premature or sickly infants, who are less likely to have the full complement of infantile facial features and who are slower to display developmental milestones such as vocalizations, eye contact, and responses contingent on adult behaviors than healthy infants, are more likely to experience child abuse (Martin, Beezely, Conway, & Kempe, 1974; Sherrod, O'Connor, Vietze, & Altemeier, 1984).

Infant Cries as a Signal for Attention

Infants' cries are also powerful stimuli to elicit adult attention. They are difficult to ignore, and many a new parent has felt a cringe of fear over an infant's aversive cry, worrying what tragedy it might portend. Although it is possible that adults learn to respond so urgently to infants' stressful cries ("I respond when the baby cries, and the aversive crying stops, and the baby is still alive"), it is also possible (and consistent with an evolutionary developmental perspective) that parents are "prepared" to respond actively to such cries and require little in the way of formal learning to make these responses. For example, Bowlby (1969) proposed that infants' cries serve as an innate releasing mechanism, similar to the way a red belly invokes aggressive responses in a stickleback fish. Over evolutionary time, adults who responded to infants' cries of distress were more apt to have babies who lived to reproduce. Alternatively, distress cries may elicit, in a similarly reflexive way, sympathy and altruism from older individuals (Murray, 1985), which result in increased care and chance of survival for the helpless infant. Likewise, crying may have indicated abandonment in prehistoric times, due to the high rate of mother–infant contact that was necessary to promote survival. Therefore, mothers would be more likely to keep their infants in close proximity to them at all times, which would forestall infanticide by conspecifics or, perhaps, predators (Lummaa et al., 1998). Crying may also indicate the hardiness and vigor of the infant.

Despite the attractiveness of these hypotheses, they require a theory concerning how distress cries may have initially elicited parental care and came eventually to be selected in infants. In one study, Furlow (1997) found evidence that suggests that the acoustic structure of infants' cries correlates with infant health. Mothers may use the cues provided by the pitch and frequency of infant cries as an indicator of infant phenotypic quality. It has been suggested that mothers have a sort of template they use to assess the status of their infants, with one factor being infant cries (Mann, 1992). Furthermore, because the act of prolonged crying expends substantial energy, infants who can afford to cry persistently may be healthier, and thus a wiser choice for parental investment, than infants with less robust cries. Mothers would hence be less likely to abandon an infant who is prone to prolonged crying bouts.

Research done with the Masai, a pastoralist society from Kenya, provides some support for this hypothesis. Famine and drought are common among the Masai, and during these times, infant and child mortality rates rise substantially. DeVries (1984) noted that of the 13 infants born during a period of famine, only six survived. Only one of the six infants who was classified as "fussy," meaning he or she cried a lot, had died; conversely, five of the seven infants classified as "easy," having few crying bouts, died. DeVries proposed that perhaps under stress from famine, the infants with easy temperaments were easier to ignore and thus perished. The infants who complained the most, through crying and fussing, were more likely to be soothed by the mothers and hence were fed. The fussy infants, of course, also could have been those with the more robust constitutions, contributing both to their heartier cries and their greater survival.

Thompson and his colleagues provided an alternative theory of the evolutionary function of infants' cries, hypothesizing that infant distress cries mimic respiratory emergency and are thus a form of deception infants use to gain attention and nurturing from parents (Soltys, Brown, Thompson, Pietrzak, Thompson, 2000; Thompson, Dessureau, & Olson, 1998; Thompson, Olson, & Dessureau, 1996). They proposed that there is, and was in the environment of evolutionary adaptedness, a basic respiratory-monitoring mechanism in adults that orients parents to the possibility of choking in infants. Babies evolved mechanisms to take advantage of this response in their parents by simulating respiratory distress in non-life-threatening situations. To test this, they examined the cries of 24-month-old infants as a function of how long they had been left alone in an unfamiliar room (Soltys et al., 2000). (This was part of a Strange Situation test, in which the mother left the room for 3 minutes and then returned.) They proposed that deviations in infants' pitch-to-cry rate ratio reflect distress, which should activate evolved caregiving responses in adults. Moreover, they proposed that these ratios should deviate more from normal the longer the baby is left alone. They found exactly this relationship over the 3-minute session, supporting their hypothesis that infants' distress cries reflect deception on the part of the infant (i.e., simulating hypo- or hyperventilation). Despite the positive findings, the authors also reported that adult judges rated the infants as becoming *less* distressed over the course of the 3-minute session, complicating the interpretation of the findings. The "deception" hypothesis by Thompson and his colleagues is intriguing and subject to empirical test. However, it is left to future research to determine how adults, particularly parents, interpret infants' cries and the relation among infant distress, the qualities of their cries, and adult response to those cries.

Is Secure Attachment the Optimal Strategy?

Although all but the most deprived of infants become attached to their mothers, infants differ in their quality of attachment. Bowlby (1969) proposed that the most adaptive attachment strategy was one in which infants learned to use their mother as a "secure base" from which to explore the surrounding environment in relative safety. Infants whose attachment styles deviated from this pattern were less likely to survive in ancient environments and are associated with psychological maladjustment in modern societies. Mary Ainsworth and her followers have done extensive research attempting to document the truth of Bowlby's claims (Ainsworth et al., 1978; Ainsworth & Wittig, 1969). Using the well-known Strange Situation, in which attachment classification of infants and toddlers is assessed by patterns of their reactions to the departure and return of caregivers in a novel environment, most infants (about two-thirds) in the United States are classified as securely attached, whereas the remaining infants fall into one of three groups of insecure attachment (avoidant, resistant/ambivalent, disorganized/disoriented). Research has found that securely attached infants are likely to have mothers (or other caregivers) who respond to them contingently and who are responsive to their signals of physical and social need (Isabella & Belsky, 1991; Egeland & Farber 1984; Teti, Nakagawa, Das, & Wirth, 1991). Longitudinal research has also reported that children and adolescents who were classified as securely attached as infants and toddlers display better social and cognitive functioning than infants classified as insecurely attached (Jacobsen, Edelstein, & Hofmann 1994; Lewis, Feiring, McGuffog, & Jaskir, 1984; Pipp, Easterbrooks, & Harmon, 1992). These relatively robust patterns led many to the conclusion, consistent with Bowlby's original proclamation, that secure attachment represents the most adaptive style, with aspects of insecure attachment being predictive of poor adjustment and psychopathology (Karen, 1990).

More recent accounts have proposed that, in addition to promoting the survival of infants, attachment systems evolved to adapt individuals to subsequent environments (Belsky, 1997; Chisholm, 1996; Hinde, 1982; Wiley & Carlin, 1999). From this perspective, different patterns of attachment should develop as a function of the ecological conditions of a child's local environment (including amount of parental investment). Moreover, attachment classifications should reflect adjustments to contemporary environments and should not necessarily be stable over time when ecological conditions vary (Lewis, Feiring, & Rosenthal, 2000). In support of this last contention is evidence of less stability of attachment classification over a 6-month period for infants living in high-stress, low-income environments relative to infants living in homes characterized by less stress (Egeland & Sroufe, 1981). A similar pattern has been observed in stability of attachment from infancy to young adulthood, with attachment classification showing stability for middle-class samples (C. E. Hamilton, 2000; Waters et al., 2000) but not for a high-risk sample (Weinfield, Sroufe, & Egeland, 2000).

In addition, research from countries outside North America has reported modal attachment classification styles that differ from that reported in most American studies. For example, 49% of infants in a sample from northern Germany displayed an avoidant attachment style, with only 33% being classified as securely attached (Grossmann, Grossman, Spangler, Suess, & Unzer, 1985), a pattern that was attributed not to a rejection of infants by their mothers but to mothers' desire to comply with cultural norms. Similarly, in a study of kibbutz-reared infants in Israel, about 50% of infants were classified as insecurely attached (Sagi et al., 1985). These patterns suggest that cultural differences in child-rearing practices produce different patterns of attachment; the alternative interpretation, that about half of northern German and kibbutz-reared children have maladaptive attachment styles, is not supported by evidence. A more parsimonious conclusion is that children are sensitive to environmental conditions and develop attachment behaviors that are best suited for those environments.

Similarly, researchers who look at older children have consistently reported that more positive child outcomes are associated with demanding but warm parenting styles (authoritative) relative to stricter, critical, and emotionally detached styles (authoritarian; Baumrind, 1967, 1973; Pettit, Dodge, & Brown, 1988). This pattern has been reported from countries around the world and for a wide range of ethnicities, social classes, and family structures (see Steinberg & Silk, 2002). But there are exceptions. For example, Baumrind (1972) reported better adjustment for a sample of inner-city African American girls whose parents practiced the harsher authoritarian, relative to the authoritative, style (see also Steinberg, Dornbusch, & Brown, 1992). Consistent with the cross-cultural attachment research, different parenting styles may be correlated with different environments to produce different adaptive patterns of behavior.

Alternative Strategies: Different Attachment Styles Adapt Children to Different Local Ecologies

Belsky, Steinberg, and Draper (1991) proposed that aspects of children's early (and later) home environment influence not only their attachment style but also important aspects of later reproductive strategies. They developed an evolutionary-based model to account for early, attachment-related experiences in the home and the subsequent timing of puberty and mating-related behaviors during adolescence. According to Belsky et al. (1991),

> a principal evolutionary function of early experience—the first 5 to 7 years—is to induce in the child an understanding of the availability and predictability of resources (broadly defined) in the environment, of the trustworthiness of others, and of the enduringness of close interpersonal relationships, all of which will affect how the developing person apportions reproductive effort. (p. 650)

Rather than positing a single "best strategy" that all humans attempt to follow, Belsky and his colleagues proposed that humans have evolved mechanisms that are sensitive to features of the early childhood environment, induce the rate of pubertal maturation, and influence reproductive strategies. Figure 9.4 presents the paths children from secure, low-stress homes and those from insecure, high-stress homes are hypothesized to take. In one case, insecure attachment; father absence; and negative, stressful family experiences are proposed to lead to early pubertal maturation, sexual promiscuity, and unstable pair bonds. In the other case, secure attachment, low-stress, and father presence result in slower pubertal timing, delayed sexual activity, and the formation of more stable pair bonds, at least for girls. Belsky

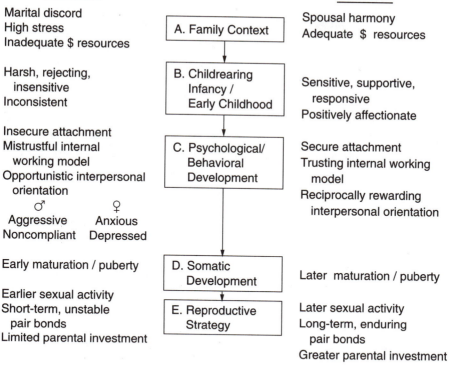

TYPE I

Marital discord
High stress
Inadequate $ resources

Harsh, rejecting,
   insensitive
Inconsistent

Insecure attachment
Mistrustful internal
   working model
Opportunistic interpersonal
   orientation

♂   ♀
Aggressive   Anxious
Noncompliant   Depressed

Early maturation / puberty

Earlier sexual activity
Short-term, unstable
   pair bonds
Limited parental investment

TYPE II

Spousal harmony
Adequate $ resources

Sensitive, supportive,
   responsive
Positively affectionate

Secure attachment
Trusting internal working
   model
Reciprocally rewarding
   interpersonal orientation

Later maturation / puberty

Later sexual activity
Long-term, enduring
   pair bonds
Greater parental investment

A. Family Context

B. Childrearing
Infancy /
Early Childhood

C. Psychological/
Behavioral
Development

D. Somatic
Development

E. Reproductive
Strategy

Figure 9.4. Two developmental pathways of reproductive strategies.
Note. From "Childhood Experience, Interpersonal Development, and Reproductive Strategy: An Evolutionary Theory of Socialization," by J. Belsky, L. Steinberg, and P. Draper, 1991, *Child Development, 62,* pp. 647–670. Copyright 1991 by Society for Research in Child Development. Reprinted with permission.

et al. argued, as have others (Chisholm, 1996), that the former strategy may be adaptive for children growing up in unpredictable environments with little expectation of social support. In such cases, both males and females invest relatively more in mating than in parenting, taking a "quantity over quality" perspective. In the latter case, in which children receive social support in a low-stress, adequately resourced environment, they invest relatively more in parenting than in mating, taking a "quality over quantity" perspective. In other words, Belsky and his colleagues proposed that children follow alternate reproductive strategies, depending on the availability of resources in their rearing environment, which results in differential investment in the next generation.

There has been substantial research from a wide range of countries investigating aspects of the Belsky, Steinberg, and Draper model since its publication in 1991, most of which confirms their basic assumptions. For example, research both before and after the publication of Belsky et al.'s paper has reported the not-surprising finding that children who reach puberty early are more apt to be sexually active than children who attain puberty later (Brooks-Gunn & Furstenberg, 1989). Girls from father-absent homes reach puberty earlier than girls living with their biological fathers (Surbey, 1990; Wierson, Long, & Forehand, 1993), as do girls who experience socioemotional stress during childhood (Chisholm, 1999; Ellis, McFadyen-Ketchum, Dodge, Pettit, & Bates, 1999; Graber, Brooks-Gunn, & Warren, 1995; Hulanicka, 1999). Effects are smaller or nonexistent for boys (Kim, Smith, & Palermiti, 1997), although boys from high-stress, father-absent homes tend to be noncompliant and aggressive (Draper & Harpending, 1987). This sex difference of enhanced effects for girls makes sense, given the differential investment in offspring by males and females. Because females' investment in any conception is greater than males', they should be more sensitive than males to environmental factors that may affect the rearing of offspring (e.g., malnutrition, stress, lack of resources; Surbey, 1998b).

However, the theory is controversial, and some modifications to the model have been proposed. For example, some researchers have suggested that most of the variance in rate of attaining puberty can be accounted for by genetic factors, specifically that fast-maturing girls have mothers who matured quickly (Surbey, 1990). Although research confirms this (Ellis & Graber, 2000; Moffitt et al., 1992), the effects of stress and father absence are still found after controlling for the age at which mothers reached puberty (Ellis & Graber, 2000; Chasiotis, Scheffer, Restmeier, & Keller, 1998), indicating that it is likely not only genetic factors are contributing to the "rate-of-maturation" effect.

One way genetic differences may contribute to the rate-of-maturation effect is to produce differences in children's receptivity to variations in rearing environments. Reviewing research from both the human and nonhuman primate literatures, Belsky (2000) proposed that some juveniles are more sensitive to individual differences in parenting. This greater plasticity is advantageous when environments are unpredictable, permitting children to adjust as well as possible to a wide range of less-than-optimal conditions (e.g., insecure attachment, father absence). Other children, however, function best in an environment that provides high levels of support and secure attachment. This is presumably the more "typical" environment, and it makes sense that some children (perhaps most children) will be adapted to this species-typical context. As with any set of traits, variability provides the stuff on which natural selection works, and parents can hedge their bets by producing some children who are receptive to change and others who will thrive in the "expected" environment.

Most recently, Ellis and Graber (2000) suggested important modifications to the Belsky et al. model. They proposed that maternal depression and the presence of a stepfather or mother's boyfriend were mitigating factors in influencing girls' maturation rate. For instance, they noted that maternal psychopathology is strongly predictive of marital discord, stressful relations with children, and divorce, making it a likely precipitating cause of the interpersonal stress and father absence that is predictive of early pubertal timing. With respect to the presence of a stepfather or mother's boyfriend, research on a variety of mammals has shown that the presence of pheromones from unrelated males results in accelerated rate of female pubertal development (Drickamer, 1988; see Sanders & Reinisch, 1990). In nonhuman mammals, this may serve to increase the likelihood of reproduction between the developing females and the unrelated adult males. Research with humans has demonstrated that women's reproductive cycles are influenced by male axillary (underarm) secretions (Cutler, Krieger, Huggins, Garcia, & Lawley, 1986). Thus, Ellis and Garber proposed, it is not father absence, per se, that is responsible for accelerated pubertal timing, but the presence of an unrelated adult male (see also Surbey, 1990).

To evaluate their hypotheses, Ellis and Garber examined pubertal timing for 87 American girls from predominately lower-middle-class to middle-class homes. In addition, they conducted structured interviews with the girls' mothers assessing mood disorder (focusing on depression); obtained measures of dysfunctional family relationships and stress in the mothers' romantic relationships, again from the girls' mothers; and collected data about the presence of biological fathers, stepfathers, or mothers' boyfriends. Consistent with previous research, they reported that girls who experienced stress in the home and father absence reached puberty earlier than girls who experienced less stress and whose biological father resided in the home. Both of these effects, however, were mediated by mothers' history of mood disorders (which also predicted age of puberty), suggesting that maternal psychopathology may be the distal cause of family stress and possibly father absence, which promotes early pubertal timing in girls. Perhaps the most interesting aspect of Ellis and Garber's results concerned the effect of the presence of a stepfather or mother's boyfriend in interaction with family stress on girls' pubertal timing. Girls with stepfathers or who had mothers with boyfriends reached puberty significantly earlier when there was high family stress than did girls living with their biological fathers (regardless of the level of stress) or girls with stepfathers or mothers' boyfriends living with low family stress (see also Chasiotis et al., 1998; Surbey, 1990). Moreover, there was a significant relation between pubertal maturation and age of the daughter when a stepfather or mother's boyfriend came into her life ($r = -.37$), such that the younger the girl was when the unrelated father figure arrived, the earlier she attained puberty. In contrast, the relation between pubertal timing and the age at which the biological father left was not significant ($r = -.13$). In related research by Ellis and his colleagues (1999), father presence and positive aspects of (especially) father–daughter interaction predicted later pubertal timing in girls, an indication that positive qualities of a father's investment affect rate of pubertal timing in their daughters.

We believe that Ellis and Graber's (2000) results (and those of Ellis et al., 1999) are intriguing. On the one hand, they provide support for the general model proposed by Belsky and his colleagues (1991). Girls are sensitive to resource-related factors in the family and follow alternate developmental pathways depending on the interaction of a host of factors (maternal mood disorder, family stress, absence of a father, quality of paternal investment). Yet, in home environments of high stress, it appears that the presence of an unrelated father figure may be the most immediate event that precipitates early pubertal timing and that this is related to the age at which this person enters the daughter's life. In nonhuman animals, the presence of unrelated males hastens the onset of puberty in females, affording greater reproductive opportunity between the older males and younger females (Sanders & Reinisch, 1990). In contemporary and probably ancient human groups, sexual activity between a stepdaughter and stepfather would likely be a source of stress, not something that would be adaptive to the family structure nor to (most, if any) individuals in the family.

One interesting speculation is that stepfather presence contributes to the family stress that is related to accelerated puberty. This interpretation suggests that the alternative developmental pathways girls take in response to some family ecological factors may not be immediately related to adaptive reproductive strategies (i.e., investing more in mating relative to parenting) but, more simply, are a reaction to the presence of an unrelated male. Daughters' earlier pubertal timing may be an atavistic response to a mechanism that promotes enhanced reproductive fitness in mammalian species requiring less parental support than humans. When, as in humans, the unrelated male serves as a father figure and is expected to provide some investment in the stepoffspring, the accelerated puberty may result in increased family stress and result in the daughter seeking experiences outside the home and contributing to her establishing sexual relationships with peers. However, the findings that positive father–daughter interactions foster slower development (Ellis et al., 1999) suggest that the picture is more complicated, perhaps more in line with the original Belsky et al. proposal. Additional research is needed to explore further the complex interactions involving girls' responses to chronic environmental conditions on their pubertal timing and related psychosexual behavior; but regardless of the eventual determination of these effects, the extant literature makes it clear that girls (and perhaps to a lesser extent boys) are sensitive to such factors and as a result follow different developmental pathways that influence their reproductive strategies.

Toddlers' Peer Relationships

Toddlers' peer relationships, too, are beginning during this period. As we noted earlier, familiarity between conspecifics, as a possible proxy for kinship, affords opportunities for sustained cooperative interaction. Research with toddlers supports this point to the extent that the social behavior between familiar, compared to nonfamiliar, children is more complex and positive (Howes, 1988, cited in Rubin et al., 1998). Indeed, "friendships" between toddlers have been observed, where friends are defined in terms of individuals who repeatedly engage in complementary interactions marked by positive affect (Rubin et al., 1998). Similar criteria for friendship among nonhuman primates have been proffered (Smuts, 1985).

Childhood

Friendships for children are definitive peer relationships. Friendships are typically defined for children as reciprocally nominated individuals considered to be "friends" (Hartup, 1996). Friends tend to resemble each other behaviorally and attitudinally (Hartup, 1996); for example, two active children are more likely to be friends than are an active and a sedentary child.

In nonhuman primates, friendship is defined in terms of close affiliative bonds. As is the case with children, nonhuman primates treat their friends differently than they do nonfriends. For example, there are cases of chimpanzees intervening in a conflict between a "friend" and another chimp and of infant chimps preferentially associating with "friends" of their mothers (Tomasello & Call, 1997, p. 203).

Friends, compared to nonfriends, also act differently with each other, at both quantitative and qualitative levels. Quantitatively, friends interact with each other more often. They also have more conflict, but they tend to resolve their conflicts more frequently (Hartup, 1996; Pellegrini, Galda, Bartini, & Charak, 1998). In fact, Piaget (1965) proposed that it is through interaction with peers that techniques for resolving conflict develop. Investigating this issue, Nelson and Aboud (1985) assessed changes in third- and fourth-grade children's social knowledge as a result of discussing interpersonal issues with a friend or a nonfriend acquaintance. Friends were more critical of each other during their discussions than were nonfriends, but they were also more likely to provide detailed reasons for their own point of view. Pairs of friends changed the most in terms of social knowledge as a result of their discussions. When friends disagreed, their ultimate solutions to the dilemmas were more mature than when nonfriends disagreed. In other words, healthy social conflict between friends resulted in greater growth in social knowledge than did conflict between nonfriends.

At a qualitative level, friends' interactions are more complex than those of nonfriends; for example, their play is more complex (Howes, 1994). In addition, their collaborative interactions are cognitively more sophisticated; for example, friends use language indicative of reflection on cognitive and linguistic processes constitutive of collaborative problem solving (Pellegrini et al., 1998). These levels of mutuality and sophistication are probably a result of the evolutionary pressure for reciprocal altruism and cooperation (Axelrod & Hamilton, 1981). Friends are particularly important to preteen children who are apt to be victimized. Children who possess many of the physical and personality features characteristic of victims are less apt to be victimized and experience less deleterious effects of victimization if they have friends (Hodges, Boivin, Vitaro, & Bukowski, 1999; Perry, Hodges, & Egan, in press).

Adolescence

Friendships continue to be important in adolescence, especially their supportive nature. In early adolescence, however, friendship is viewed in exclusive terms, later changing to show less jealousy and possessiveness (Rubin et al., 1998). Particularly important for the period of adolescence, relative to childhood, is the friendship quality of "intimacy" as is evidenced by self-disclosure and closeness (Newcomb & Bagwell, 1995). In keeping with the movement from family to peer orientation, friends, rather than family, become sources of advice and support during this time of rapid and abrupt transition. The result of this close association is that friends are both models for and reinforcers of group behavior (Hartup, 1996).

Although friends generally are positive sources of information during adolescence (Brown, Clasen, & Eicher, 1986), the opposite also has been well documented. For example, when aggressive adolescent boys are together, they use more deviant forms of talk and are more excited by it than are nonaggressive boys (Dishion, Andrews, & Crosby, 1995). It is probably the case that youngsters with similar values and reputations choose to associate with each other (homophilly) and they become more alike with time (Hartup, 1996; Pellegrini et al., 1999).

The importance of similarity between friends may have evolutionary significance. Attraction toward similar conspecifics may have its origins in the attraction of kin to each other. As noted above, it is in the context of interactions with kin that cooperation and reciprocity develops (W. D. Hamilton, 1964). By extension, cooperative and reciprocal interactions occur with close and familiar non-kin because of the likelihood that they will be reciprocated (Trivers, 1971). Further, self-disclosures, which are a hallmark of adolescents' friendships, are also reciprocated in friendships, and these reciprocal self-disclosures between friends further reinforce the relationship (Youniss, 1996). Friendships typified by closeness, self-disclosure, and reciprocity may be especially important for adolescence as it marks a point when youngsters are entering a new social world with new social relationships. That adolescence is marked by fewer but more intense friendships suggests that they may help individuals negotiate this new territory.

GROUPS

Infancy and Toddler Years

By the end of their first year, human infants are able to make categorical distinctions between men and women (Leinbach & Fagot, 1993) and between adults and children (Brooks & Lewis, 1976) and are attracted to other, even unfamiliar, infants (Brooks & Lewis, 1976). They show a preference for children of their own sex as early as 2 years of age (Fagot, 1985), and 2-year-old boys (but not girls) demonstrate a preference for imitating same-sex as opposed to opposite-sex activities (Bauer, 1993). This all suggests that infants and toddlers are able to classify people according to potentially important social categories (sex and age) and show an early preference for affiliating with peers.

Even for preverbal toddlers, the group level of social complexity is expressed in terms of dominance. As noted above, dominance does not reside in an individual but is a result of the particular social and ecological constraints in a specific niche (Dunbar, 1988; Hinde, 1980). Dominance is usually expressed in terms of hierarchies, such that individuals are differentially ordered, transitively (Archer, 1988), for access to resources. From this view, dominance is not an end, but a means to an end—access to resources. Once a dominance hierarchy is established, rates of aggression decline as individuals know their place in the social order, recognizing the high costs and low benefits associated with challenges to dominant individuals (de-Waal, 1982; Dunbar, 1988; Strayer & Noel, 1986).

Resources can be expressed along a number of dimensions, such as mates, props, and food. It is probably the case that resources are differently defined according to different developmental periods. During the period of infancy and toddlerhood and probably childhood as well, access to props, such as toys, are a valued resource in peer interaction.

Dominance, even at this young age, can best be conceptualized in terms of "leadership," as discussed above, and can be specifically defined in terms of agonistic and affiliative behaviors. Dominant individuals are leaders of their groups, and they use a variety of strategies, both cooperative and aggressive, to establish and maintain dominance (Hawley, 1999). A variety of strategies may be used at different periods in development by individuals in different niches. For infants and toddlers, like some nonhuman primates (Chance, 1967), dominance can be recognized by attention structures, or the extent to which individuals attend to, or look at, other individuals. This definition of dominance has also been used by child developmentalists (Vaughn & Waters, 1981). Attention structure as a measure of dominance, however, has been criticized because it redefines dominance in terms of attention rather than access to resources, and it does not reflect the seemingly contradictory roles of aggression and affiliation in dominance (Hinde, 1974). Despite the meaning of attention structures, it seems that some individuals within toddler groups receive more attention than others.

Childhood

Even during the preschool years, children tend to segregate themselves into same-sex groups, but this tendency increases during the school years (Maccoby & Jacklin, 1987). Preference for same-sex friends and play groups is not unique to Western culture but is found universally (Edwards & Whiting, 1988). Parents and teachers seem to foster same-sex interaction early (M. Lewis, Young, Brooks, Michalson, 1975), but boys and girls also differ in how they play: Boys engage in more R&T play and center their interactions on movable toys, whereas girls are more apt to engage in dramatic play and table activities (see chapter 10). In fact, some suggest that girls actively avoid contact with boys because of their roughness (Haskett, 1971). Also, from the preschool years through adolescence, girls are more apt to play in smaller groups than boys, whereas boys' play groups tend to be larger, rougher, and more competitive than girls' (Lever, 1976, 1978), in keeping with coalitional male–male competition (Geary, 1998). Such observations led Maccoby (1988) to suggest that biologically based differences in play style may contribute to sex-segregated play groups.

One of the functions of same-sex groups during childhood seems to be the development of social skills and the establishment of social hierarchies. That children acquire proper social skills in sex-segregated groups is illustrated by what happens when children's primary friendships are with children of the opposite sex. In one study, third- and fourth-grade children's social skills were evaluated as a function of whether or not they had cross-sex friends (Kovacs, Parker, & Hoffman, 1996). The primary finding was that children with primarily opposite-sex friends were less well adjusted socially than children with only (or primarily) same-sex friends, although they were better adjusted than children without friends.

Social categorization, involving in-group favoritism and out-group discrimination, is readily apparent during childhood (Bigler, Jones, & Lobliner, 1997; Powlishta, 1995). As we have mentioned, the most pervasive social group difference among preadolescents is gender, with children playing primarily in same-sex groups. For example, in one study, 8- to 10-year-old children viewed videos of unfamiliar boys and girls and rated them on a variety of dimensions (e.g., masculinity, femininity, liking). Similar to the in-group favoritism found among adults, children rated same-sex targets more positively than opposite-sex targets (Powlishta, 1995; see also Egan & Perry, 2001).

As with toddlers, dominance structures are an important dimension of children's social structure. A variety of prosocial and aggressive strategies are used (especially by boys) to establish and maintain leadership in their peer groups. It is usually the case that boys, in the initial phases of group formation, use aggressive strategies in their competition with peers over resources (Strayer & Noel, 1986). That is, they tend to use aggression selectively and effectively to acquire resources. After dominance hierarchies are established, rates of aggression decrease and leaders use prosocial and cooperative strategies more frequently. Interestingly, rates of aggression for preschool children are positively correlated with popularity (Vollenweider, Vaughn, Azria, Bost, & Krzysik, in press). It may be the case that dominant children are those who are able to use aggression in an effective and Machiavellian way; they do not use aggression indiscriminately and reactively. For example, they may use it to help friends or allies (Strayer & Noel, 1986).

Primatologist Franz de Waal (1982, 1989), in his discussion of aggression among nonhuman primates, suggested that, in some ecologies, aggression leads to affiliation, rather than dispersal, when interactants reconcile after an aggressive bout. In cases where relationships between dominant and subordinate members are important and subordinates are free to leave the area, reconciliation cements the social order and enables group members to continue their interaction. This mixture of aggression, reconciliation, and cooperation probably come together in the formation and maintenance of dominance hierarchies.

The functions and dynamics of peer-group structure during childhood are nicely illustrated in a classic study by social psychologist Muzafer Sherif and his colleagues (1961). In the Robbers Cave experiment, 22 unacquainted fifth-grade boys attending summer camp were divided into two groups. Over the course of several weeks, each group participated in enjoyable activities such as doing crafts, building hideouts, and playing organized games, with each group of boys not being aware of the other. Group cohesiveness was emphasized, in part by organizing activities that required cooperation. For example, one evening the staff failed to cook dinner, and the boys had to divide responsibilities and prepare the meal themselves. Over time, clear positions of status emerged, with some children becoming recognized leaders and others followers. Both groups even adopted names: Rattlers and Eagles.

Once group cohesion had been established, the two groups were brought together and a series of "friendly" competitions was arranged (e.g., baseball, tug-of-war). Although the boys did not know it, the camp counselors arranged the games so that each group won and lost equally. When a group lost a competition, within-group conflict arose, often including threats of physical attack against one another or a change in leadership. As competition continued, however, within-group conflict decreased and group solidarity increased, often expressed by hostility toward the other group. The groups would abuse each other verbally ("You're not Eagles, you're pigeons"); they engaged in raids on the other campsite and theft or destruction of property; and physical violence (rock throwing) had to be stopped by counselor intervention. Thus, competition between groups led, after a brief period of disharmony, to greater within-group cohesion and overt hostility toward the other group. The findings from Sherif's study illustrate at least three of the four "evolutionary adaptations" proposed in J. R. Harris's (1995) group selection theory: (a) social affiliation and within-group favoritism, (b) status-seeking behaviors and establishment of social hierarchies, and (c) between-group hostilities.

Adolescence

Dominance for adolescence, like childhood, is an important dimension of group structure. Also like childhood, dominance in adolescence, at least for boys, involves both prosocial and aggressive strategies (Pellegrini & Bartini, 2000). The resources for which individuals are competing are heterosexual relationships. This period is marked by sexual maturity and the onset of sexual activity (Brooks-Gunn & Furstenberg, 1989). The separate affiliative (in the form of popularity) and aggressive (in the forms of observed and self-reported aggression) components of dominance each predict unique variance in boys' heterosexual desirability (in the form of being invited to a hypothetical party by girls in their class; Pellegrini & Bartini, 2000).

Dominance in early adolescence does, in some ways, differ from childhood (Hawley, 1999; Pellegrini & Bartini, 2000). These differences are probably caused by the corresponding changes associated with moving into puberty (which witnesses rapid change in body size) and moving from primary school to middle school. Each of these changes mean that youngsters must renegotiate status in their peer groups as they move from being the physically largest individuals in established relationships in their primary school to being the smallest in a new and larger group. In a recent longitudinal study (Pellegrini & Bartini, 2000), boys' dominance status decreased as they moved from primary to middle school. Boys then seemed to use aggressive strategies as a way to establish dominance in these new settings as the frequency of aggression increased from primary to middle school. By mid-year of the first year of middle school, however, aggression declined again.

This developmental trend in aggression and dominance is consistent with the logic of dominance. That is, dominance relationships are established when individuals compete with each other for resources. Based on a series of these encounters, individuals know their status in relation to other individuals. Consequently, they do not challenge other, more dominant individuals as they recognize the high probability of defeat. In short, the costs associated with such a challenge outweigh possible benefits.

The use of unbridled aggression, even by dominant individuals, is very costly. Costs associated with injury and possible defeat often result in alternative strategies being used. For example, we see in our longitudinal work (Pellegrini & Bartini, 2000) that after dominance relationships are established by using agonistic strategies, the affiliative dimension of dominance becomes more salient. That is, after dominance is established, dominant individuals use strategies such as reconciliation to re-integrate their defeated peers into the group (de Waal, 1989). This sort of strategy benefits both subordinate and dominant individuals to the extent that the former often gain access to resources held by the latter and the latter minimize the possibility that subordinates will form alliances to defeat him. It may be in such contexts that previously victimized children form affiliative networks and possible alliances. In short, dominance during adolescence, as in earlier periods, is related to the ability to use both prosocial and aggressive strategies. The goal toward which dominance is aimed is related to one's desirability as a mate.

The question that begs to be asked in all this is: How do subordinate individuals get access to resources? An alternative-strategy perspective suggests that it would be foolish for a low-ranking member to threaten or fight a higher-ranking individual. Instead, in nonhuman primates, the lower ranking individual uses an alternative strategy, such as stealing food or clandestinely copulating with a desirable female (Nunn, 1999). The female may show an interest in such a strategy because it is beneficial to her and her offspring to have a number of males think the offspring is theirs and thereby invest in it by providing food and protection. Further, as discussed earlier in this chapter, a promiscuous strategy with low investment in offspring may be used in niches low in resources (Belsky et al., 1991).

Our discussion of dominance thus far has centered on strategies that males use to attain dominance and leadership in groups of males. Males' dominance, as we noted above, orders them in terms of their access to resources, generally, and mating partners, specifically. Darwin (1871) also noted that females' choice of mates influences mating. Our knowledge of the ways in which females accomplish this goal is extremely limited in both the human and animal literatures (Gowaty, 1992). Ethologists who have studied this (Gowaty, 1992; Nunn, 1999; Smuts, 1985, 1995) have found that female primates use alliances with conspecifics for defense against unwanted sexual overtures and access to desired males. In gaining access to males, females often compete, through alliances and deception, with other females. We also know that females use relational aggression at different rates from males, and Geary (2001) has suggested that females use relational aggression to disrupt interpersonal networks that are so important to them. In short, and consistent with Darwin's original formulations, there is within- and between-sex competition for mates.

These differences in the ways in which females compete with each other for access to males affect the sorts of aggression used by males and females. First, it is well known that males use more verbal and physical aggression than females (Maccoby, 1998). Further, we know that girls, more than boys, use relational or indirect aggression, or aggression used in the service of peer relations (Crick & Grotpeter, 1995), for example, excluding one girl from a social group or damaging her reputation. With this said, it should still be noted that girls' use of relational aggression is still at lower levels than boys' use of other forms of aggression (Maccoby, 1998).

Differences between boys and girls for all types of aggression are due to a complex pattern of differences associated with socialization, hormonal, and evolutionary histories (Maccoby, 1998). For example, because males were historically hunters meant that they developed stronger and bigger bodies. The physical action patterns used to hunt could also be used against conspecifics to attain goals (get access to females). These differences in evolutionary history are also reflected in associated biological (hormonal, arousal, and self-regulation) and socialization histories.

How does this debate inform the child development literature on the ways in which females establish and maintain dominance? Females' use of interpersonal alliances to get and maintain access to resources is a starting point. From this view, it is important to explore further females' use of relational aggression (Crick & Grotpeter, 1995). We know that girls use it more than boys, but it would be helpful to know the distribution of its use in both boys and girls across the life span. Further, it would be helpful to know the goal of within- and between-sex aggression. That is, relational aggression is used to manipulate social relations, but we do not know for certain what these relationships are the vehicle for. Is it used by preschool girls, like preschool boys, to gain access to favored props? In adolescence, is it used against other girls to gain access to potential mates?

EVOLUTIONARY CONSIDERATIONS OF THE DEVELOPMENT OF SOCIAL BEHAVIOR

One of the themes running through this book is that human infants have been prepared by evolution to "expect" a species-typical environment.

Most critical to a human-typical environment are other people, particularly a mother, father, and other kin, as well as small numbers of unrelated (or less-directly related) individuals. Some of these unrelated individuals will be peers, with whom one will grow up. Beyond this, human social environments are highly diverse, and children must develop effective ways of dealing with these seemingly unpredictable environments. In a previous chapter, we discussed some of the social–cognitive processes that would be useful in dealing with conspecifics. In this chapter, we saw how those processes might be put into action, looking at various forms of "real" social behavior and how it changes over childhood.

Despite the diversity of social environments children around the world and throughout history (and prehistory) have faced and the range of behavioral solutions that children and adults must develop to cope effectively, there are many commonalities and several general, evolutionary mechanisms that can describe, explain, and predict social development. For instance, successful human social interaction is comprised of affiliative behaviors, including cooperation, and agonistic behaviors, particularly aggression. These forms of interaction, which become more complex with age, present themselves across the life span and are also evidenced in nonhuman primates. For the most part, the value of these behaviors can be understood in terms of economic models of evolution, balancing the costs and the benefits of particular behaviors in certain situations. However, other evolutionary theories, such as parental investment, can also help predict under what circumstances some people will behave aggressively, even violently, and how such seemingly counterproductive behavior (from a modern perspective) can be viewed as stemming from evolved mechanisms that produced more average benefits than costs in ancient environments.

We continue to emphasize that, despite evolved biases toward dyadic affiliation, status seeking, and in-group favoritism and out-group avoidance, specific social behaviors are acquired as a result of the dynamic transaction between children and their environments. As we commented in this chapter, much of this transaction and much of what children learn about social development, particularly past the early childhood years, is with peers. Perhaps the "mechanism" by which most socially relevant information is learned with peers is through play, and it is to this topic that we now turn.

10

HOMO LUDENS:
THE IMPORTANCE OF PLAY

Play is not unique to humans, but we do take it very seriously. In fact, historian Johan Huizinga (1950) referred to humans as *Homo ludens*, or "playful man." According to Huizinga, play is older than culture, has not been changed much by civilization, and has permeated from the beginning "the great archetypal activities of human society" (p. 4). Play is indeed old, phylogenetically speaking; it is observed in most mammals and even some reptiles (Burghardt, 1998). Further, it is characteristic of the juvenile period (Fagen, 1981) and is observed much less frequently in adults (Biben, 1979; Breuggeman, 1978).

How important is play to human development, and how might an evolutionary perspective enhance our understanding of the role of play in human social, emotional, and cognitive development? As we see in this chapter, play is often defined as having "no apparent function." Yet, evolutionists (and historians such as Huizinga) rarely have viewed play as functionless, although the nature of its function often has been hotly debated (Bekoff & Byers, 1998; Martin & Caro, 1985; Pellegrini & Smith, 1998; Power, 2000; P. K. Smith, 1982; Spinka, Newbury, & Bekoff, 2001). Indeed, some ethologists (Hinde, 1974) and developmental psychologists (Power, 2000) suggested that animals and children spend significant portions of their time and energy budgets in play; therefore it should, from an evolutionary perspective, serve an important function. As we discuss in this chapter, play has typically been interpreted as having deferred, or delayed, functions—as preparation for later adult behavior (Bruner, 1972). However, play also has immediate benefits, a position consistent with the perspective that natural selection exerts functional pressure during childhood (Bjorklund & Pellegrini, 2000; Hamilton, 1966; Spinka et al., 2001), as it is necessary to survive childhood to reach sexual maturity.

Play has held particular allure for sociobiologists, ethologists, and child developmentalists for years, despite associated difficulties such as its definition and functions (Hinde, 1980; E. O. Wilson, 1975). One level of response to these difficulties has been to define play as anything a juvenile does and to list numerous corresponding functions (see Burghardt, 1999, 2001; Martin & Caro, 1985, for a discussion of these problems). Starting at least with Groos (1898, 1901), play has been considered a hallmark of the juvenile period; he and many other scholars subsequently (Piaget, 1962; Vygotsky, 1978) have suggested that play serves numerous important developmental functions. Play during childhood, according to this view, enables children to learn the skills necessary for successful functioning in adulthood.

Some students of animals' and children's play have seen it as a source of creativity or behavioral flexibility that may eventually lead to discovering new ways to solve old problems (Bekoff, 1997; Fagen, 1981; Oppenheim, 1981; Sutton-Smith, 1997); and because of the youthful tendency toward play and curiosity in animals, it is likely that innovations will be introduced by the young rather than by adults. Support for this contention comes from the controversial observations (see Galef, 1996, for a critique) of potato washing in Japanese macaque monkeys (Kawai, 1965). A group of Japanese scientists provisioned a troop of wild monkeys with sweet potatoes, which were often sandy. One juvenile monkey learned to wash potatoes in seawater before eating them, and this was subsequently learned by other juveniles and then some adult females (few adult males ever learned this). This innovation was then passed on to infants as part of the "culture." Although it is unlikely that important cultural innovations will be made through the play of human children, the discoveries children make through play may serve as the basis of later innovations or true creativity, which become important later in life. These individual variations in behavior, in turn, may open the door to the "Baldwin effect," in which the phenotypic behavioral variability within populations becomes genetically fixed (Spinka et al., 2001; see chapter 3).

In this chapter we define play from an evolutionary developmental psychological perspective. Play lends itself rather nicely to such a broad-based view as expressed in ethological and sociobiological research (Bekoff, 1997; E. O. Wilson, 1975). Indeed, E. O. Wilson (1975) nominated play as one of five areas of animal behavior (along with kin selection, parent–offspring conflict, territoriality, and homosexuality) that warrant a sociobiological explanation. We define play in terms of its ontogeny, phylogenetic comparisons, and proximal factors influencing its occurrence (Tinbergen, 1963). We argue that, whereas some aspects of play do indeed appear to have delayed benefits, serving to prepare juveniles for adulthood, others have immediate benefits, consistent with the idea that aspects of childhood are adaptive for that time in development only and not as preparations for adulthood (Bjorklund, 1997b). It is surely the case that the nature and timing of benefits vary considerably, with, for example, species and ecologies (Bekoff, 1997; Burghardt, 1999; Spinka et al., 2001). We then discuss possible functions of play using cost–benefit analyses (see chapter 9), an examination of design features of play, and experimental deprivation and enrichment studies. The cost–benefit analysis method, although controversial (Allen & Bekoff, 1995), is helpful not only in determining possible functions of play but also in locating that period in development when benefits may be accrued. As part of our discussion of functions, we examine sex differences, particularly as they related to life in the environment of evolutionary adaptedness.

WHAT IS PLAY? GENERAL DEFINITIONAL ISSUES

An important step in understanding play is defining it, most globally, in terms of the ways in which it is different from behavior not considered playful and more specifically in terms of its design features. Based on these features, we can begin to make inferences about possible functions. That is, we use structural features to make functional inferences. We discuss the following forms of play: locomotor, object, social, and fantasy.

Each of these forms of play is a lot like art; we all know it when we see it but find it difficult to define. Given the complexity of the phenomenon, it is generally considered that no one definition of play is sufficient (Burghardt, 1999; Martin & Caro, 1985; Pellegrini & Smith, 1998), and thus definitions are typically multidimensional. For example, in Rubin, Fein, and Vandenberg's (1983) review of the child development literature, children's play is defined according to three dimensions: psychological disposition, behavior, and contexts supporting (or preceding) play. Ethologists also advocate defining play along dimensions (Burghardt, 1999; Martin & Caro, 1985; Pellegrini & Smith, 1998), suggesting structural (behavioral), consequential (those behaviors following play), and relational (where it occurs) criteria.

Perhaps the most commonly agreed on criterion for play is that is does not seem to serve any immediate purpose (although see Burghardt, 1998, for an exception). In Rubin et al.'s (1983) scheme, this immediate "purposelessness" criterion is related to the "means-over-ends" dispositional criterion. Attending to means over ends assumes that children are less concerned with the outcome of their behavior than with the behavioral processes per se. The importance of means over ends for a definition of animal play also has primacy for ethologists (Martin & Caro, 1985).

Embedded in the purposeless criterion of play is that players often self-handicap; that is, stronger or more facile players give weaker or smaller players an advantage in play (Spinka et al., 2001). For example, in a chase bout, a faster squirrel may initially outrun a conspecific, only to return more closely and begin the chase again. With children, older children often fall or stay on their backs so a younger player can assume a superordinate play position. These self-handicapping behaviors seem purposeless because they do not seem to result in the superordinate player gaining any resources.

In addition to dispositional dimensions, play has also been defined according to phenomena that precede and succeed the play behaviors themselves. Rubin et al. (1983) defined antecedents to play in terms of context, or those circumstances that elicit and support play. Those aspects of context that afford play include a familiar environment (in terms of props and people) that is safe and friendly; a minimally intrusive adult; and children who are free from stress, hunger, and fatigue. Martin and Bateson (1993) invoked a similar method in their use of spatial relations to categorize behaviors, including play; that is, behaviors can belong to the same category if they co-occur in a specific setting or among a certain group, such as juveniles. For example, all behavior observed in the playground could generally be considered play.

Correspondingly, play can be defined in terms of its consequences, or behaviors that follow play. This strategy is often used by ethologists (Hinde, 1980) and has been used in the child development literature to define specific dimensions of play, such as rough-and-tumble play (R&T; Pellegrini, 1988) and parallel play (Bakeman & Brownlee, 1980). For example, R&T refers to vigorous behaviors such as wrestling, grappling, kicking, and tumbling that would in other contexts be regarded as aggressive (Pellegrini & Smith, 1998). A behavior is categorized as R&T, not aggression, if children stay together after the conclusion of the bout; if they separate, it is defined as aggression (although see de Waal, 1989, for an alternative view). When classifying play, it can be useful to complement structural definitions with antecedent and consequential dimensions of play.

Such a multidimensional approach to defining play is useful, and some studies have looked at how observers may combine information from several cues to decide whether an activity is playful (Costabile et al., 1991; Smith & Vollstedt, 1985). Although the antecedents and consequences of behaviors may be useful in identifying a phenomenon, they should be considered as elicitors and outcomes, respectively, of those behaviors, rather than as components of the behavior per se. Thus, it is important to keep these dimensions conceptually distinct, especially when we consider the role of play in serving a developmental function.

ONTOGENY, PHYLOGENETIC COMPARISONS, AND PROXIMAL INFLUENCES

That play is a multidimensional construct is evident when we examine its different forms during childhood. Play is easy to recognize and ubiquitous in mammals, which suggests that it may have a common phylogenetic history or that common selection pressures may have acted on these species (Spinka et al., 2001). The forms of play that occur primarily during the juvenile period have been carved up differently by ethologists and child developmentalists.

Ethologists generally consider three forms of play, all with the attribute of apparent purposelessness: locomotor play, social play, and object play (Bekoff & Byers, 1981). Child developmentalists have also studied these forms of play, but they have spent much of their time studying pretend, or fantasy, play (Fein, 1981). Indeed, fantasy play is often cited by child psychologists as the paradigm example of play (McCune-Nicolich & Fenson, 1984; Smith & Vollstedt, 1985). The extent to which nonhuman animals engage in fantasy play in their natural ecologies, however, is controversial (Bekoff, 1997; see also discussion below).

Locomotor play involves exaggerated and repetitious movements, which often occur in novel sequences. These movements are the sort of behaviors described by Piaget (1962) and Bruner (1972) that very young children engage in as they are beginning to master the workings of their bodies and the objects that furnish their worlds. *Social play* refers to the seemingly goalless behavior between a juvenile and another juvenile or an adult. In the nonhuman literature, most social play occurs among juveniles, whereas in the human literature social play first appears in adult–infant interaction and later in peer play (see Pellegrini & Smith, 1998, for review). *Object play* can be solitary or social and involves manipulation of the material environment in ways that are not epistemic. *Fantasy play* can be social or nonsocial, is imaginary, and has an "as-if" orientation to objects or other participants. These types of play are not considered to be hierarchic, and they often co-occur with each other; for example, R&T involving two children can be social, locomotor, and fantastic (e.g., the enacting of a superhero battle).

Most mammalian species engage in social, object, and locomotor play (Bekoff & Byers, 1981; Burghardt, 1998; Fagen, 1981; Power, 2000). Extensive reviews of the animal literature suggest that play accounts for about 10% of animals' time and energy budgets (Byers & Walker, 1995; Fagen, 1981; Martin & Caro, 1985; P. K. Smith, 1982). Children's play, on the other hand, is more differentiated than that of nonhumans, to the extent that developmental progressions of different forms of play have been documented (Fein, 1981; Pellegrini & Smith, 1998). As part of this differentiation, exploration and play have been parsed, where exploration is epistemic—in which objects are manipulated so as to gain information. Play, by definition, has no goal as such.

Although understanding children's play is surely relevant to a proper understanding of the social, emotional, and intellectual development of all members of the species, some researchers have been particularly interested in sex differences in play. Evolutionary psychologists have focused on sex differences in behavior, following predictions from parental investment theory (Trivers, 1972). According to this theory, the sex that invests most in offspring, which in most mammals is the female, is more particular in selecting a mate and thus is more cautious in consenting to sex (see chapter 8). The least investing sex, males in mammals, are less selective in choosing a mate and compete among themselves for access to the higher investing sex. Based on these simple differences, according to the theory, males and females evolved different adaptive strategies for behaviors related to mating and parenting, and humans are no exception (Bjorklund & Shackelford, 1999; Buss & Schmidt, 1993). From an evolutionary developmental perspective, sex-stereotypic adult behaviors should not arise fully formed, but immature versions of them should be observable in childhood, serving as preparations for adulthood (Bjorklund & Pellegrini, 2000). If play is an important mechanism whereby children acquire information about their social and material worlds that has been shaped by natural selection, sex differences in play may reflect biases in males and females that orient them to different experiences and prepare them for different juvenile and adult lifestyles. We examine sex differences in play throughout this chapter and interpret these differences in terms of their potential adaptive value to ancestral men and women.

We reiterate that not all presumably adaptive characteristics of infancy and childhood are necessarily adaptations for later adulthood. Rather, according to evolutionary developmental psychological theory, some aspects of childhood behavior and cognition have been selected over the course of evolution to adapt individuals to their current environment rather than to prepare them for a future one (see Bjorklund, 1997b; Bjorklund & Pellegrini, 2000; see also chapters 2 and 6). We propose that there are immediate and deferred benefits of play, both for characteristics of play that are species universal and for characteristics that differentiate males and females.

Exploration

Especially during infancy, much time is spent exploring the environment rather than playing with it, although the exact amounts of time and energy spent in exploration have not been documented. *Exploration* is an information-gathering venture and is evidenced, in its earliest forms, by mouthing and simple manipulation of objects (Belsky & Most, 1981; Hutt, 1966). When exploring, infants are thought to be guided by the question "What can it do?" This object orientation is different from the more person-centered orientation guiding play, "What can *I* do with it?" (Hutt, 1966). Through exploration, children come to know their environments. This knowledge provides the basis for play. As such, exploration must be considered separate from play.

A similar pattern exists in nonhuman juveniles. For example, with domestic piglets, exploration of novel stimuli increases and then wanes; as it wanes, play with the objects increases (Wood-Gush & Vestergaard 1991; Wood-Gush, Vestergaard, & Petersen, 1990). Further, when animals habituate to objects or conspecifics in their immediate niches, they often travel to other areas to seek out novelty. In chimpanzees, exploration appears in the first two months of life and then gives way to more playful behaviors (Loizos, 1967).

The value, or function, of exploration may be that it affords animals the opportunity to learn about the properties of their immediate environments (Eibl-Eibesfeldt, 1967) and about new and different niches. Exploration may be especially useful for animals that colonize or forage into new areas because it enables them to learn about those new places. Consequently, it has been suggested that the history of exploration in evolution may be less dependent on an animal's phylogenetic place than on its ecology (Wood-Gush & Vestergaard, 1991). Counter to this argument, however, exploration does appear only among animals with differentiated cortexes (Suomi, 1982). The positive relation between exploration and brain size (and its correlate play) is probably due to the dialectic relation between seeking out stimuli and brain complexity.

In humans, exploration precedes play ontogenetically and microgenetically. Based on limited time samples of infants interacting with their mothers in a laboratory, exploration dominates behavior for the first 9 months of life (Belsky & Most, 1981). By age 12 months, play and exploration co-occur. By 18 months, play accounts for more of the child's interactions with the environment than does exploration (Belsky & Most, 1981). During this period of late infancy and early childhood, boys engage in more exploration than girls (Bornstein et al., 1996). In addition, boys, as compared to girls, tend to explore larger areas. Proximally, this is probably related to boys' higher levels of androgens, which relate to physical activity. More distally, in the environment of evolutionary adaptedness, males likely took on the role of hunter, which involved exploration of and familiarity with diverse environments. Microgenetically, exploration also precedes play to the extent that children of all ages must explore an object, or know its properties, before they can play with it (Hutt, 1966). Thus, novel objects or circumstances elicit exploration.

Locomotor Play

Locomotor play is physically vigorous, for example, chasing, climbing, and wrestling. There are three distinctive forms of locomotor play: rhythmic stereotypies, exercise play, and R&T (see Pellegrini & Smith, 1998, for a more thorough review). Each of these forms of play has a distinct inverted-U age curve and possibly different functions. We discuss each in turn, as well as associated sex differences.

Rhythmic Stereotypies

Rhythmic stereotypies are gross motor movements without any apparent function, for example, body rocking and foot kicking, and occur in the first year of life (Thelen, 1979, 1980). The onset of these behaviors is probably controlled by neuromuscular maturation. They are first observed at birth, peak around the mid-point of the first year, and account for about 40% of a 1-hour observational period (Thelen, 1980). There are no apparent sex differences in rhythmic stereotypies.

Rhythmic stereotypies can also occur in the context of adult–child interaction. For example, Roopnarine and colleagues (1993) described instances of parents bouncing children on their knees and throwing them in the air. This sort of activity accounted for 13% of all the play activity of 1-year-olds; object play with adults accounted for the other 87%. Thelen (1980) reported similar levels of parent–infant physical play in the form of vestibular stimulation.

Exercise Play

Exercise play is gross locomotor movement in the context of play. It is physically vigorous and may or may not be social. Exercise play can start at the end of the first year, and much of it, like later aspects of rhythmic stereotypies, takes place in the context of adult–child interaction (e.g., a parent chasing a child around the yard). Although adult–child exercise play peaks at around 4 years of age (MacDonald & Parke, 1986), there are cases where the adult role is to encourage young children to engage in exercise. For example, there are observations of adults in a Botswana foraging group encouraging infants to chase after and catch large insects (Konner, 1972).

Exercise play is very common during the preschool period, although many studies do not differentiate exercise play from R&T or pretend play, with which it co-occurs (Pellegrini & Perlmutter, 1987; P. K. Smith, 1973; P. K. Smith & Connolly, 1980). Thus, it may be underreported in the literature. Where it is reported, it, like other forms of play, follows an inverted-U developmental curve, peaking at around ages 4 to 5 years (Eaton & Yu, 1989; Routh, Schroeder, & O'Tuama, 1974). Specifically, at 2 years of age, exercise play accounts for about 7% of children's observed behavior in a daycare center setting (Rosenthal, 1994) and increases to about 10% for 2- to 4-year-olds in daycare settings (Field, 1994) and at home (Bloch, 1989). At 5 to 6 years of age, exercise play accounts for about 13% of in-home behavior (Bloch, 1989). Two ethological studies of children's behavior in preschools suggest that exercise play (defined as chase, jump, climb, and so forth) accounts for about 20% of all observed behavior in school (McGrew, 1972; P. K. Smith & Connolly, 1980).

As children move into primary school, the rate of exercise play declines, although there are fewer studies on which to base this judgment. For 6-year-olds, exercise play accounted for only 13% of all their outdoor behavior during school recess (Pellegrini, 1995b). This relative decrease from the preschool period may, however, be an underestimate. Specifically, in this study, observations were conducted on the school playground, a venue that affords this sort of play. Most preschool studies, on the other hand, occur in classrooms which, due to spatial and social policy constraints (many schools discourage this sort of behavior indoors), are less conducive to exercise play (Smith & Connolly, 1980).

A clear limitation of this work, as with much of child development, is that children's exercise play has been studied for the most part in schools and in preschools specifically. However, if we want to understand the circumstances in which children develop and make accurate estimates of the time and energy they spend in play, expanded efforts must be made to study children in their homes and communities (Barker, 1968; Bronfenbrenner, 1979; McCall, 1977). One such effort was mounted by Simons-Morton and colleagues (1990) in a study of 9- to 10-year-olds' exercise and physical fitness. Using self-reported frequencies of moderate to vigorous physical activity across a 3-day period, they found that children exercised more before and after school than during school. During the course of the day, they engaged in one or two bouts of exercise play (of 10 minutes or longer) each day.

Several proximal factors affect exercise play. As noted above, spatial density is a major determinant, in that more exercise play is observed in more, as compared to less spacious environments (P. K. Smith & Connolly, 1980). It is obviously easier to move around quickly in a spacious area than in a restricted one. Other less obvious factors are also important. For example, when children (Pellegrini, Huberty, & Jones, 1995; P. K. Smith & Hagan, 1980) and other young animals (Müller-Schwarze, 1984) are deprived of opportunities to exercise and then given opportunity, levels and durations of exercise increase as a function of deprivation. It may be the case that during the period of childhood, when skeletal and muscular systems are maturing rapidly, the body overcompensates for lost opportunity to exercise those rapidly developing systems. Relatedly, malnourishment inhibits exercise play (Burghardt, 1998, 1999). During such stressful times, the body probably uses valuable but scarce nutrients for physical growth rather than exercise play.

Last, ambient temperature also affects exercise play. For example, low levels of exercise play are observed in tropical climates (Cullumbine, 1950), and high levels of exercise are observed during cool, compared to warm, periods (Pellegrini, Horvat, & Huberty, 1998). These trends may be related to the thermoregulative role of exercise (Barber, 1991; Burghardt, 1988). In short, exercise play is relatively common in early childhood, accounting for around 20% of school activity and about 10% of home behavior and peaks in the early primary school years.

Reliable sex differences exist in children's exercise play, with boys engaging in it more than girls (Pellegrini & Smith, 1998). This difference is probably due, in part, to the more general sex difference in physical activity, with differences increasing from infancy to mid-adolescence (Eaton & Enns, 1986).

There may be cognitive implications, too, of exercise play and general locomotion. Locomotion probably aids in environmental information gathering. In support of this claim is research showing that animals living in changeable environments, relative to those living in more stable environments, engage in more locomotor play (Spinka et al., 2001). Similar findings exist for children. In a meta-analysis of experimental studies of the effects of locomotor experiences on children's spatial cognition, locomotor experience was found to be related to spatial cognition, but the effects were mediated by age such that experience was more important during early childhood than infancy and the toddler period (Yan et al., 1998). It is probably the case that with maturity, children are able to remember and integrate more information gathered during their movements and then use this information to solve spatial problems. As we noted in chapter 6, sex differences in locomotor activity may be responsible, in part, for the male advantage that is reported for some types of spatial cognition (Bjorklund & Brown, 1998).

The importance of locomotion in spatial cognition is highlighted in studies of hormonal differences between the sexes. These differences may be partially responsible for differences in both locomotion and spatial cognition. Male hormones are related to increased levels of locomotion across several species (see Quadagno, Briscoe, & Quadagno, 1977, for a review). In a natural experiment involving children, Resnick and colleagues (1986) compared "normal" males and females with females who had been prenatally exposed to abnormally high levels of male hormones (congenital adrenal hyperplastia; CAH). Results were consistent with the hypothesis: CAH females had play styles and spatial cognition skills (card and mental rotation tasks) that were higher than female controls and closer to those of males, suggesting that locomotion may be especially important in boys' spatial cognition.

The finding of sex differences in spatial-orientation abilities, favoring males, is consistent with the theory that males need enhanced skills of this type for tracking game and traveling away from the home base during hunting and perhaps in warfare, although the latter possibility has not been explored. Females, on the other hand, may develop greater object location and memory skills, necessary for a life as a gatherer. Research has confirmed this pattern of spatial abilities beginning in childhood (Eals & Silverman, 1994; Silverman & Eals, 1992; Silverman et al., 2000; see also chapter 6). Differences in experience, related to differences in physical play styles between girls and boys, may be one proximal basis for this pattern.

Rough-and-Tumble Play

R&T, probably the most thoroughly studied aspect of play by scientists with an evolutionary perspective, is commonly observed in most mammals and accounts for about 10% of their time and energy budgets (Fagen, 1981). That said, we must note that R&T, like exercise play, is probably underestimated in the child development literature, as most scholars do not include it in their coding schemes, and when they do, they often subordinate it to fantasy play, with which it co-occurs (Pellegrini & Perlmutter, 1987; P. K. Smith & Connolly, 1980). R&T is necessarily social, involving roughhousing with another person, usually a peer.

Most observations of R&T with peers, like other forms of play, have been collected in school settings. For the preschool period, observations occur primarily in classrooms, whereas primary and middle school observations take place on school playgrounds. As such, these estimates represent a limited portion of children's day and are confounded by school policy that tends to discourage R&T. R&T with peers during the preschool period accounts for 3%–5% of observed classroom behavior (Pellegrini, 1984) and increases to 7%–10% at ages 6–10 years (Boulton, 1992; Humphreys & Smith, 1987; Pellegrini, 1988). It declines to 5% at 11–13 years (Boulton, 1992; Pellegrini, 1995a) and to 3% at 14 years (Pellegrini, 1995b). The patterns are similar in nonhuman animals such that R&T increases from infancy, peaks during the juvenile period, and then declines, almost disappearing from the behavioral repertoire of adult animals (Fagen, 1981).

There are also robust sex differences in R&T. Specifically, males engage in R&T more than females in virtually all human cultures (DiPietro, 1981; Whiting & Whiting, 1975) and in many mammalian species (Meaney, Stewart, & Beatty, 1985; P. K. Smith, 1982). These differences hold for parent–child play (Carson et al., 1993; Roopnarine et al., 1993) as well as peer play (DiPietro, 1981; Pellegrini, 1989). The sex differences are more robust for play fighting than for chasing (Smith & Connolly, 1980), perhaps reflecting the degree to which society models and reinforces this sort of behavior for males. For example, fathers spend more time than mothers in R&T, and they do so with sons more than daughters (Carson et al., 1993; Parke & Suomi, 1981). Further, in schools, girls' play is supervised more closely than that of boys (Fagot, 1974), and this may inhibit physically vigorous play, which many teachers consider inappropriate for girls (Pellegrini, 1989).

The vigorous and rough nature of boys' play has been hypothesized as one reason for the sex segregation of play groups that is found universally (Edwards & Whiting, 1988; Maccoby, 1998). Segregation in play groups and males' play being rougher than females' are also typical of nonhuman primate play (Biben, 1998), as well as that of other mammals, such as ungulates (Clutton-Brock, Iason, & Guiness, 1987).

Social learning must be considered in conjunction with hormonal events to obtain a full understanding of sex differences in R&T and exercise play as well. Hormonal influences on physical activity and play typically center on the role of androgens on neural organization and behavior (Meaney et al., 1985). Normal exposure to androgens during fetal developmental predisposes males, compared to females, toward physical activity generally and toward exercise play and R&T more specifically. Excessive amounts of these hormones during fetal development are hypothesized to "masculinize" females' play behavior (Collaer & Hines, 1995). The experimental literature, using mice, hamsters, rats, and monkeys, supports the androgenization hypothesis (Collaer & Hines, 1995; Quadagno et al., 1977).

Of course, such experiments are impossible to perform with humans. Yet we do have "natural experiments" where human fetuses are exposed to abnormally high does of androgens, for example, CAH, as mentioned above. The CAH studies support the androgenization hypothesis. Using a combination of questionnaires on which parents and children are asked to report on behaviors and limited behavioral observations, androgenized girls, relative to controls, prefer male activities and male toys (Berenbaum & Hines, 1992; Berenbaum & Snyder, 1995; Hines & Kaufman, 1994; Money & Erhardt, 1972).

In one study (Berenbaum & Hines, 1992), boys and girls who had been exposed to high levels of androgen before birth were observed when they were between ages 3–8 years, and their toy preferences were contrasted with groups of nonaffected children. Children were brought into a playroom where stereotypically masculine, feminine, and neutral toys were available. The results of this study are shown in Figure 10.1. As expected, unaffected boys played more often with the masculine toys and unaffected girls played more often with the feminine toys. The difference in toy preference was not significant between the CAH and nonaffected boys. In contrast, the CAH girls spent much more time playing with the masculine toys than did control girls, such that masculine toy preferences were comparable with those of the boys. This pattern of toy preference was replicated in later research (Berenbaum & Snyder, 1995).

These robust sex differences in exercise and R&T play should be related to functional benefits of these forms of play and are discussed later. The proximal factors affecting R&T are similar to those affecting exercise play, as they are both physically vigorous.

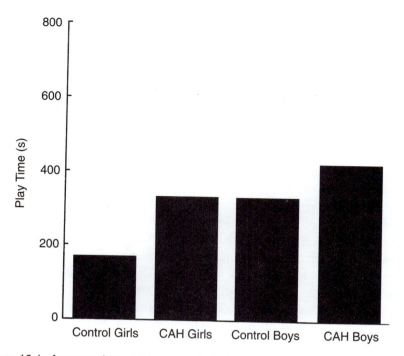

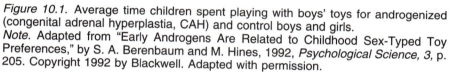

Figure 10.1. Average time children spent playing with boys' toys for androgenized (congenital adrenal hyperplastia, CAH) and control boys and girls.
Note. Adapted from "Early Androgens Are Related to Childhood Sex-Typed Toy Preferences," by S. A. Berenbaum and M. Hines, 1992, *Psychological Science, 3*, p. 205. Copyright 1992 by Blackwell. Adapted with permission.

Physical play, in the form of both exercise play and R&T, constitutes a significant portion of school children's waking behavior (up to 20% by some estimates) and is more prevalent in boys than in girls. Such "high-energy" behavior is often at odds with "proper" school behavior. Contemporary schools require children to sit quietly for protracted periods of time, something that is incompatible with physical play. It should not be surprising, therefore, that schools are more likely to identify boys as "behavior problems" than girls. In fact, as we noted in chapter 2, the rapidly rising incidence of cases of attention deficit hyperactivity disorder (ADHD) may be related less to a rise in neurological problems than to a misdiagnosis of "normal" play behavior—which may be adaptive in some environments—although such behavior is maladaptive in a school setting (Panksepp, 1998; Pellegrini & Horvat, 1995). The widespread use of psychostimulant drugs to reduce the hyperactivity and increase the attention focusing of some children diagnosed with ADHD may reduce the desire and opportunity to play, and may have unintended consequences, such as reduced neural and behavioral plasticity (Panksepp, 1998; but see Goldstein & Barkley, 1998).

Object-Oriented Play

One form of usually solitary play that displays a sex difference is object-oriented play. *Object-oriented play* is not found in mammals other than humans and the great apes (McGrew, 1992; Power, 2000; Tomasello & Call, 1997) and involves the nonfunctional manipulation of objects, such as banging or throwing them. These descriptions, however, must be tempered by the general paucity of comparative research in this area (M. Bekoff, personal communication, 2001). Object-oriented play would seem to be valuable in emergent tool use, as individuals learn the properties of objects and their potential functions.

Boys are more likely to build things and to experimentally manipulate objects, taking them apart and attempting to put them together again (Hutt, 1972). This pattern is not limited to contemporary America but has been reported in a U.S. sample in the 1960s (Sutton-Smith, Rosenberg, & Morgan, 1963) and in a variety of industrial and pre-industrial cultures (Eibl-Eibesfeldt, 1989). It also has been found that, with children ages 47–69 months old, not only do males play more with blocks and other toys that provide practice with manipulations and transformations, but they also build more structures as compared with females (Caldera et al., 1999). Object-oriented play is influenced by sex hormones. CAH girls, exposed to high levels of androgen prenatally, engage in more object-oriented play than unaffected girls (Berenbaum & Hines, 1992; Collaer & Hines, 1995).

Not all studies report sex differences in object-oriented play, and some studies actually find more such play in girls than boys (see Christie & Johnsen, 1987). Moreover, sex differences in tool use in chimpanzees also favor females (see McGrew, 1992). Termite fishing and nut cracking, activities that both involve the use of simple tools, are most frequently performed by females. Male chimpanzees will occasionally use tools when hunting (e.g., throwing rocks), but they use tools less frequently than females do. Thus, although sex differences in object-oriented play seem to show a male advantage in humans, the nature of the differences is complex, and further research to specify the behaviors involved, the conditions in which they are most likely to be found, and their consequences, if any, is warranted (see Pellegrini & Bjorklund, in press).

Some recent evidence exists that preschool boys are more likely to successfully use simple tools to solve problems than preschool girls (Chen & Siegler, 2000). Children between ages 1.5 and 3 years sat at a table on which an attractive toy was placed out of their reach. Immediately in front of the children was a set of six objects, only one of which could be used to pull the toy to the child. Boys were more apt to use the proper tool and to successfully retrieve the toy than girls, although the sex difference was smaller following demonstration of the tool use strategy (see also Gredlein,

2001). Similar sex differences were reported for 9- and 10-month-old infants (Bates, Carlson-Luden, & Bretherton, 1980). Chen and Siegler suggested that the primary locus of the sex difference in their study was boys' greater eagerness to use the tool. This is consistent with the differences in object-oriented play observed by other researchers (Sandberg & Meyer-Bahlburg, 1994) and suggests that such an orientation may result in enhanced tool use and problem solving for males.

It is difficult to get an accurate estimate of the amount of time children engage in object-oriented play, because various researchers have established substantially different criteria for classifying object-oriented play. When simple contact with and manipulation of objects is the criterion, more than one third of preschool children's time is involved in such "play" (Rubin & Maioni, 1975; Rubin et al., 1978), with this figure being similar for observations made among groups of hunter–gatherers (Bakeman et al., 1990; Sigman et al., 1988). However, when more differentiated coding schemes for object "play" are used, the percentage observed drops substantially. Specifically, P. K. Smith and Connolly's (1980) study of children's behavior in preschools found, across three samples, that object play accounted for 10%–15% of all behavior.

In short, the amount of time children spend in play with objects seems to vary considerably depending on the coding scheme used as well as the venue of the observations. Conservative estimates of object play, at 10%–15%, are derived from more differentiated coding schemes and are consistent with other time budget studies of human (Pellegrini, Horvat, & Huberty, 1998; Pellegrini & Smith, 1998) and nonhuman (Fagen, 1981) play. More liberal estimates, derived from more global coding schemes, suggest that more than a third of all behavior, at least in preschool settings, is object-oriented. These higher figures are similar to estimates of object manipulation, not play. Of course, these higher estimates may reflect preschool organization that supports this sort of activity as well as the global nature of the coding scheme typically used to derive these figures. When a similar coding scheme is used outside of school, the estimates drop precipitously, to approximately 10% (P. K. Smith & Connolly, 1980). Future research should, however, differentiate object play from object manipulation because estimates of occurrence help researchers make functional inferences.

These dramatic differences in time spent in object play in different venues also point to the importance of describing children's behavior in a variety of settings. Although the call to study children in settings other than preschools is hardly new (Barker, 1968; Bronfenbrenner, 1979; McCall, 1977), we repeat the call. Such descriptive information is imperative if we are to understand the role of play, or indeed any other behavior, in individual and species development. Time or energy spent (as a measure of the cost) in an activity is a necessary first step in documenting possible function (or benefit). From this view, a behavior is functional when benefits outweigh costs. More will be said about function and cost–benefit analyses below.

Fantasy Play

Fantasy play, or symbolic play or pretense, is the paradigm case of play during childhood. Fantasy includes an as-if orientation to actions, objects, and peers. Often, it involves playing a distinct pretend role such as mommy, firefighter, or doctor. Most generally, fantasy involves a player taking a stance that is different from reality (Lillard, 1993a, 1993b) and using a mental representation of a situation as part of an enactment (Lillard, in press).

Commonly agreed-on components of fantasy include decontextualized, sequential combinations, self–other relations, and object and role substitutions (Fein, 1981; Mathews, 1977). Decontextualization refers to using an action pattern or prop outside its functional context. For example, an aggressive act can be used in a nonaggressive context. Decontextualized acts are sequentially combined into play bouts of varying length. For example, a doll can be fed and then rocked, or a play face (initiating play) can be followed by a play bite. Self–other relations have players moving from solitary to social play. Last, object and role substitutions have objects or players representing other things or roles. For example, a doll can be used as a baby (object substitution) or a dominant individual can assume the role of a subordinate (role substitution; self-handicapping). In the animal play literature, self-handicapping is treated as a form of role reversal (Rosenberg, 1990).

Is Fantasy Play Limited to Homo Sapiens? Groos (1898) suggested that fantasy play occurred in many species, but most recently scholars have suggested that fantasy play is limited to humans (Tomasello & Call, 1997), so that, unlike social, object, and locomotor play, fantasy play may represent a "phylogenetic discontinuity."

That fantasy play is limited to humans is based on the premise that it involves having a "metarepresentation" (Leslie, 1987). Metarepresentation involves children having an awareness that they are pretending and consciously realizing that a pretend situation is mentally represented (Taylor & Carlson, 1997). When children come to this realization about fantasy, they also realize that minds represent the world in a variety of ways. This is when they are said to possess a theory of mind (see chapter 7), as indicated by passing false-belief tasks. From a metarepresentational view, fantasy play can be viewed as requiring mental representation and as an indication of children's general cognitive development (Piaget, 1962). When children pretend, they recognize that individual minds mentally represent the world, a capacity not developed until around 4 years of age.

Alternatively, some theorists have proposed that fantasy does not necessarily involve metarepresentation. This view holds that in early stages of fantasy, children may be aware that they are pretending and enact scripts or behavioral routines in social fantasy play (P. L. Harris, 1990; 1991; Jarrold et al., 1994; Lillard, 1993a, in press; P. K. Smith, in press). According to P.L. Harris's (1991) simulation theory, young children's ability to recognize and take part in fantasy play does not involve representing mental states. Fantasy, from this view, has the young child imagining a pretend world through simulation. In simulation, the young child initially assumes that his or her mental simulations are the same as his or her playmate's. Repeated social encounters with peers are central to the model, because with repeated experiences children (at about 3 years of age) recognize that this assumption is not often valid and that their views and mental states often differ from others. Simulation, at this point, involves children putting their beliefs and desires to one side and temporarily imagining those of their peers. With this simulation done, children will then go back to their own mental functions and use the simulated information to appreciate what the others think or feel. In the early stages (at about age 3 years), understanding others is a mixture of the imagined views of others and children's own egocentric perspective. For example, the child realizes that he or she and a peer share different beliefs about a real apple and a plastic apple. By 4–5 years of age, children are able to imagine that they have different beliefs from others with respect to the two, mutually exclusive objects. So understanding others' points of view depends on accuracy of simulation, not on the nature of the representation. With repeated encounters, children's simulations of others becomes more accurate, an accuracy not dependent on metarepresentation of others' mental states.

From this view, children realize that fantasy is real and that they and their peers can have different views of different objects. The social dimension of fantasy play results in children's views conflicting with that of their peers, because they recognize that fantasy is inherently ambiguous: There is a limited correspondence between the play theme and reality. To sustain playful interaction in the face of ambiguity, children must accommodate others' perspectives and clarify ambiguity (Pellegrini, 1982). The use of explicit play–frame markers such as "let's pretend" or more implicit play frame markers such as exaggerated voice or movements informs children that the interactions are playful. The use of these markers, in turn, indicates an intent to communicate a fantasy theme. Only after maturation and repeated interactions with others do children come to understand others' mental states. Thus, understanding others is not dependent on having a theory of others' minds but, instead, is a result of repeated negotiations and simulations with others.

There is some evidence, admittedly controversial, for aspects of fantasy play, or pretense, in great apes, as well as social carnivores (Bekoff, 1997). There are numerous reports of home-reared (enculturated) chimpanzees engaging in elaborate symbolic play (Gardner & Gardner, 1971; C. Hayes, 1951; Jensvold & Fouts, 1993; Temerlin, 1975). The sign language trained chimpanzee, Washoe, for instance, reportedly bathed and dried her dolls (Gardner & Gardner, 1971), and Lucy, another human-reared chimpanzee, signed to her dolls (Temerlin, 1975). There are also several observations of wild chimpanzees displaying behavior that has been interpreted as fantasy play (e.g., treating a log as a baby; see Wrangham & Peterson, 1996). Such observations in wild apes are rare, and observations in enculturated apes, although presumably reflecting symbolic representation, are open to alternative explanations.

The evidence for the existence of fantasy play among nonhumans, according to Bekoff (1997), is as follows. First, both great apes and canids (wild dogs) have a relatively high encephalization quotient (the ratio of actual brain size to brain weight expected from a regression equation for the relationship between brain size and body size; see chapter 4). Consequently, they have the cognitive capacity required for fantasy. Second, and like the fantasy play of children, nonhuman primate and canid social play is characterized by "decontextualized" behaviors and behavioral sequences of routines: They are taken from one context and used differently in another context (Fein, 1981). For example, bites in social play for wolves are gentler than those in real fighting. These individual behaviors are also embedded into predictable sequential combinations where initiations are marked by explicit negotiations between players.

This sort of play is consistent with the view that play is simulation or action based and seems consistent with simulation theory's descriptions of 3-year-old humans' fantasy. These interactions do not depend on animals having a theory of others' minds. Animals take behaviors from functional contexts (behaviors used in a real fight context) and simulate them in a social play context. Their sustained and playful interactions are evidence that animals recognize that these behaviors have different aims from these same behaviors in serious contexts (Rosenberg, 1990). Further, animals use of "metacommunicative" play signals (Bekoff, 1977) seems to be a way in which animals announce "This is play" and suggests that animals recognize that conspecifics may have a view of the situation that is different from theirs. Canids use play "bows," and nonhuman primates use play faces to mark the beginning of a bout. In addition, animals use play signals to clarify possible ambiguity, or "honest mistakes" in play (Fagen, 1981). Play markers, such as bows or play faces, reliably follow play behaviors that might be interpreted as too rough. Following these markers (or behavioral "punctuations"), animals reliably resume play (Bekoff, 1997).

There are also examples of animals substituting roles in play. For example, animals in several species, such as big horn sheep, chimpanzees, and sable antelopes (Fagen, 1981), like children, take on different roles by self-handicapping in play; dominant individuals often take on subordinate roles in play fighting. Self-handicapping is often punctuated with play signals to further clarify the playful tenor of the bout. That some nonhuman primates (e.g., Diana monkeys, *Cercopithecus diana*) may have mental representations of predators (Züberbühler, 2000) suggests that the reversals commonly observed in play fighting may be indicative of a mental representation of predator. In this naturalistic playback experiment, Züberbühler (2000) tested the extent to which monkeys responded to acoustically different alarm calls depending on predator's distance (close or far), elevation (above or below), or category (eagle or leopard). The vocal signals of males and females in response to experimentally manipulated alarm calls indicated that they responded to the predator category. These results suggest that the vocalizations may be "linked to mental representations of external objects or events, such as a concept of a predator" (p. 926). Similar results have been reported with captive ringtailed lemurs, *Lemur catta* (Pereira & Macedonia, 1991). Role reversals and self-handicapping in play seem to indicate that animals may substitute one role for another and that these roles may be based on conceptual categories. That is, in self-handicapping, nonhuman primates may be "pretending" to be a predator.

In short, some social, nonhuman animals may engage in fantasy according to criteria used to categorize children's fantasy. Descriptions of animal play seem consistent with the view that early forms of fantasy are action-based simulations and are not dependent on players having a model of others' minds. We recognize that this is a controversial claim; students of fantasy suggest that fantasy has players making attributions about each others' mental states. To our knowledge, however, the question of children's attributions of other mental states does not typically enter into differentiating fantasy play from other forms of behavior in most behavioral studies. Research simply uses an as-if criterion. The issues of intent and levels of abstraction are separate.

The mechanism by which animals come to understand others' intentionality, according to Bekoff (1997), may be similar to that suggested by Gopnik and Meltzoff (1997) for the bases of understanding intent in human infants. From this view, there is a cross-modal representational system that connects the self to others. Body movements of others are mapped onto one's own kinesthetic sensations, much as newborn infants are proposed to do when they imitate. This may be a basis by which individuals unify acts of others with their own.

Although the suggestion that some social mammals other than humans may engage in aspects of fantasy play is provocative, we reiterate that it is speculative but deserves consideration. Further, we do not believe that there is solid evidence to propose that such behavior reflects an ability to make attributions about conspecifics' mental states. When fantasy play is defined by the display of decontextualized behavior and role substitution in relation to conspecifics, we feel comfortable attributing such play to great apes in some contexts (perhaps primarily to those reared much as human children), and perhaps even to canids, following Bekoff (1997). However, fantasy play in humans typically involves more than decontextualized behavior and role substitution. For example, Bering (2001) proposed that symbolic play be differentiated between *feature-dependent make-believe*, in which imaginative behavior is directed toward an object that resembles another object (a child treating a shoe as if it were a telephone or Washoe bathing her doll as if it were a baby), and *true symbolic play*, in which individuals follow social scripts in the absence of any perceptually eliciting stimuli while in the context of play. There is some evidence that most of the fantasy play of human children until nearly their third birthday can best be described as feature-dependent make-believe and that only after this time does children's fantasy play clearly involve representations of objects that are perceptually dissimilar from the objects they purport to represent (Ungerer et al., 1981). Bering (2001) argued that the incidence of fantasy play observed in great apes reflects only feature-dependent make-believe and does not require advanced forms of symbolic representation, as is seemingly required for attributing beliefs and knowledge to others (theory of mind). We concur and propose that the fantasy play observed in nonhuman animals, particularly canids, may reflect a species-specific adaptation and may not be governed by the same type of cognitive systems as the fantasy play observed in humans (Burghardt, 1999). In other words, the similarity between canid and human fantasy play likely reflects phenotypically similar but phylogenetically independent (convergent) solutions to the problems associated with living in socially complex groups. Because of the nearness of our last shared common ancestor, it seems more likely that the cognitive abilities for feature-dependent make-believe displayed by some great apes may have been the basis for the evolution of humanlike symbolic abilities, such as those seen in true symbolic play.

Fantasy Play Among Human Foragers. The hypothesized continuity from nonhuman to human fantasy is further supported by the numerous systematic accounts of fantasy play among human foragers. Evolutionary psychologists and ethologists find foragers interesting to study to the extent that they represent what life might have been like in the environment of evolutionary adaptedness. Two groups of African foragers that have been systematically studied are the !Kung, a Kalahari hunter–gatherer group (Blurton Jones, 1993; Blurton Jones & Konner, 1973; Konner, 1972), and the Hadza, foragers who live in northern Tanzania (Blurton Jones, 1993).

In both groups, the children exhibit fantasy play. For example, Hadza children play with dolls they make out of rags and play at being predators. Variations in the fantasy play of each group, however, do exist. Specifically, time spent by children in play generally varies according to costs and benefits involved in play versus foraging and the impact of each on their survival and reproduction. In cases where children can efficiently and safely engage in hunting and gathering, as with the Hadza, less time is spent in play and more time is spent in production. In the case of the !Kung, hunting and gathering are relatively dangerous (there is a high risk of children getting lost) and food is relatively available for adults to forage, so they have a longer period of childhood (i.e., time free from hunting and gathering) and spend more time in camp playing with other juveniles (Blurton Jones, 1993).

Development of Fantasy Play in the Industrialized World. Fantasy is, of course, commonly observed among children in the industrialized world. The vast majority of the research studies of children's fantasy play in industrialized countries is done in preschool classrooms. Thus, estimates of time and energy budgets are limited. Fantasy begins during the second year of life, peaks during the late preschool years, and declines during the primary school years. It accounts for more than 15% of the total time budget (Field, 1994) and for 10%–17% of preschoolers' and 33% of kindergartners' play behaviors, and then it declines (Fein, 1981).

That children beyond preschool are discouraged from playing may contribute significantly to age trends in fantasy. Many parents report, anecdotally, that their children continue engaging in fantasy well in the elementary school years. One of us (AP) has a 12-year-old son, who along with two or three of his friends, spend hours engaging in pretend using Lego creations. More in-home observational and diary studies documenting the amount of time spent in fantasy are clearly necessary.

One such example involving young children was carried out by Haight and Miller (1993). This intensive observational study in homes of young children shows the important supportive role of mothers in early pretend play interactions. Based on these in-home observations, rates of pretend began at .06 minutes/hour for 12–14-month-old children, increased to 3.3 minutes/hour at 24 months, and reached 12.4 minutes/hour at 48 months. Thus, there is convergence between the preschool and home observational data that pretend play accounts for 12%–15% of children's time budget.

Fantasy play is influenced by social variables as well as play props. For example, research has shown that young children are more likely to engage in fantasy play when they are playing with someone else than when they are alone and that mothers in particular bring out high levels of symbolic play in their children (Bornstein et al., 1999; Youngblood & Dunn, 1995). Consistent with Vygotsky's (1978) idea of a zone of proximal development, young children who interact with a more skilled partner who structures the situation appropriately for them advance in their skills faster than when such support is not provided. For instance, in one study examining the play of mothers with their 21-month-old children, many mothers were observed to adjust their level of play to that of their children's or slightly above the level their child was displaying (Damast, Tamis-LeMonda, & Bornstein, 1996). Moreover, mothers who knew more about play development were more likely to increase the level of play with their children than were less knowledgeable mothers and thus provided appropriately challenging play interactions.

That children's fantasy play is supported by older individuals, in the form of play with older children or siblings, is also consistent with the data from foraging societies and assumptions about the environment of evolutionary adaptedness. In foraging societies, children spend much of their time with other children, but these children are in mixed, not homogeneous, age groups (Draper, 1976; P. K. Smith, 1998). Young children's fantasy, in these cases, is supported by older conspecifics.

Related research has shown that children who are securely, as compared to insecurely, attached to their mothers engage in more sophisticated pretend play. For example, securely attached children initiate more play interactions, and the tenor of their interactions with their mothers while playing is more positive than that of insecurely attached youngsters (Roggman & Langlois, 1987). Further, mothers tend to engage in symbolic play with their daughters more frequently than with sons (Bornstein et al., 1999). Also, children's pretense is more sustained and complex when they are playing with friends compared with acquaintances (Howes, 1994). The mutuality and emotional commitment of friends may motivate children to sustain cooperative interaction (Hartup, 1996).

Regarding social fantasy play, children typically talk about mental and cognitive states in the process of negotiating roles; for example, "Doctors can't say that." Social pretend play facilitates children's reflection on language, as they must render meaning linguistically explicit; this happens in play more than in realistic discourse, as the former is more abstract and more easily results in ambiguity unless explicated. The ability to talk about language is related specifically to children's phonemic awareness—awareness of the rules governing the sound system of their native language—and to subsequent school-based reading. An important component in learning to read involves children learning letter–sound correspondence (Pellegrini, Galda, Bartini, & Charak, 1998; Pellegrini, Galda, Shockley, & Stahl, 1995), making fantasy play an important (but, of course, not the sole) factor in facilitating metalinguistic and early reading ability. Sex differences in fantasy play, favoring girls, are consistent with the finding that girls also outperform boys on linguistic tasks (Maccoby & Jacklin, 1974; see further discussion of sex differences in fantasy play below).

Play themes generally follow the themes inherent in the props available (Pellegrini & Perlmutter, 1989; P. K. Smith & Connolly, 1980). Not surprisingly, when children play with props suggesting a medical theme, their play follows that theme. When props have less explicit themes, however, such as blocks or scraps of plastic foam, themes are more varied (Pellegrini & Perlmutter, 1989). Further, the sex role stereotypicality of the props influences the rates and levels of sophistication of pretend play. Boys' play with female-preferred toys, such as dolls, is less sophisticated than it is with male-preferred toys, such as blocks. Indeed, the play of older preschool boys is less sophisticated than that of younger boys when they play with female-preferred props (Pellegrini & Perlmutter, 1989). This suggests that children's pretending begins conforming to sex role stereotypes during this period. That is, boys come to view playing with female-preferred props as something to be avoided. Consistent with this argument, girls' play with male-preferred props is not less sophisticated than their play with female-preferred props.

Other research has documented the greater avoidance of opposite-sex activities and play materials for male than for female toddlers (Bauer, 1993; O'Brien & Huston, 1985). For example, 25-month-old girls showed comparable imitation of stereotypically male, female, and neutral activities; boys, in contrast, displayed greater imitation for the masculine and neutral activities than for the feminine activities (Bauer, 1993). Moreover, boys spent more time interacting with the neutral and masculine props and demonstrating a knowledge of sex stereotypes, and avoided female activities, as young as 2 years of age. Although the greater social value afforded male roles may indeed contribute to this sex difference, the very early age at which it appears suggests that boys may be more predisposed than girls toward early and strong sex-stereotypic behavior.

By 3 years of age onward, pretend play involves quite sophisticated social role-playing skills with peers (Howes, 1994). Extended social role playing was described by Smilansky (1968) as "sociodramatic play" and involves acting out quite complex narrative sequences. For example, two 3-year-old children may play with doctor props and a doll such that they string together a brief series of interrelated conversational turns.

Joan: Ooo, the baby has goo on her.
Susan: Ooo, goo.
Joan: Clean it.
Susan: All done.
Joan: The baby's sick.

In the above example, the first four utterances centered on a common theme: a dirty baby. By the fifth utterance, however, the theme changed. With age, these themes become more involved and less dependent on props. Props with explicit themes, such as doctor props, act as a "scaffold" or support for pretend interaction by making clear what the interaction should be about; interactions are less dependent on verbal explication of meaning and pretend transformations (Pellegrini, 1985). By the end of the preschool period, most children are very capable of initiating and sustaining social pretend play bouts with ambiguous props with minimal adult support (McLoyd, 1980).

There are also reliable sex differences in fantasy play during the preschool period. Girls engage in fantasy play both more frequently and at more sophisticated levels than do boys. Although the pretense of girls tends to revolve around domestic, dramatic themes, the pretense of boys tends to be more fantastic and physically vigorous, often co-occurring with play fighting and superhero themes (Pellegrini & Perlmutter, 1987; P. K. Smith, 1974). The potential functional nature of such sex differences is discussed later.

In summary, fantasy play accounts for 12%–15% of preschool children's time budget at home and at school. Both sex and social partners in play, in turn, affect these rates. Sex differences in time spent in different forms of pretend play as well as social partners in play should also have implications for function.

To conclude this section, we have presented an overview of the prevalence of different forms of play in infants' and children's time and energy budgets. The figures we cited suggest that play of all forms follows an inverted-U developmental curve, with different peaks for different forms of play (see Table 10.1). For example, physically vigorous play and fantasy play both peak around ages 5–7 years and account for 10% and 25% of children's time budgets, respectively. This level of time investment leads us to posit that play does indeed serve some function. Given this level of cost, it would not have been naturally selected if it did not. To further explore this argument, we consider below the costs in relations to possible benefits of play.

TABLE 10.1
Approximate Percentage of Waking Time Children of Different Ages
Engage in Exercise, Rough-and-Tumble, Fantasy, and Object Play

	Toddlers (2 years)	Preschool (3–5 years)	Early School (6–10 years)	Adolescence (11–15 years)
Exercise	7	10	15	—
R&T	—	5	10	5
Fantasy	5	25	15	—
Object Play	73[a]	15	—	—

Note. Figures do not include all forms of play, and the various forms of play are not exclusive (e.g., R&T and fantasy play may co-occur). Percentages based on data presented in text. R&T = rough-and-tumble play. Dashes (—) indicate insufficient data.
[a]Percentage of total amount of object play with parents relative to other forms of play with parents.

FUNCTION AND THE DEVELOPMENT OF PLAY

As we have seen, behaviors are often classified as play if they appear to have no apparent immediate benefit to the actor. Yet, perhaps paradoxically, play is typically seen as serving an important function in children's development. At one level, the paradox can be reconciled by supposing that the apparent lack of immediate function actually conceals either a function delayed in development or an immediate function of which the player (and even the observer) may be unaware. Researchers in animal play (Bekoff & Byers, 1981; Martin & Caro, 1985) and child play (Pellegrini & Smith, 1998) have discussed its numerous hypothesized developmental functions. For the most part, these are delayed functions; consideration of immediate functions of play has been largely absent from the child development literature (Spinka et al., 2001).

What Is Meant by Function?

Function has at least two meanings in the social and behavioral sciences, following from Tinbergen's (1963) classic "Four Whys" (see chapter 2). In its ultimate sense, *function* can be defined in terms of an evolutionary history of biological adaptation for that species. A behavior is functional in this "ultimate" sense (Hinde, 1980) if it typically has added to the survival or reproductive success of individuals over many generations. Functions can also be defined in terms of beneficial consequences during the life span of any individual irrespective of the evolutionary history of selection. Thus, this meaning of function refers to outcomes during the life of the individual player (Hinde, 1980; Symons, 1978). Benefits may be accrued immediately, during the period of childhood, or deferred until adulthood. For example, social physical-activity play may have immediate value in terms of affiliation with a peer group, or it may have deferred benefits and relate to later ability to encode and decode social signals (Pellegrini & Smith, 1998).

Kagan (1996) has suggested that most developmental theorists, including Piaget (1962) and Vygotsky (1978), proposed that the benefits of play are understood in terms of what good it does in later development and not during childhood. Kagan (1996) suggested that the deferred-benefits view may be due to bias among developmental psychologists toward the importance of early experience in human development. From this perspective, play serves as preparation for later adult functioning. Consequently, play is viewed as an imperfect version of adult behavior. As we mentioned earlier, sex differences in play have often been interpreted as preparing boys and girls for the different roles they will likely assume in adulthood (see Geary, 1998, 1999). Bateson (1976) has referred to this as the *scaffolding* view of play: Play functions in the assembly of skills and is disassembled when the skill is complete (Bruner, 1972). The classic example of play serving deferred benefits is the play fighting characteristic of juvenile males seen as practice for adult hunting and fighting skills (P. K. Smith, 1982).

An alternative view is that play is not an incomplete or imperfect version of adult behavior but a benefit to the specific niche of childhood. Bateson's (1976) term for the immediate functions of play is *metamorphic*. Specifically, play behavior and its beneficial consequences are unique to a specific period of development, and the advantages of those consequences for later development may not be readily apparent, for example, because of discontinuity in development. Immediate function may reflect the usefulness of play to a specific period. In this way, we consider play as a specific adjustment to the context of childhood (Bateson, 1976, 1981b; Bjorklund, 1997b; Bjorklund & Green, 1992). For example, the sense of mastery and self-efficacy associated with children's play probably relates to children experimenting with new and different activities. Once activities are chosen, they should be sustained; sustenance, in turn, affords opportunities for learning specific skills (Bjorklund & Green, 1992). In a similar vein, boys' R&T may serve as a way to learn and practice social signaling (Martin & Caro, 1985), with exaggerated movements and a play face communicating playful intent. Further, it is a way in which boys establish leadership in their peer group and assess others' strength (Pellegrini & Smith, 1998). R&T also has immediate nonsocial benefits; it provides opportunities for the vigorous physical exercise that is important for skeletal and muscle development (Bruner, 1972; Dolhinow & Bishop, 1970).

How Are Functional Inferences Made?

Function can be inferred in several ways, including cost–benefit analyses, consideration of design and contextual features, and experimental enhancement or deprivation of play components.

Cost–Benefit Analyses

Cost–benefit analyses can be used to establish a prima facie case for functional significance and are commonly used in the ethological literature (Caro, 1995; Martin, 1982; Pellegrini et al., 1998; see also chapter 9). This method assumes that for a construct to be naturally selected, benefits should outweigh costs. Thus, we assume a correspondence between costs and benefits: High costs should have high benefits; low costs imply either low or high benefits (Fagen, 1981).

Costs with respect to play can be defined in terms of caloric expenditure, time, or survivorship. For example, the research evidence indicates that childhood injuries sometimes occur in a play context (Cataldo et al., 1986);

many childhood deaths due to injury occur while children are riding bicycles (Peterson et al., 1997). There are also caloric costs associated with play. Playing, especially physical play, requires energy, and the energy demands of play must be considered in light of other caloric requirements, for example, calories required for basic metabolism, growth, and more direct learning tutorials. The extensive animal literature (Fagen, 1981; Martin & Caro, 1985) and the more limited child literature (Pellegrini et al., 1998; Pellegrini & Smith, 1998) on caloric costs of play suggest that play accounts for a modest portion, less than 10%, of total caloric budget.

Cost–benefit analyses represent a promising, although challenging (Allen & Bekoff, 1995), method for studying play, primarily because children can be studied in natural ecologies. Although studies of cost are seemingly basic to our understanding of play, there are surprisingly few of them (Pellegrini et al., 1998); consequently, we cannot catalogue the size of likely benefits of play according to this method. Basic time–budget information on play, as we described earlier, is scant and is generally limited to school contexts (see Pellegrini & Smith, 1998, for a review). Further, we do not know the extent to which childhood accidents are embedded in play or nonplay contexts. For example, do drownings occur while children are playing or bathing? This sort of information is certainly important for providing adequate services for children.

Some hints about probable functions of play according to this model can be hypothesized from the animal play research. Generally, play accounts for a limited portion of an animal's time and energy budgets—around 10% (Martin & Caro, 1985; Pellegrini et al., 1998). Based on these rather low levels, cost–benefit analyses suggest that benefits of physical play for animals are probably modest and immediate. They should be immediate rather than deferred, because of the risks encountered during the time from infancy to maturation. Specifically, there is a high probability of prereproductive mortality, 38%–42% among contemporary foragers (Worthman, 1993). Given these risks, the probability of reaping these benefits is maximized if they are accrued during the juvenile period rather than at maturation. This does not preclude, however, the possibility that some aspects of physical play in humans do not also have some deferred benefits (e.g., R&T as preparation for adult fighting in boys); and fantasy play, possibly unique to our species, may be particularly beneficial for practicing roles that will characterize children as adults (see Geary, 1998).

Design Features

A design features argument involves explication of the similarities between aspects of play behavior and similar features in functional behavior. Take one feature of social play—reciprocal role taking. Role taking in play at Period 1 could be related (conceptually and empirically) to the ability to take different, more mature roles later in Period 2, such as the ability to play cooperative games (Pellegrini, 1995b). Establishing function, then, involves making logical and empirical connections between the features at Periods 1 and 2 and between play and functional behaviors.

There are, however, some problems with this design features approach to the study of function. Any one play behavior may serve more than one function. Further, there may not be a behavioral similarity between play and later functional behaviors (discontinuity in development). Similarly, a mature behavior may develop from many different antecedents (i.e., "equifinality"; Martin & Caro, 1985).

Design feature arguments have been made frequently for R&T and for fantasy play. Regarding R&T, the social and physically vigorous dimensions of play, respectively, have been linked theoretically to later aggressive/fighting status and cardiovascular fitness (Biben, 1998; Humphreys & Smith, 1987; Pellegrini, 1993, 1995a; P. K. Smith, 1982). Empirically, R&T relates proximally to children's vigorous games (Pellegrini, 1988). The role alternation and chasing in R&T are similar to the turn taking and running in games such as tag. As children approach adolescence, R&T, especially for boys, appears to be increasingly incorporated into assertion or establishing and maintaining social dominance because dominance involves use of agonistic and cooperative strategies (Pellegrini & Bartini, 2000).

Further, experiencing both superordinate and subordinate role characteristics of R&T probably relates to social competence. Evidence from non-human primates, for example, suggests that juvenile squirrel monkeys, *Saimiri sciureus*, deprived of opportunities to engage in play fighting where they are in superordinate (the one who pins) or subordinate (the one who is pinned) roles, are later bullies and "sissies," respectively (Biben, 1998). Thus, although boys' R&T may have some immediate benefits, such as facilitating skeletal and muscle development, it also appears to teach boys something about aggression, fighting, and social competition. Further, by engaging in the role alternation characteristic of play fighting, boys are gaining experience in superordinate and subordinate roles (Biben, 1998), something that is useful in competitive interactions of all sorts but especially useful in their encounters with other males on matters of dominance and eventual mating choices.

These benefits of R&T are consistent with the fact that boys engage in R&T more than girls. Specifically, the forms of behaviors that boys use to establish and maintain dominance (see chapter 9) are similar to those they exhibit in R&T and mirror the activities associated with male–male competition (primitive warfare) in hunter–gatherer societies (Keeley, 1996). Further, and more distally, any hunting or fighting benefits associated with

R&T also reflect males' roles in the environment of evolutionary adaptedness (P. K. Smith, 1982).

Another sex difference can be found in play parenting. Boys and girls show a comparable interest in infants until about age 6; from this time onward, girls are, on average, more responsive to babies than are boys, and this is also reflected in the incidence of play parenting (see Geary, 1998). This pattern has been reported cross-culturally, including in traditional hunter–gatherer societies (Eibl-Eibesfeldt, 1989) and across cohorts in the United States (see Geary, 1998), suggesting that it is not limited to the particular values of contemporary Western culture. Part of play parenting often involves the construction of complex and reciprocal family relations, such as mother, father, and baby. According to Pitcher and Schultz (1983), "even from the youngest age, girls are quite knowledgeable about the details and subtleties in these roles" (p. 79). In contrast, boys' fantasy play is more likely to focus on power, dominance, and aggression, often being part of R&T. In brief, characteristics of the fantasy play of boys and girls can be seen as antecedents for later roles (e.g., male–male competition, parenting) they will play (or would have surely played in the environment of evolutionary adaptedness) in adulthood (see Pellegrini & Bjorklund, in press).

Some aspects of girls' greater interests in infants and play parenting are related to sex hormones. For example, girls' interest in infants and young children increases after menarche (Goldberg, Blumberg, & Kriger, 1982), and girls exposed prenatally to high levels of androgen (CAH) display significantly less play parenting and interest in infants than unaffected girls (Berenbaum & Snyder, 1995). Hormones are not likely the entire story, however. In most cultures, girls, more than boys, are assigned the task of child care (Edwards & Whiting, 1988). Yet, even without being assigned such responsibilities, girls engage in more play parenting than boys. From an evolutionary perspective, such an orientation would facilitate girls being attentive to activities related to caring for children and practicing skills that would serve them well when they become mothers. Women in all cultures of the world spend more time interacting with and caring for children than men, and this distribution of childcare responsibilities was likely similar in our evolutionary past (see chapter 8).

We noted earlier that boys are more likely to engage in object-oriented play than girls. From an evolutionary perspective, the sex difference in object-oriented play may be related to traditional adult males' greater involvement with a wider variety of tools than females. Although in traditional cultures both males and females use and make tools (and this surely would have been the case for our ancestors), men make most of the tools, including weapons, musical instruments, and ceremonial objects (Daly & Wilson, 1983). Males' greater orientation toward manipulating objects may thus reflect an ontogenetic adaptation related to an adult life in which tool use and construction play an important role (see further discussion of sex differences in object-oriented play and its possible relation to tool use below).

The two sexes may have evolved near identical brains, biologies, and aptitudes, but with different biases, based on inherited genes expressed in a species-typical environment (including prenatal environment) that orient them to slightly different aspects and interpretations of the world. When these different orientations receive support from the environment, boys and girls have different experiences and develop different expertise. Play thus serves as a vehicle for acquiring and mastering important sex-specific aspects of behavior. Sex differences in play, and the experiences that such differences afford, may serve as a proximal mechanism for adult sex differences in behavior. Such sex-specific expertise would have been adaptive, we argue, for our ancestors. Such biases in boys' and girls' play orientation need not receive strong environmental support, either prenatally (as when the prenatal environment contains excessive androgen for female fetuses) or postnatally. Children can acquire through play only what the environment has to offer. But we believe that such biases were selected over evolutionary time because they yielded adaptive behavior and that these biases persist today. Their benefit to modern people, in which sex differences in competition for resources and tool use have been minimized, are questionable but nonetheless part of our mammalian and primate heritage.

Experimental Deprivation Studies

Deprivation studies are more common in the animal play literature (Müller-Schwarze, 1984; especially with rats, Pellis, Field, Smith, & Pellis, 1996; Potegal & Einon, 1989) than in the children's play literature. In these studies, children (or young animals) are deprived of an aspect of play, typically social play and physical-activity play. The assumption behind these studies is that deprivation of an aspect of play, if it is developmentally important (i.e., if it serves a beneficial consequence), should result in a "rebound effect" (Burghardt, 1984) when children are given the opportunity to play again; children compensate for the deprivation by engaging in high levels of play for longer time. Most deprivation studies can be criticized on the grounds that more than one thing is involved when we deprive children of play (although see Potegal & Einon, 1989, for a rare exception). So when children are deprived of social play, they are often deprived, simultaneously, of other forms of social interaction as well.

The effects of depriving children of play opportunities have largely been limited to studies of locomotor play. In three sets of field experiments, P.K. Smith and Hagan (1980) and Pellegrini and colleagues (Pellegrini &

Davis, 1993; Pellegrini, Huberty, & Jones, 1995) experimentally deprived a sample of British preschoolers and American primary school children, respectively, of opportunities for locomotor play. Results were consistent across all experiments in showing that increased deprivation led to increased levels of play when opportunities for play were resumed. Further, with the primary school sample, break time maximized children's attention to school tasks when they returned to the classroom.

It was argued that children engaged in longer and more intense bouts of locomotor play after deprivation so as to compensate for lost opportunities for physical exercise, which serves important training functions (Byers & Walker, 1995). Further, the relatively transient effects of physical training suggest that they are immediate and can occur at any point during the life span (Byers & Walker, 1995). For example, in studies of both older and younger athletes, exercise training improves performance rather rapidly. However, the effects of the training also wane rapidly. In cases specific to children who were not athletes, the results were similar. For example, one study reported that distance-running training decreased heart rate and increased maximum oxygen uptake in 8- to 12-year-old children (Lussier & Buskirk, 1977).

Regarding the effect of break time on attention to school tasks, we argue, consistent with the cognitive immaturity hypothesis (Bjorklund, 1997b; Bjorklund & Green, 1992; Pellegrini & Bjorklund, 1997), that non-focused play activities, such as those found at break time, provide a release from more focused school work. This recognition for the need of nonfocused play activities for optimal learning in young children is apparently realized by many Asian school systems. Although it is well known, for example, that Japanese children spend more days in school than American children, it is not as well known that the Japanese school system provides young children more breaks from rigorous work than American schools. For example, first- and fifth-grade children in both Japan and Taiwan have more breaks and recesses than American children do. In both countries, the school day is highly structured, with 40- to 45-minute classes followed by 10- to 15-minute breaks (Stevenson & Lee, 1990). A common observation by Westerners in Asian schools is how attentive the children are. Such intense concentration is attributed partly to the frequent opportunities children have for vigorous play during between-class breaks. Stevenson and Lee reported that first- and fifth-grade children in Minneapolis schools had no more than two recesses per day, whereas first-grade children in schools in Sendai, Japan, and Taipei, Taiwan, had four, and fifth-grade children had between five and eight (see Table 10.2). It is also worth noting that although the Taiwanese and Japanese fifth-graders spent more hours per week in school than the American children, the school day for first-graders

TABLE 10.2
Number of Days Spent in School per Year, Number of Hours Spent
in School per Day, and the Number of Recesses per Day for First-
and Fifth-Grade Children in Minneapolis, United States; Taipei,
Taiwan; and Sendai, Japan

City	Grade	Days/year	Hours/day	Recesses/Day
Minneapolis	First	174	6	2
	Fifth	174	6	2 (at most)
Taipei	First	230	4	4
	Fifth	230	8.3	4
Sendai	First	234	5	4
	Fifth	234	6.5	5

Note. Adapted from "Context of Achievement," by H. W. Stevenson and S. Y. Lee, 1990, *Monographs of the Society for Research in Child Development, 55*(1-2), Serial No. 221. Copyright 1990 by Society for Research in Child Development. Adapted with permission.

in both Taipei (4 hours) and Sendai (5 hours) was shorter than in Minneapolis (6 hours), reflecting an awareness of the attentional limitations of these young children.

Experimental Enrichment Studies

Many experimental enrichment studies involve the effects of object play generally and pretend or sociodramatic play specifically on various dimensions of children's social cognitive status, such as role taking (Burns & Brainerd, 1979), narrative competence (Pellegrini, 1984; Pellegrini & Galda, 1982), associative fluency (Dansky & Silverman, 1973, 1975), and problem solving (Sylva, Bruner, & Genova, 1976). The effect on some ability, perspective taking, for example, is then examined. Most of the criterion measures in these studies are taken in close temporal proximity to the training, thus assuming that play has immediate rather than deferred benefits.

By way of illustration, we consider experimental studies of object play. Object play has typically been studied in the context of using objects to solve problems. The classic study in this area was conducted by primatologist Wolfgang Köhler (1925), in which a chimpanzee solved a problem (assembling a combination of separate sticks to reach bananas hanging out of reach). The chimpanzee "played" with the sticks, seemingly with no goal in mind and, serendipitously, discovered that the sticks went together. The combined and longer stick could then be used to reach the bananas.

Two sets of studies are representative of this approach in child development. Sylva, Bruner, and Genova (1976) tested young children in a Köhler-like paradigm: Separated sticks, clamps, and a latched reward box were situated at the end of a table. The task for children was to assemble the three sticks, reach for the latch, open it, and retrieve the colorful piece of chalk. There were three conditions: a play condition, a control condition, and an observational learning condition in which children observed an adult assemble the sticks. The basic premise was that play provided the behavioral flexibility to solve a novel problem. The results indicated that children in the play condition, relative to the other conditions, generated solutions that moved from simple to complex and were more goal directed. Sylva and colleagues argued that play afforded opportunities for spontaneous problem solving with minimal frustration, as play stresses means over ends. This sort of serial ordering of activities and subroutines with objects, they argued, is necessary for tool invention that may have been evident in our phylogenetic history. Unfortunately, sex differences were not reported in the results.

In a similar vein, Dansky and Silverman (1973, 1975) examined the effects of object play on children's associative fluency, or novel uses of conventional objects. Children in the play conditions were encouraged to "play" with the experimental objects for 10–15 minutes. Other children participated in control and observational-learning conditions. Following training, children were asked (typically by the same experimenter who did the training) to provide novel uses for experimental objects or for new objects. In each case, children in the play conditions provided more novel uses than those in other conditions.

The research by Sylva and her colleagues and by Dansky and Silverman demonstrated that children who played with objects were more likely to use those objects later to solve a new problem. Thus, children may have learned something specific about the objects through play that increased the likelihood of the objects being used as tools in a subsequent situation. In more recent research, Gredlein (2001) asked whether a greater tendency toward object-oriented play, in general, would relate to subsequent tool use. In his study, 3-year-old children participated, one at a time, in a free-play session with sets of toys, some of which promoted object-oriented play (e.g., Legos, Lincoln Logs). In a separate session, children sat in front of a table on which were placed objects (none of which they had seen in the play session), only one of which could be used to successfully retrieve an out-of-reach toy. (This was based on the procedure used by Chen and Siegler, 2000, described earlier.) Gredlein reported that boys displayed more object-oriented play and a greater tendency to spontaneously use tools to retrieve a toy than did girls, suggesting that boys' greater experience with object-oriented play was responsible, in part, for their better performance on the tool-use task. As with Chen and Siegler (2000), Gredlein reported that girls' performance on the tool-use task increased to the level of the boys after being given only a simple hint, suggesting that there were no sex differences in the actual competence in using tools. Rather, the primary locus of the sex difference was likely boys' greater motivations for object-oriented play and for using tools, with their greater object-oriented play possibly enhancing their tendency to see objects as potential tools.

Although these studies showed positive benefits of play, they must be interpreted cautiously, primarily because laboratory-based experimental manipulations of play tell us only how certain variables may affect behavior; they do not necessarily tell us about the ways in which these behaviors develop in nature (Bronfenbrenner, 1979; McCall, 1977). Further and more basically, many play enrichment experiments have serious internal validity problems (P. K. Smith, 1988). For example, many experiments confound play treatment with adults tutoring children in play (Pellegrini, 1984; Smith & Syddall, 1978). In addition, lack of control for experimenter bias in much of this work invalidates many conclusions regarding function (Simon & Smith, 1985; P. K. Smith & Whitney, 1987). Even if internal validity issues had been controlled, the treatments have often been of such short duration, for example, 10 minutes in some cases (Dansky & Silverman, 1973), that we must question the efficacy of such short treatments to change otherwise stable behavior in a child ages 3–5 years.

Discussion of these limitations should not lead to the conclusion that object play is unimportant. Studies in natural habitats, such as those of nonhuman primates washing potatoes and fishing for termites, suggest that animals use objects in novel and creative ways, and play affords opportunities to be flexible with objects. Consequently, research designs must provide ample opportunities to support this sort of flexibility. For example, repeated observations in relatively familiar circumstances should support youngsters' exhibition of competence in a variety of tasks (Bronfenbrenner, 1979).

PLAY, EVOLUTION, AND DEVELOPMENT

Juvenile members of most mammal species play and, from a life history perspective, humans spend more time as juveniles than does any other large animal. In fact, if play were to be used as a defining characteristic of the juvenile period relative to life span, it would extend well into the reproductive years (and perhaps into senescence) in humans. Although play is characterized by its seeming "purposelessness," both animal and human theorists of play concur that it does indeed have a function, perhaps numerous ones. Some functions of play appear to be immediate, such as exercise and the establishment and maintenance of social relations during childhood. Others appear to be deferred, including the practicing of adult roles in a nonthreatening context. Play serves as a way for youths to explore both their physical and social worlds and to modify their neural circuitry in the meantime. As the comparative work of Blurton Jones (1993) with two groups of foragers showed, these patterns are embedded in the contingencies of the local economies.

We argue, as have others (Bruner, 1972; Fagen, 1981; Martin & Caro, 1985; Pellegrini & Smith, 1998), that play has been subjected to selection pressures over the course of mammalian evolution. Play in humans is extended, because human children have so much to learn that they require not only a long time to learn it, but also safe environments in which to master their eventual adult roles. The interactions and lessons acquired during play among peers, perhaps more than any other single socialization agent, afford children the opportunity and flexibility to learn what it means to be a man or woman in their society (cf., J. R. Harris, 1995).

Children through history (and prehistory), as well as juveniles from other social mammal species, could "expect" environments that afforded the opportunity to play. Although many of the environments that children grow up in today may be safer and less physically demanding than ancient environments must have been, the requirements of formal schooling, adult supervision of "organized" play (e.g., Little League), and the lure of sedentary activities such as television viewing, modify the quantity and quality of play experienced by many contemporary children, which may, in turn, have subtle effects on their development. It is a hackneyed phrase that "play is the work of children," but play seems to have been especially adapted for the period of childhood and is what children are "intended" to do. Remembering this may cause us to think twice before modifying children's environments to achieve one goal (e.g., more focused learning opportunities in schools) at the expense of play.

11

EPILOGUE: EVOLUTION AND DEVELOPMENT

AN EVOLUTIONARY ACCOUNT OF DEVELOPMENT AND A DEVELOPMENTAL ACCOUNT OF EVOLUTION

A Common Ground

Developmental psychology has been a fractured field, with diverse subdisciplines and a variety of theoretical perspectives that often seem at odds. Communication among practitioners, even in seemingly highly related fields such as specialties within cognitive development, is often poor, with developmental scientists of one ilk (e.g., those who study basic processes such as working memory) failing to appreciate or understand what other developmentalists (e.g., those who study metacognition) do. For those who study higher level processes, such as metacognition or theory of mind, to reduce analysis much below the level of "meaning" yields much detail but little understanding; for those who study basic-level processes, spending one's time examining such ill-defined notions as self-awareness and metamemory may be philosophically pleasing but scientifically vacuous. And even if these hypothetical practitioners were to see the merit in the work of the others, the levels of analysis are so different that there is often little room for common interest and nothing really to talk about. The divide is even greater when communication between "major" subdisciplines within developmental psychology is considered.

Perhaps this poor communication is only the result of scientists' need to specialize when basic knowledge is expanding at an exponential rate. Perhaps the day of the generalist is past. There is surely some truth to this, but we believe in the potential unification of knowledge (E. O. Wilson, 1998), at least within the domain of developmental psychology, and that it can be achieved only when some common ground for understanding development is found.

In this book we have presented data from a broad range of topics in developmental psychology and believe that, despite the diversity of phenomena, we have found that common ground (cf., Buss, 1995). What relates them is a developmental perspective that is based on the principles of evolution. Evolutionary developmental psychology, we believe, serves as a metatheory for developmental psychology (Bjorklund, 1997a), bringing together under a single tent aspects of development as diverse as a fetus's response to teratogens, sensory development, spatial cognition, aggression, play, and theory of mind. This metatheory does not preclude the need for more domain-specific theories to account for the particular aspect of development under question. What it provides, we believe, is an overarching set of principles that can be applied to all aspects of development and, if shared by enough people in the field, can serve as the ground for communication among all developmental scientists.

What It Means to Be Human

Evolutionary developmental psychology also contributes significantly, we believe, to answering the central question of psychology, "What does it means to be human?" This question has intrigued humankind at least since the time of the ancient Greeks and likely further back to the cave painters of Lascaux and their hunter–gatherer brethren. There is no canonical approach to answering this "meaning-of-life" question, but we believe, as we suspect most of our readers do, that reason rather than revelation is the best course to follow. But what sort of data does one look for to reason about the nature of humankind? Obviously, one needs to examine carefully adult humans to get an idea of what a "finished" product looks like. However, human nature is mediated through culture, and so one needs to evaluate how people living in vastly different social circumstances understand and solve the problems that face all humans. Evolutionary psychology has done a good job, we believe, constructing clever experiments performed on college sophomores and using cross-cultural data (both original and archival) to generate a database for assessing this central question (Barkow et al., 1992; Buss, 1995).

But other sources of evidence exist that evolutionary psychologists have not necessarily ignored but have not fully embraced. First there is comparative psychology, particularly similarities and differences between *Homo sapiens* and our close genetic relatives the great apes. We have made much use of comparative data in this book, believing in the basic continuity of cognitive evolution, although recognizing that each species has evolved unique solutions to the problems it faced and thus being ever on the lookout for phylogenetic discontinuities. Second, and most important for us, is an examination of infants, children, and the process of development. Because of the behavioral and cognitive plasticity of our species, looking only at "completed" adults can give a distorted picture of original human nature. What "knowledge" are children born with (and where did that come from?), and how do primitive psychological biases get translated into mature forms of thought and behavior? We firmly believe that a developmental perspective provides the best venue for understanding human nature. A developmental perspective, of course, is not sufficient to answer the "big questions," but it is necessary. Moreover, from our perspective, human nature is not simply something one is born with. Rather, human nature develops, being expressed as evolved dispositions that interact with a species-typical environment over the course of ontogeny.

BASIC PRINCIPLES OF EVOLUTIONARY DEVELOPMENTAL PSYCHOLOGY

Evolutionary developmental psychology assumes that not only are the behaviors and cognitions that characterize adults the product of natural selection, but so are characteristics of children's behaviors and minds. Evolutionary developmental psychology is concerned not only with universals—patterns that characterize all members of a species—but also with how individuals adapt to their particular life circumstances. Below we present some of the basic principles of evolutionary developmental psychology as they have been articulated throughout this book.

Evolutionary Developmental Psychology Involves the Expression of Evolved, Epigenetic Programs

Although many psychologists are not aware of it, evolutionary psychologists do not argue for a form of genetic determinism but rather hold that evolved psychological mechanisms interact with the local environment to produce a particular pattern of behavior. However, as a rule, evolutionary psychologists concerned with adult functioning have seen no need to specify the nature of the organism–environment interaction. Evolutionary developmental psychology addresses this shortcoming by providing a well-defined model of the way in which nature and nurture are proposed to interact following the precepts of the developmental systems approach (Gottlieb, 2000; Oyama, 2000a). Developmental psychologists have come to rephrase the nature–nurture issue, asking not "how much" of any characteristic is due to nature or nurture but rather "How do nature and nurture interact to produce a particular pattern of development?" But simply restating the questions in this way advances the argument little. The developmental systems approach specifies how biological and environmental factors at multiple levels of organization transact to produce a particular pattern of ontogeny.

From this perspective, new morphological structures or behaviors do not simply arise as a result of the reading of the genetic blueprint but emerge as a result of the continuous and bidirectional transaction between all levels of biological and experiential factors, from the genetic through the cultural. The first important point here is that "everything develops." Nothing appears in the phenotype de novo, or fully formed; new structures and functions emanate from earlier ones in a probabilistic fashion. The second important point is that "experience" is broadly defined and encompasses events both exogenous and endogenous to the organism, including self-produced activity, such as the firing of a nerve cell. Functioning at one level influences functioning at adjacent levels, with constant feedback between levels. From the developmental systems perspective, there are no pure genetic effects and no pure environmental effects. Everything develops as a function of the bidirectional relationship between structure and functions, continuously across development, from conception until death.

What, then, is one to make of the evolved psychological mechanisms postulated by evolutionary psychologists? Such mechanisms can be thought of as genetically coded "messages" or "rules of thumb" (E. O. Wilson, 1998) that, following epigenetic rules, interact with the environment (again, broadly defined) over time to produce behavior. The unique experiences of every individual ensure substantial variability in developmental outcome; because the experiences of no two individuals are identical, no two phenotypes, of any species, will be exactly the same. However, the developmental systems approach proposes that individuals inherit not only a species-typical genome but also a species-typical environment. For humans, this includes a womb, a lactating mother, and a social group including both kin and non-kin. It is the similarity of the species' environment, as well as the similarities of a species' genome, that is responsible for the universal physical, behavioral, and psychological features that humans (or ducks or chimpanzees) share. To the extent that individuals grow up in environments similar to those of their ancestors, development should follow a species-typical pattern.

The concept of the human newborn as a tabula rasa is outdated and nearly universally rejected by developmental psychologists. What, then, are infants born with? The answer is epigenetic programs that have evolved over eons and are responsive to the general types of environments that our ancient ancestors experienced. It is these programs, in constant transaction with the environment (broadly defined), that produce the patterns of development and eventual adult behavior that define us as a species.

An Extended Childhood Is Needed in Which to Learn the Complexities of Human Social Communities

Many things make *Homo sapiens* distinct from other species, but one that is particularly pertinent to a developmental perspective is our extended juvenile period. Humans spend more time as "prereproductives" than any other mammal, although the timing of our ontogeny is merely an extension of a trend observed in primate phylogeny. There are clearly costs to the prolongation of youth. Although death rates decline sharply following infancy, many children in the prehistoric past surely died from infection, disease, and accidents before reaching sexual maturity. There must have been some powerful benefits associated with delayed development to outweigh the substantial costs. Those benefits can be seen in mastering the complexities of a human social community. A currently popular account of human cognitive evolution holds that the single most potent pressure on human intellectual evolution was the need to cooperate and compete with conspecifics. As hominid groups became more complex, individuals needed a better awareness of themselves and the thoughts, desires, and knowledge of conspecifics to better understand, and perhaps manipulate, others. Those who could better master their social world reaped more of the benefits in terms of quality mates and resources and passed those cognitive characteristics along to their offspring. But, despite the similarity of human groups across the globe, there is also much variability, requiring a flexible and plastic intelligence to master the vagaries of group living. This requires not only a large brain, but also a long time to accomplishment. It was the confluence of a large brain, social complexity, and an extended juvenile period that set the stage for the modern human mind.

Many Aspects of Childhood Serve as Preparations for Adulthood and Were Selected Over the Course of Evolution

It is perhaps a canon of developmental psychology that experiences in infancy and childhood prepare the individual for life as an adult. Evolutionary developmental psychology takes a similar view, and proposes that certain aspects of infancy and childhood were selected because of the opportunities they afforded children for learning the ways of their social group that would be useful in adulthood. Aspects of play are good candidates for preparatory mechanisms; children learn much about their physical environment and social world through play that will serve them well later in life.

Although there are many aspects of childhood that may fit this description, the "childhood-as-preparation-for-adulthood" hypothesis is probably best reflected in sex differences. Evolutionary psychologists dealing with adult functioning have often focused on sex differences, postulating that, because men and women have different self-interests, they have evolved different psychologies. This is reflected especially in sex differences with regard to mating, child-rearing, and intrasex competition. However, these behaviors, dispositions, and thought processes do not appear with the first burst of pubertal hormones or on hearing the cries of one's newborn child but have developmental histories, with children adapting their gender-specific behavior to the local norms, based on evolved predispositions.

We have documented repeatedly throughout this book examples of sex differences in childhood that appear to prepare us for the adult roles of men and women or, more appropriately, for the "traditional" sex roles that typified our ancestors in the environment of evolutionary adaptedness. Different styles of play serve as one example. Beginning during the preschool years, boys in all cultures (and males in many nonhuman mammalian species) display higher rates of rough-and-tumble play (R&T) than girls. One function proposed for R&T play is preparation for adult fighting and hunting in males, based on the similarity between such play and adult behaviors. Girls, on the other hand, engage in more play parenting (e.g., doll play) than boys, and their play is less apt to be centered on physically based dominance relationships, a difference that has been viewed as an evolved tendency related to females being the primary caretakers for their offspring (and that adult male relationships are more likely to be based on physical dominance). Sex differences during childhood in aggression, competition, risk taking, spatial cognition, and behavioral inhibition, among others, have similarly been hypothesized to reflect differential preparations for boys and girls for adult life.

From an evolutionary developmental perspective, such specieswide sex differences do not reflect a form of biological determinism. Rather, evolved epigenetic rules bias boys and girls toward different environments and experiences and, to the extent that one's culture supports those biases, will lead children in the "right" direction (i.e., a form of adult behavior that has, over many generations, been associated with reproductive success). But, as we have emphasized throughout this book, such evolved biases per se are not sufficient to produce a developmental pattern. Human behavior is highly flexible, and although some outcomes are more likely than others, all require environmental support to be realized.

Some Characteristics of Infants and Children Were Selected to Serve an Adaptive Function at Specific Times in Development and Not as Preparations for Adulthood

Although many aspects of infancy and childhood can be seen as preparations for adulthood, this is not the entire picture. Many features of infancy and childhood have been selected in evolution to serve an adaptive function at that time in development only and not to prepare the child for later life. In the previous section, we identified play as a means by which children learn information that will serve them well as adults—play as preparation for adulthood. Yet, we also believe that many aspects of play serve an immediate function. Play provides exercise and knowledge of one's current environment and serves as a safe venue for establishing a social hierarchy and learning "one's place" in the cohort. A prolonged childhood clearly affords children the time to learn and become prepared for life in a complicated, adult social group; however, providing children with experience that will aid them as reproductive adults will do no good unless they make it through infancy and childhood. Evolution, we propose, has endowed children (and the youth of other species) with many characteristics that adapt them well to their immediate environments rather than prepare them for a future one. In addition to play, we suggested several social–cognitive phenomena, such as neonatal imitation and overestimation of one's abilities, that may function primarily to adapt infants and children to life at a particular time in ontogeny rather than to prepare them for a qualitatively different life as an adult.

Many, But Not All, Evolved Psychological Mechanisms Are Domain-Specific in Nature

The basic tenet of evolutionary psychology is that what has evolved are domain-specific information processing programs, selected to deal with relatively specific types of recurring problems that our ancestors faced in the environment of evolutionary adaptedness. The mind is not a general-purpose device that can be applied equally well to a vast range of problems but consists of a set of independent, specialized modules. From the point of view of the developmentalist, infants enter the world "prepared" to process and learn some information more readily than others, and these preparations serve as the foundation for social and cognitive development across childhood and adolescence. Modules have been proposed for physical knowledge (object permanence), mathematics, language, and theory of mind, among many others.

Despite our basic agreement with mainstream evolutionary psychology about the importance of domain-specific mechanisms, we also believe that people possess domain-general abilities, that these abilities have been selected over phylogeny, and that they interact with domain-specific mechanisms over the course of ontogeny. We do not believe, of course, that a single, domain-general mechanism can account for all learning and cognition; however, we propose that there may be several domain-general mechanisms (working-memory capacity, speed of processing) that influence performance on many tasks and that such mechanisms experienced selective pressure over evolutionary time. This proposal seems to describe the extant data and must be considered in any fully articulated evolutionary psychological theory.

Evolved Mechanisms Are Not Always Adaptive for Contemporary People

Just because some social, behavioral, or cognitive tendency was adaptive for our prehistoric ancestors, does not mean that it is adaptive for modern humans. Similarly, just because some tendencies (e.g., violence among young adult males) are "natural" based on evolutionary examination, does not mean that they are morally "good" or inevitable. Concerning child development in modern cultures, formal schooling may represent the best example of the "evolved-mechanisms-are-not-always-currently-adaptive" principle. From the perspective of evolutionary developmental psychology, much of what we teach children in school is "natural" in that teaching involves tasks never encountered by our ancestors, and some "normal" individual differences in behaviors (such as those shown by some children with high levels of activity) may be especially maladaptive in contemporary environments.

THE EVOLUTION OF EVOLUTIONARY DEVELOPMENTAL PSYCHOLOGY

Developmental psychology has its roots in late-19th-century biological thinking, in which evolutionary theory played a major role (Cairns, 1998). Evolutionary theory disappeared from mainstream developmental explication throughout most of the 20th century, although it was kept alive in the thinking of mainly developmental psychobiologists whose species of choice were usually infrahuman mammals and birds (Mason, 1968a; Oppenheim, 1981; Spear, 1984). Evolutionary theory is back, and respectable members of the academic community can profess a belief in natural selection, evolved psychological mechanisms, and a commonality with the great apes and be taken seriously.

In some respects, developmental psychologists are relative latecomers to this bandwagon. This is due, in part, to the history and subject matter of our field. Because mainstream evolutionary psychologists talked about "Darwinian algorithms" that seemingly are expressed in the adult given the proper environmental context, there seemed little need for development. This was also true in the dominant account of biological evolution, the modern synthesis, which essentially made no room for ontogeny in influencing phylogeny. But developmentalists produced their own theories, some of which, such as those of the neonativists, behavioral geneticists, and people looking at the role of ontogeny in evolution, included, implicitly or explicitly, assumptions about evolution. These theorists did not always view the processes of evolution or ontogenesis similarly, however, making any unified developmental theory unlikely.

We believe that the zeitgeist has changed, and we are pleased to be part of growing group of developmental psychologists who see the possibility of an evolution-based theory of ontogeny that will encompass all who think seriously about development. Although it is possible that the current emphasis on evolutionary accounts of development will fade, we doubt it. It will surely change in character over time as more researchers take evolution seriously in formulating their experiments and theories; but we believe that the theory that is the basis of modern biology should also be the basis of modern psychology and developmental psychology, and that it will not quickly be relegated to history.

REFERENCES

Abravanel, E., & Gingold, H. (1985). Learning via observation during the second year of life. *Developmental Psychology, 21,* 614–623.

Adler, A. (1927). *Understanding human nature.* New York: Greenberg.

Agrawal, A. A., Laforsch, C., & Tollrian, R. (1999). Transgenerational induction of defences in animals and plants. *Nature, 401,* 60–63.

Ahmed, A., & Ruffman, T. (1998). Why do infants make A not B errors in a search task, yet show memory for the location of hidden objects in a nonsearch task? *Developmental Psychology, 34,* 441–453.

Ainsworth, M. D. S., Blehar, M. C., Waters, E., & Wall, S. (1978). *Patterns of attachment: A psychological study of the strange situation.* Hillsdale, NJ: Erlbaum.

Ainsworth, M. D. S., & Wittig, D. S. (1969). Attachment and exploratory behavior of one-year-olds in a strange situation. In B. M. Foss (Ed.), *Determinants of infant behavior* (Vol. 4, pp. 113–136). London: Methuen.

Akazawa, T., Muhesen, S., Dodo, Y., Kondo, O., & Mizouguchi, Y. (1995). Neanderthal infant burial. *Nature, 377,* 585–586.

Alexander, R. D. (1974). The evolution of social behavior. *Annual Review of Ecology and Systematics, 5,* 325–384.

Alexander, R. D. (1989). Evolution of the human psyche. In P. Mellers & C. Stringer (Eds.), *The human revolution: Behavioural and biological perspectives on the origins of modern humans* (pp. 455–513). Princeton, NJ: Princeton University Press.

Allen, C., & Bekoff, M. (1995). Function, natural design, and animal behavior: Philosophical and ethological considerations. *Perspectives in Ethology, 11,* 1–46.

Allman, J. M. (1999). *Evolving brains.* New York: Scientific American Library.

Allman, J., & Hasenstaub, A. (1999). Brains, maturation times, and parenting. *Neurobiology of Aging, 20,* 447–454.

Allman, J., Rosin, A., Kumar, R., & Hasenstaub, A. (1998). Parenting and survival in anthropoid primates: Caretakers live longer. *Proceedings of the National Academy of Sciences USA, 95,* 6866–6869.

Als, H. (1995). The preterm infant: A model for the study of fetal brain expectation. In J-P. Lecanuet, W. P. Fifer, N. A. Krasnegor, & W. P. Smotherman (Eds.),

Fetal development: A psychobiological perspective (pp. 439–471). Hillsdale, NJ: Erlbaum.

Anderson, K. G., Kaplan, H., & Lancaster J. (1999a). Paternal care by genetic fathers and stepfathers I: Reports from Albuquerque men. *Evolution and Human Behavior, 20,* 405–431.

Anderson, K. G., Kaplan, H., Lam, D., & Lancaster J. (1999b). Paternal care by genetic fathers and stepfathers II: Reports by Xhosa high school students. *Evolution and Human Behavior, 20,* 433–451.

Antell, S. E., & Keating, D. P. (1983). Perception of numerical invariance in neonates. *Child Development, 54,* 695–701.

Antinucci, F. (Ed.). (1989). *Cognitive structure and development in nonhuman primates.* Hillsdale, NJ: Erlbaum.

Apanius, V., Penn, D., Slev, P. R., Ruff, L. R., & Potts, W. (1997). The nature of selection on the major histocompatibility complex. *Critical Reviews in Immunology, 17,* 179–224.

Archer, J. (1988). *The behavioural biology of aggression.* London: Cambridge University Press.

Archer, J. (1992). *Ethology and human development.* Hemel Hempstead, UK: Wheatsheaf.

Arterberry, M., Yonas, A., & Bensen, A. S. (1989). Self-produced locomotion and the development of responsiveness to linear perspective and texture gradients. *Developmental Psychology, 25,* 976–982.

Asfaw, B., White, T., Lovejoy, O., Latimer, B., Simpson, S., & Suwa, G. (1999). *Australopithecus garhi:* A new species of early hominid from Ethiopia. *Science, 284,* 629–635.

Astington, J. W., & Jenkins, J. M. (1995). Theory of mind development and social understanding. *Cognition and Emotion, 9,* 151–165.

Austad, S. N. (1997). *Why we age: What science is discovering about the body's journey through life.* New York: Wiley.

Avis, J., & Harris, P. L. (1991). Belief-desire reasoning among Baka children: Evidence for a universal conception of mind. *Child Development, 62,* 460–467.

Axelrod, R., & Hamilton. W. D. (1981). The evolution of cooperation. *Science, 211,* 1390–1396.

Azmitia, M. (1992). Expertise, private speech, and the development of self-regulation. In R. M. Diaz & L. E. Berk (Eds.), *Private speech: From social interaction to self-regulation* (pp. 101–122). Hillsdale, NJ: Erlbaum.

Baenninger, M., & Newcombe, N. (1995). Environmental input to the development of sex-related differences in spatial and mathematical ability. *Learning and Individual Differences, 7,* 363–379.

Baghurst, P. A., McMichael, A. J., Wigg, N. R., Vimpani, G. V., Robertson, E. F., Roberts, R. J., & Tong, S-L. (1992). Environmental exposure to lead and children's intelligence at the age of seven years. *New England Journal of Medicine, 327,* 1279–1284.

Bahrick, L. E., & Pickens, J. N. (1995). Infant memory for object motion across a period of three months: Implications for a four-phase attention function. *Journal of Experimental Child Psychology, 59,* 343–371.

Bai, D. L., & Bertenthal, B. I. (1992). Locomotor status and the development of spatial skills. *Child Development, 63,* 215–226.

Baillargeon, R. (1987). Object permanence in 3 1/2- and 4 1/2-month-old infants. *Developmental Psychology, 23,* 655–664.

Baillargeon, R. (1994). How do infants learn about the physical world? *Current Directions in Psychological Science, 3,* 133–140.

Baillargeon, R., Kotovsky, L., & Needham, A. (1995). The acquisition of physical knowledge in infancy. In G. Lewis, D. Premack, & D. Sperber (Eds.), *Casual understandings in cognition and culture* (pp. 79–116). Oxford, England: Oxford University Press.

Bakeman, R., Adamson, L. B., Konner, M., & Barr, R. G. (1990). !Kung infancy: The social context of object exploration. *Child Development, 61,* 794–809.

Bakeman, R., & Brownlee, J. (1980). The strategic use of parallel play: A sequential analysis. *Child Development, 51,* 873–875.

Baldwin, J. M. (1902). *Development and evolution.* New York: McMillan.

Balter, M. (2001). Scientists spar over claims of earliest human ancestor. *Science, 291,* 1460–1461.

Baltes, P. B. (1997). On the incomplete architecture of human ontogeny: Selection, optimization, and compensation as foundation of developmental theory. *American Psychologist, 52,* 366–380.

Bandura, A. (1986). *Social foundations of thought and action: A social cognitive theory.* Englewood Cliffs, NJ: Prentice-Hall.

Bandura, A. (1989). Social cognitive theory. In R. Vasta (Ed.), *Annals of child development* (pp. 1–60). Greenwich, CT: JAI.

Bandura, A., & Walters, R. H. (1963). *Social learning theory and personality development.* New York: Holt, Rinehart, & Winston.

Barber, N. (1991). Play and energy regulation in mammals. *Quarterly Review of Biology, 66,* 129–147.

Bard, K. A., Fragaszy, D., & Visalberghi, E. (1995). Acquisition and comprehension of a tool-using behavior by young chimpanzees (*Pan troglodytes*): Effects of age and modeling. *International Journal of Comparative Psychology, 8,* 47–68.

Barker, R. G. (1968). *Ecological psychology: Concepts and methods for studying the environment of human behavior.* Stanford, CA: Stanford University Press.

Barkow, J. H., Cosmides, L., & Tooby, J. (Eds.). (1992). *The adapted mind: Evolutionary psychology and the generation of culture.* New York: Oxford University Press.

Barnes, K. E (1971). Preschool play norms: A replication. *Developmental Psychology, 5,* 99–103.

Baron-Cohen, S. (1989). The autistic child's theory of mind: A case of specific developmental delay. *Journal of Child Psychology and Psychiatry, 30,* 285–298.

Baron-Cohen, S. (1995). *Mindblindness: An essay on autism and theory of mind.* Cambridge, MA: MIT Press.

Baron-Cohen, S., Leslie, A., & Frith, U. (1985). Does the autistic child have a "theory of mind"? *Cognition, 21,* 37–46.

Baron-Cohen, S., Leslie, A., & Frith, U. (1986). Mechanical, behavioral, and intentional understanding of pictures and tools in autistic children. *British Journal of Developmental Psychology, 4,* 113–125.

Baron-Cohen, S., Wheelwright, S., Stone, V., & Rutherford, M. (1999). A mathematician, a physicist and a computer scientist with Asperger syndrome: Performance on folk psychology and folk physics tests. *Neurocase, 5,* 475–483.

Barton, R. A., & Harvey, P. H. (2000). Mosaic evolution of brain structure in mammals. *Nature, 405,* 1055–1058.

Bates, E. (1999). On the nature of language. In R. Levi-Montalcini, D. Baltimore, R. Dulbecco, & F. Jacob (Series Eds.) & E. Bizzi, P. Calissano, & V. Volterra (Vol. Eds.), *Frontiere della biologia [Frontiers of biology]: The brain of Homo sapiens* (pp. 241–265). Rome: Giovanni Trecanni.

Bates, E., Carlson-Luden, V., & Bretherton, I. (1980). Perceptual aspects of tool use in infancy. *Infant Behavior and Development, 3,* 127–140.

Bates, E., Dale, P. S., & Thal, D. (1994). Individual differences and their implications for theories of language development. In P. Fletcher & B. MacWhinney (Eds.), *Handbook of child language* (pp. 96–151). Oxford, England: Blackwell.

Bates, E., & Goodman, J. C. (1997). On the inseparability of grammar and the lexicon: Evidence from acquisition, aphasia and real-time processing. *Language and Cognitive Processes, 12,* 507–586.

Bateson, P. P. G. (1976). Rules and reciprocity in behavioural development. In P. P. G. Bateson & R. A. Hinde (Eds.), *Growing points in ethology* (pp. 401–421). Cambridge, England: Cambridge University Press.

Bateson, P. P. G. (1981a). Control of sensitivity to the environment during development. In K. Immelman, G. Barlow, L. Petrinovich, & M. Main, (Eds.), *Behavioral Development* (pp. 432–455). London: Cambridge University Press.

Bateson, P. P. G. (1981b). Discontinuities in the development and changes in the organization of play in cats. In K. Immelsmann, G. W. Barlow, L. Petrinovich, & M. Main (Eds.), *Behavioral development* (pp. 281–295). New York: Cambridge University Press.

Bateson, P. P. G. (1988). The active role of behavior in evolution. In M.-W. Ho & S. Fox (Eds.), *Process and metaphors in evolution* (pp. 191–207). Wiley, Chichester.

Bauer, P. J. (1993). Memory for gender-consistent and gender-inconsistent event sequences by twenty-five-month-old children. *Child Development, 64,* 285–297.

Bauer, P. J. (1997). Development of memory in early childhood. In N. Cowan (Ed.), *The development of memory in childhood* (pp. 83–111). Hove East Essex, England: Psychology Press.

Bauer, P. J., Wenner, J. A., Dropik, P. L., & Wewerka, S. S. (2000). Parameters of remembering and forgetting in the transition from infancy to early childhood. *Monographs of the Society for Research in Child Development*, 65(4, Serial No. 263).

Bauer, P. J., & Wewerka, S. S. (1995). One- and two-year-olds recall events: Factors facilitating immediate and long-term memory in 13.5 and 16.5-month-old children. *Child Development*, 64, 1204–1223.

Baumrind, D. (1967). Child care practices anteceding three patterns of preschool behavior. *Genetic Psychology Monographs*, 75, 43–88.

Baumrind, D. (1972). An exploratory study of socialization effects on black children: Some black–white comparisons. *Child Development*, 43, 261–267.

Baumrind, D. (1973). The development of instrumental competence through socialization. In A. D. Pick (Ed.), *Minnesota symposium on child psychology* (Vol. 7, pp. 3–46). Minneapolis: University of Minnesota Press.

Baumrind, D. (1993). The average expectable environment is not good enough: A response to Scarr. *Child Development*, 64, 1299–1317.

Bayley, N. (1949). Consistency and variability in the growth of intelligence from birth to eighteen years. *Journal of Genetic Psychology*, 75, 165–196.

Bekoff, M. (1977). Social communication in canids: Evidence for the evolution of a stereotyped mammalian behavior. *Science*, 197, 1097–1099.

Bekoff, M. (1997). Playing with play: What can we learn about cognition, negotiation, and evolution. In D. Cummins & C. Allen (Eds.), *The evolution of mind* (pp. 162–182). New York: Oxford University Press.

Bekoff, M., & Byers, J. A. (1981). A critical re-analysis of the ontogeny and phylogeny of mammalian social and locomotor play. In K. Immelmann, G. Barlow, L. Petronovich, & M. Main (Eds.), *Behavioural development* (pp. 296–337). Cambridge, England: Cambridge University Press.

Bekoff, M., & Byers, J. A. (Eds.). (1998). *Animal play: Evolutionary, comparative, and ecological perspectives*. New York: Cambridge University Press.

Bellis, M. A., & Baker, R. R. (1990). Do females promote sperm competition? Data for humans. *Animal Behaviour*, 40, 997–999.

Belsky, J. (1997). Attachment, mating, and parenting: An evolutionary interpretation. *Human Nature*, 8, 361–381.

Belsky, J. (2000). Conditional and alternative reproductive strategies: Individual differences in susceptibility to rearing experience. In J. Rodgers & D. Rowe (Eds.), *Genetic influences on fertility and sexuality* (pp. 127–146). Boston: Kluwer.

Belsky, J., & Most, R. (1981). From exploration to play. *Developmental Psychology*, 17, 630–639.

Belsky, J., Steinberg, L., & Draper, P. (1991). Childhood experience, interpersonal development, and reproductive strategy: An evolutionary theory of socialization. *Child Development*, 62, 647–670.

Berenbaum, S. A., & Hines, M. (1992). Early androgens are related to childhood sex-typed toy preferences. *Psychological Science*, 3, 203–206.

Berenbaum, S. A., & Snyder, E. (1995). Early hormonal influences on childhood sex-typed activity and playmate preferences: Implications for the development of sexual orientation. *Developmental Psychology, 31*, 31–42.

Bering, J. M. (2001). Theistic percepts in other species: Can chimpanzees represent the minds of non-natural agents? *Journal of Cognition and Culture, 1*, 107–137.

Bering, J. M., Bjorklund, D. F., & Ragan, P. (2000). Deferred imitation of object-related actions in human-reared juvenile chimpanzees and orangutans. *Developmental Psychobiology, 36*, 218–232.

Bertenthal, B. I., Campos, J., & Barrett, L. (1984). Self-produced locomotion: An organizer of emotional, cognitive, and social development in infancy. In R. Emde & R. Harmon (Eds.), *Continuities and discontinuities in development* (pp. 175–209). New York: Plenum.

Bertenthal, B. I., Campos, J. J., & Kermoian, R. (1994). An epigenetic perspective on the development of self-produced locomotion and its consequences. *Current Directions in Psychological Science, 3*, 140–145.

Berthier, N. E., DeBois, S., Poirier, C. R., Novak, M. A., & Clifton, R. K. (2000). Where's the ball? Two- and three-year-olds reason about unseen events. *Developmental Psychology, 36*, 384–401.

Bevc, I., & Silverman, I. (1993). Early proximity and intimacy between siblings and incestuous behavior: A test of the Westermark theory. *Ethology and Sociobiology, 14*, 171–181.

Bevc, I., & Silverman, I. (2000). Early separation and sibling incest: A test of the revised Westermark theory. *Evolution and Human Behavior, 21*, 151–161.

Biben, M. (1979). Predation and predatory play behaviour of domestic cats. *Animal Behaviour, 27*, 81–94.

Biben, M. (1998). Squirrel monkey play fighting: Making a case for a cognitive training function for play. In M. Bekoff & J. A. Byers (Eds.), *Animal play* (pp. 161–182). New York: Cambridge University Press.

Bickerton, D. (1990). *Language and species*. Chicago: University of Chicago Press.

Bigler, R. S., Jones, L. C., & Lobliner, D. B. (1997). Social categorization and the formation of intergroup attitudes in children. *Child Development, 68*, 530–543.

Bjorklund, D. F. (1987). A note on neonatal imitation. *Developmental Review, 7*, 86–92.

Bjorklund, D. F. (1997a). In search of a metatheory for cognitive development (or, Piaget's dead and I don't feel so good myself). *Child Development, 68*, 142–146.

Bjorklund, D. F. (1997b). The role of immaturity in human development. *Psychological Bulletin, 122*, 153–169.

Bjorklund, D. F. (2000). *Children's thinking: Developmental function and individual differences* (3rd ed.). Belmont, CA: Wadsworth.

Bjorklund, D. F., Bering, J., & Ragan, P. (2000). A two-year longitudinal study of deferred imitation of object manipulation in an enculturated juvenile chimpanzee (*Pan troglodytes*) and orangutan (*Pongo pygmaeus*). *Developmental Psychobiology, 37*, 229–237.

Bjorklund, D. F., & Brown, R. D. (1998). Physical play and cognitive development: Integrating activity, cognition, and education. *Child Development, 69,* 604–606.

Bjorklund, D. F., Gaultney, J. F., & Green, B. L. (1993). "I watch therefore I can do": The development of meta-imitation over the preschool years and the advantage of optimism in one's imitative skills. In R. Pasnak & M. L. Howe (Eds.), *Emerging themes in cognitive development: Vol. II. Competencies* (pp. 79–102). New York: Springer-Verlag.

Bjorklund, D. F., & Green, B. L. (1992). The adaptive nature of cognitive immaturity. *American Psychologist, 47,* 46–54.

Bjorklund, D. F., & Harnishfeger, K. K. (1990). Children's strategies: Their definition and origins. In D. F. Bjorklund (Ed.), *Children's strategies: Contemporary views of cognitive development* (pp. 309–323). Hillsdale, NJ: Erlbaum.

Bjorklund, D. F., & Harnishfeger, K. K. (1995). The role of inhibition mechanisms in the evolution of human cognition and behavior. In F. N. Dempster & C. J. Brainerd (Eds.), *New perspectives on interference and inhibition in cognition* (pp. 141–173). New York: Academic Press.

Bjorklund, D. F., & Kipp, K. (1996). Parental investment theory and gender differences in the evolution of inhibition mechanisms. *Psychological Bulletin, 120,* 163–188.

Bjorklund, D. F., & Kipp, K. (2001). Social cognition, inhibition, and theory of mind: The evolution of human intelligence. In R. J. Sternberg & J. C. Kaufman (Eds.), *The evolution of intelligence* (pp. 27–53). Mahwah, NJ: Erlbaum.

Bjorklund, D. F., & Pellegrini, A. D. (2000). Child development and evolutionary psychology. *Child Development, 71,* 1687–1798.

Bjorklund, D. F., & Pellegrini, A. D. (2002). Evolutionary perspectives on social development. In P. K. Smith & C. Hart, *Handbook of social development.* London: Blackwell.

Bjorklund, D. F., & Schwartz, R. (1996). The adaptive nature of developmental immaturity: Implications for language acquisition and language disabilities. In M. Smith & J. Damico (Eds.), *Childhood language disorders* (pp. 17–40). New York: Thieme Medical.

Bjorklund, D. F., & Shackelford, T. K. (1999). Differences in parental investment contribute to important differences between men and women. *Current Directions in Psychological Science, 8,* 86–89.

Bjorklund, D. F., Yunger, J. L., Bering, J. M., & Ragan, P. (in press). The generalization of deferred imitation in enculturated chimpanzees (*Pan troglodytes*). *Animal Cognition.*

Bjorklund, D. F., Yunger, J. L., & Pellegrini, A. D. (2002). The evolution of parenting and evolutionary approaches to childrearing. In M. Bornstein (Ed.), *Handbook of parenting* (2nd ed.; pp. 3–30). Mahwah, NJ: Erlbaum.

Blasi, H. C., & Bjorklund, D. F. (2001). El desarrollo de la memoria: Avances significativos y nuevos desafíos. (Memory development: Accomplishments of the past and directions for the future). *Infancia y Aprendizaje, 24,* 233–254.

Bloch, M. N. (1989). Young boys' and girls' play at home and in the community: A cultural ecological framework. In M. Bloch & A. Pellegrini (Eds.), *The ecological context of children's play* (pp. 120–154). Norwood, NJ: Ablex.

Bloom, L., Lightbown, P., & Hood, L. (1975). Structure and variation in child language. *Monographs of the Society for Research in Child Development, 40*(2, Serial No. 160).

Blurton Jones, N. (1993). The lives of hunter–gatherer children: Effects of parental behavior and parental reproductive strategy. In M. F. Pereira & L. A. Fairbanks (Eds.), *Juvenile primates: Life history, development, and behaviors* (pp. 309–326). New York: Oxford University Press.

Blurton Jones, N. G., Hawkes, K., & O'Connell, J. F. (1997). Why do Hadza children forage? In N. L. Segal, G. E. Weisfeld, & C. C. Weisfeld (Eds.), *Uniting psychology and biology: Integrative perspectives on human development* (pp. 279–313). Washington, DC: American Psychological Association.

Blurton Jones, N., & Konner, M. (1973). Sex differences in behaviours of London and Bushman children. In R. Michaels & J. Crook (Eds.), *Comparative ecology and the behaviours of primates* (pp. 690–750). London: Academic.

Boesch, C. (1991). Teaching among wild chimpanzees. *Animal Behaviour, 41*, 530–532.

Boesch, C. (1993). Toward a new image of culture in chimpanzees. *Behavioral and Brain Sciences, 16*, 514–515.

Boesch, C., & Shinskey, J. L. (1998). On perception of a partially occluded object in 6-month olds. *Cognitive Development, 13*, 141–163.

Boesch, C., & Tomasello, M. (1998). Chimpanzee and human cultures. *Current Anthropology, 39*, 591–604.

Bogartz, R. S., Shinskey, J. L., & Speaker, C. (1997). Interpreting infant looking: The event set x event set design. *Developmental Psychology, 33*, 408–422.

Bogin, B. (1997). Evolutionary hypotheses for human childhood. *Yearbook of Physical Anthropology, 40*, 63–89.

Bogin, B. (1999). *Patterns of human growth* (2nd ed.). Cambridge, England: Cambridge University Press.

Bolk, L. (1926). On the problem of anthropogenesis. Proc. Section Sciences Kon. Akad. Wetens. *Amsterdam, 29*, 465–475.

Bonner, J. T. (1980). *The evolution of culture in animals*. Princeton University Press: Princeton, NJ.

Bonner, J. T. (1988). *The evolution of complexity by means of natural selection*. Princeton University Press: Princeton, NJ.

Bornstein, M. H. (Ed.) (1995). *Handbook of parenting*. Mahwah, NJ: Erlbaum.

Bornstein, M. H., Ferdinandsen, K., & Gross, C. G. (1981). Perception of symmetry in infancy. *Developmental Psychology, 17*, 82–86.

Bornstein, M. H., Haynes, O. M., O'Reilly, A. W., & Painter, K. M. (1996). Solitary and collaborative pretense play in early childhood: Sources of individual variation in the development of representational competence. *Child Development,* *67,* 2910–2929.

Bornstein, M., Haynes, O. M., Pascual, L., Painter, K. M., & Galperin, C. (1999). Play in two societies. *Child Development, 70,* 317–331.

Bouchard, T. J., Jr., Lykken, D. T., McGue, M., Segal, N. L., & Tellegen, A. (1990). Sources of human psychological differences: The Minnesota study of twins reared apart. *Science, 250,* 223–228.

Bouchard, T. J., Jr., & McGue, M. (1981). Familial studies of intelligence: A review. *Science, 212,* 1055–1059.

Boulton, M. J. (1992). Participation in playground activities in middle school. *Educational Research, 34,* 167–182.

Boulton, M. J., & Underwood, K. (1992). Bully/victim problems among middle school children. *British Journal of Educational Psychology, 62,* 73–87.

Bowlby, J. (1969). *Attachment and loss: Vol. 1. Attachment.* London: Hogarth.

Boyer, P. (1998). Cognitive tracks of cultural inheritance: How evolved intuitive ontology governs cultural transmission. *American Anthropologist, 100,* 876–889.

Boysen, S. T. (1993). Counting in chimpanzees: Nonhuman principles and emergent properties of number. In S. T. Boysen & E. J. Capaldi (Eds.), *The development of numerical competence: Animal and human models* (pp. 39–59). Hillsdale, NJ: Erlbaum.

Boysen, S. T., & Berntson, G. G. (1989). Numerical competence in a chimpanzee (*Pan troglodytes*). *Journal of Comparative Psychology, 103,* 23–31.

Brainerd, C. J. (1978). *Piaget's theory of intelligence.* Englewood Cliffs, NJ: Prentice Hall.

Brédart, S., & French, R. M. (1999). Do babies resemble their fathers more than their mothers? A failure to replicate Christenfeld and Hill (1995). *Evolution and Human Behavior, 20,* 129–135.

Bredekamp, S., & Copple, C. (Eds.). (1997). *Developmentally appropriate practice in early childhood programs* (Rev. ed.). Washington, DC: National Association for the Education of Young Children.

Breuggeman, J. A. (1978). The function of adult play in free-ranging *Macca mulatta.* In E. O. Smith (Ed.), *Social play in primates* (pp. 169–191). New York: Academic Press.

Briars, D., & Siegler, R. S. (1984). A featural analysis of preschoolers' counting knowledge. *Developmental Psychology, 20,* 607–618.

Brigham, C. C. (1923). *A study of American intelligence.* Princeton, NJ: Princeton University Press.

Bronfenbrenner, U. (1979). *The ecology of human development.* Cambridge, MA: Harvard University Press.

Bronfenbrenner, U., & Ceci, S. J. (1994). Nature–nurture reconceptualized in developmental perspective: A bioecological model. *Psychological Review, 101,* 568–586.

Brooks, J., & Lewis, M. (1976). Infants' responses to strangers: Midget, adult, and child. *Child Development, 47*, 323–332.

Brooks-Gunn, J., & Furstenberg, F. F. (1989). Adolescent sexual behavior. *American Psychologist, 44*, 249–257.

Brown, B. B., Clasen, D. R., & Eicher, S. A. (1986). Perceptions of peer pressure, peer conformity dispositions, and self-reported behavior among adolescents. *Developmental Psychology, 22*, 521–530.

Brown, J., & Eklund, A. (1994). Kin recognition and the major histocompatibility complex: An integrative review. *American Naturalist, 143*, 435–461.

Brown, R. D., & Bjorklund, D. F. (1998). The biologizing of cognition, development, and education: Approach with cautious enthusiasm. *Educational Psychology Review, 10*, 355–373.

Bruner, J. S. (1972). The nature and uses of immaturity. *American Psychologist, 27*, 687–708.

Bruner, J. S. (1983). *Child's talk: Learning to use language.* New York: Norton.

Bruner, J. S. (1986). *Actual minds, possible worlds.* Cambridge, MA: Harvard University Press.

Bruner, J. S., Olver, R. R., & Greenfield, P. M. (Eds.). (1966). *Studies in cognitive growth.* New York: Wiley.

Bugental, D. B. (2000). Acquisition of the algorithms of social life: A domain-based approach. *Psychological Bulletin, 126*, 187–219.

Bugos, P. E., & McCarthy, L. M. (1984). Ayoreo infanticide: A case study. In G. Hausfater & S. B. Hrdy (Eds.), *Infanticide: Comparative and evolutionary perspectives* (pp. 503–520). New York: Aldine de Gruyter.

Burch, R. L., & Gallup, G. G., Jr. (2000). Perceptions of paternal resemblance predict family violence. *Evolution and Human Behavior, 21*, 429–435.

Burgess, R. L., & Molenaar, P. C. M. (1995). Commentary to Gottlieb's "Some Conceptual Deficiencies in 'Developmental' Behavior Genetics." *Human Development, 38*, 159–164.

Burghardt, G. (1984). On the origins of play. In P. K. Smith (Ed.), *Play in animals and humans* (pp. 5–42). Oxford, England: Blackwell.

Burghardt, G. M. (1988). Precocity, play, and the ectotherm–endotherm transition. In E. Blass (Ed.), *Handbook of behavioral neurobiology: Vol. 9.* (pp. 107–148). New York: Plenum.

Burghardt, G. M. (1998). Play. In G. Greenberg & M. M. Haraway (Eds.), *Comparative psychology: A handbook* (pp. 725–733). New York: Garland.

Burghardt, G. M. (1999). Conceptions of play and the evolution of animal minds. *Evolution and Cognition, 5*, 115–123.

Burghardt, G. M. (2001). *The genesis of animal play.* Cambridge: MIT Press.

Burns, S., & Brainerd, C. (1979). Effects of constructive and dramatic play on perspective taking in very young children. *Developmental Psychology, 15*, 512–521.

Bushnell, I. W. R., Sai, F., & Mullin, J. T. (1989). Neonatal recognition of the mother's face. *British Journal of Developmental Psychology, 7*, 3–15.

Buss, D. M. (1989). Sex differences in human mate preferences: Evolutionary hypotheses tested in 37 cultures. *Behavioral and Brain Sciences, 12*, 1–49.

Buss, D. M. (1995). Evolutionary psychology. *Psychological Inquiry, 6*, 1–30.

Buss, D. M. (1999). *Evolutionary psychology: The new science of the mind.* Boston: Allyn & Bacon.

Buss, D. M., Haselton, M. G., Shalelford, T. K., Bleske, A. L., & Wakefield, J. C. (1998). Adaptations, exaptations, and spandrels. *American Psychologist, 53*, 533–548.

Buss, D. M., Larsen, R. J., Westen, D., & Semmelroth, J. (1992). Sex differences in jealousy: Evolution, physiology, and psychology. *Psychological Science, 3*, 251–255.

Buss, D. M., & Schmidt, D. P. (1993). Sexual strategies theory: An evolutionary perspective on human mating. *Psychological Review, 100*, 204–232.

Byers, J. A., & Walker, C. (1995). Refining the motor training hypothesis for the evolution of play. *American Naturalist, 146*, 25–40.

Byrne, R. (1995). *The thinking ape: Evolutionary origins of intelligence.* Oxford, England: Oxford University Press.

Byrne, R., & Whiten, A. (Eds.). (1988). *Machiavellian intelligence: Social expertise and the evolution of intellect in monkeys, apes, and humans.* Oxford, England: Clarendon.

Byrnes, J. P., Miller, D. C., & Schafer, W. D. (1999). Gender differences in risk taking: A meta-analysis. *Psychological Bulletin, 125*, 367–383.

Cabrera, N. J., Tamis-LeMonda, C. S., Bradley, R. H., Hofferth, S., & Lamb, M. E. (2000). Fatherhood in the twenty-first century. *Child Development, 71*, 127–136.

Cairns, R. B. (1976). The ontogeny and phylogeny of social interactions. In M. E. Hahn & E. C. Simmel (Eds.), *Communicative behavior and evolution* (pp. 115–139). New York: Academic Press.

Cairns, R. B. (1979). *Social development: The origins and plasticity of interchanges.* San Francisco: Freeman.

Cairns, R. B. (1983). The emergence of developmental psychology. In P. H. Mussen (Gen. Ed.) & W. Kessen (Vol. Ed.), *Handbook of child psychology: Vol. 1. History, theory, and methods* (4th ed., pp. 41–102). New York: Wiley.

Cairns, R. B. (1998). The making of developmental psychology. In W. Damon (Gen. Ed.) & R. M. Lerner (Ed.), *Handbook of child psychology: Vol. 1. Theoretical models of human development* (5th ed., pp. 25–105). New York: Wiley.

Cairns, R. B., & Cairns, B. D. (1994). *Lifelines and risks: Pathways of youth in our time.* New York: Cambridge University Press.

Cairns, R. B., Gariepy, J-L., & Hood, K. E. (1990). Development, microevolution, and social behavior. *Psychological Review, 97*, 49–65.

Cairns, R. B., MacCombie, D. J., & Hood, K. E. (1983). A developmental-genetic analysis of aggressive behavior in mice: I. Behavioral outcomes. *Journal of Comparative Psychology, 97,* 69–89.

Cairns, R. B, Nemhauser, J. B., & Neiman, J. T. (1991). Aggressive and violent injuries among children and young adults. *Annals of Emergency Medicine, 20,* 951.

Caldera, Y. M., O'Brien, M., Truglio, R. T., Alvarez, M., & Huston, A. C. (1999). Children's play preferences, construction play with blocks, and visual-spatial skills: Are they related? *International Journal of Behavioral Development, 23,* 855–872.

Call, J., & Tomasello, M. (1994). Production and comprehension of referential pointing by orangutans (*Pongo pygmaeus*). *Journal of Comparative Psychology, 108,* 315–329.

Call, J., & Tomasello, M. (1996). The effects of humans on the cognitive development of apes. In A. E. Russon, K. A. Bard, & S. T. Parker (Eds.), *Reaching into thought: The minds of the great apes* (pp. 371–403). New York: Cambridge University Press.

Call, J., & Tomasello, M. (1999). A nonverbal false belief task: The performance of children and great apes. *Child Development, 70,* 381–395.

Campbell, A. (1999). Staying alive: Evolution, culture, and women's intrasexual aggression. *Behavioral and Brain Sciences, 22,* 203–252.

Canfield, R. L., Smith, E. G., Brezsnyak, M. P., & Snow, K. L. (1997). Information processing through the first year of life. *Monographs of the Society for Research in Child Development, 62*(4, Serial No. 250).

Caretta, C. M., Caretta, A., & Cavaggioni, A. (1995). Pheromonally accelerated puberty is enhanced by previous experience of the same stimulus. *Physiology and Behavior, 57,* 901–903.

Caro, T. M. (1995). Short-term costs and correlates of play in cheetahs. *Animal Behaviour, 49,* 333–345.

Carpenter, M., Akhtar, N., & Tomasello, M. (1998). 14- through 18-month-old infants differentially imitate intentional and accidental actions. *Infant Behavior and Development, 21,* 315–330.

Carpenter, M., Nagell, K., & Tomasello, M. (1998). Social cognition, joint attention, and communicative competence from 9 to 15 months of age. *Monographs of the Society for Research in Child Development, 63*(4, Serial No. 255).

Carson, J., Burks, V., & Parke, R. D. (1993). Parent–child physical play: Determinants and consequences. In K. MacDonald (Ed.). *Parent–child play* (pp. 197–220). Albany: State University of New York Press.

Carver, L. J., & Bauer, P. J. (1999). When the event is more than the sum of its parts: Individual differences in 9-month-olds long-term ordered recall. *Memory, 7,* 147–174.

Case, A., Lin, I-F., & McLanahan, S. (in press). How hungry is the selfish gene? *Economic Journal.*

Case, R. (1992). *The mind's staircase: Exploring the conceptual underpinnings of children's thought and knowledge*. Hillsdale, NJ: Erlbaum.

Case, R., Kurland, M., & Goldberg, J. (1982). Operational efficiency and the growth of short-term memory span. *Journal of Experimental Child Psychology, 33*, 386–404.

Casey, M. B. (1996). Understanding individual differences in spatial ability within females: A nature/nurture interactionist framework. *Developmental Review, 16*, 241–260.

Casey, M. B., Nuttall, R. L., & Peraris, E. (1999). Evidence in support of a model that predicts how biological and environmental factors interact to influence spatial skills. *Developmental Psychology, 35*, 1237–1247.

Cataldo, M. F., Dershewitz, R. Wilson, M., Christophersen, E., Finney, J., Fawcett, S., & Seekins, T. (1986). Childhood injury control. In N. A. Krasnegor, J. Arateh, & M. Cataldo (Eds.), *Child health behavior* (pp. 217–253). New York: Wiley.

Cerny, J. A. (1978). Biofeedback and the voluntary control of sexual arousal in women. *Behavior Therapy, 9*, 847–855.

Chance, M. R. A. (1967). Attention structure as a basis for primate rank order. *Man, 2*, 503–518.

Changeux, J. P. (1985). *Neuronal man: The biology of mind*. Princeton, NJ: Princeton University Press.

Changeux, J. P., & Dehaene, S. (1989). Neural models of cognitive function. *Cognition, 33*, 63–109.

Charlesworth, W. R. (1992). Darwin and developmental psychology: Past and present. *Developmental Psychology, 28*, 5–16.

Chasiotis, A., Scheffer, D., Restmeier, R., & Keller, H. (1998). Intergenerational context discontinuity affects the onset of puberty: A comparison of parent–child dyads in West and East Germany. *Human Nature, 9*, 321–339.

Chen, Z., & Siegler, R. S. (2000). Across the great divide: Bridging the gap between understanding of toddlers' and older children's thinking. *Monographs of the Society for Research in Child Development, 65*(2, Serial No. 261).

Cheyne, D., & Seyfarth, R. (1990). *How monkeys see the world*. Chicago: University of Chicago Press.

Chisholm, J. S. (1996). The evolutionary ecology of attachment organization. *Human Nature, 7*, 1–37.

Chisholm, J. S. (1999). Attachment theory and time preference: Relations between early stress and sexual behavior in a sample of American university women. *Human Nature, 10*, 51–83.

Chomsky, N. (1957). *Syntactic structures*. The Hague: Mouton.

Chomksy, N. (1965). *Aspects of a theory of syntax*. Cambridge, MA: MIT Press.

Chorney, M. J., Chorney, K., Seese, N., Owen, M. J., Daniels, J., McGuffin, P., Thompson, L. A., Detterman, D. K., Benbow, C., Lubinski, D., Eley, T., &

Plomin, R. (1998). A qualitative trait locus associated with cognitive ability in children. *Psychological Science, 9,* 159–166.

Christenfeld, N., & Hill, E. (1995). Whose baby are you? *Nature, 378,* 669.

Christie, J. F., & Johnsen, E. P. (1987). Reconceptualizing constructive play: A review of the empirical literature. *Merrill–Palmer Quarterly, 33,* 439–452.

Clark, E. A., & Hanisee, J. (1982). Intellectual and adaptive performance of Asian children in adoptive American settings. *Developmental Psychology, 18,* 595–599.

Clements, W. A., & Perner, J. (1994). Implicit understanding of belief. *Cognitive Development, 9,* 377–395.

Clements, W. A., Rustin, C. L., & McCallum, S. (2000). Promoting the transition from implicit to explicit understanding: A training study of false belief. *Developmental Science, 3,* 81–92.

Clinton, H. R. (1996). *It takes a village: And other lessons children teach us.* New York: Simon & Schuster.

Clutton-Brock, T. H. (1991). *The evolution of parental care.* Princeton, NJ: Princeton University Press.

Clutton-Brock, T. H., Iason, G. R., & Guiness, F. E. (1987). Sexual segregation and density related changes in habitat use in male and female red deer. *Journal of Zoology, 211,* 275–289.

Clutton-Brock, T. H., & Vincent, A. C. J. (1991). Sexual selection and the potential reproductive rates of males and females. *Nature, 351,* 58–60.

Coie, J. D., & Dodge, K. A. (1998). Aggression and antisocial behavior. In W. Damon (Gen. ed.) & N. Eisenberg (Ed.). *Handbook of child psychology: Vol. 3. Social, emotional, and personality development* (5th ed., pp. 779–862). New York: Wiley.

Cole, P. M. (1986). Children's spontaneous control of facial expression. *Child Development, 57,* 1309–1321.

Collaer, M. L., & Hines, M. (1995). Human behavioral sex differences: A role for gonadal hormones during early development? *Psychological Bulletin, 118,* 55–107.

Collins, W. A. (1990). Parent–child relationships in the transition to adolescence. In R. Montemayor, G. Adams, & T. Gullotta (Eds.), *Advances in adolescent development* (Vol. 2, pp. 85–106). Newbury Park, CA: Sage.

Collins, W. A., Maccoby, E. E., Steinberg, L., Hetherington, E. M., & Bornstein, M. H. (2000). Contemporary research on parenting: The case for nature *and* nurture. *American Psychologist, 55,* 218–232.

Connor, J. M., & Serbin, L. A. (1977). Behaviorally based masculine and feminine activity-preference scales for preschoolers: Correlates with other classroom behaviors and cognitive tests. *Child Development, 48,* 1411–1416.

Cook, E., Hodes, R., & Lang, P. (1986). Preparedness and phobia: Effects of stimulus content on human visceral conditioning. *Journal of Abnormal Psychology, 95,* 195–207.

Cook, M., & Mineka, S. (1989). Observational conditioning of fear to fear-relevant versus fear-irrelevant stimuli in rhesus monkeys. *Journal of Abnormal Psychology*, 98, 448–459.

Cook, M., & Mineka, S. (1990). Selective associations in the observational conditioning of fear in rhesus monkeys. *Journal of Experimental Psychology: Animal Behavior Processes*, 16, 372–389.

Cook, M., Mineka, S., Wolkenstein, B., & Laitsch, K. (1985). Observational conditioning of snake fear in unrelated rhesus monkeys. *Journal of Abnormal Psychology*, 94, 591–610.

Cooper, R. P., & Aslin, R. N. (1990). Preference for infant-directed speech in the first month after birth. *Child Development*, 61, 1584–1595.

Cooper, R. P., & Aslin, R. N. (1994). Developmental differences in infant attention to the spectral properties of infant-directed speech. *Child Development*, 65, 1663–1677.

Cosmides, L., & Tooby, J. (1987). From evolution to behavior: Evolutionary psychology as the missing link. In J. Dupre (Ed.), *The latest on the best essays on evolution and optimality* (pp. 277–306). Cambridge, MA: MIT Press.

Cosmides, L., & Tooby, J. (1992). Cognitive adaptations for social exchange. In J. H. Barkow, L. Cosmides, & J. Tooby (Eds.), *The adapted mind: Evolutionary psychology and the generation of culture* (pp. 163–228). New York: Oxford University Press.

Costabile, A., Smith, P.K., Matheson, L., Aston, J., Hunter, T., & Boulton, M. (1991). Cross-national comparison of how children distinguish playful and serious fighting. *Developmental Psychology*, 27, 881–887.

Cowan, N., Nugent, L. D., Elliott, E. M., Ponomarev, I., & Saults, J. S. (1999). The role of attention in the development of short-term memory: Age differences in the verbal span of apprehension. *Child Development*, 70, 1082–1097.

Crick, N. R., & Dodge, K. A. (1994). A review and reformulation of social information-processing mechanisms in children's social adjustment. *Psychological Bulletin*, 115, 74–101.

Crick, N. R., & Grotpeter, J. K. (1995). Relational aggression, gender, and social psychological adjustment. *Child Development*, 66, 710–722.

Crook, J. M. (1980). *The evolution of human consciousness*. Oxford: Clarendon Press.

Crouch, M. (1999). The evolutionary context of postnatal depression. *Human Nature*, 10, 163–182.

Cullumbine, H. (1950). Heat production and energy requirements of tropical people. *Journal of Applied Physiology*, 2, 201–210.

Cummins, D. D. (1996a). Evidence for the innateness of deontic reasoning. *Mind & Language*, 11, 160–190.

Cummins, D. D. (1996b). Evidence of deontic reasoning in 3- and 4-year-old children. *Memory and Cognition*, 24, 823–829.

Cummins, D. D. (1998a). Cheater detection is modified by social rank: The impact of dominance on the evolution of cognitive functions. *Evolution and Human Behavior, 20,* 229–248.

Cummins, D. D. (1998b). Social norms and other minds: The evolutionary roots of higher cognition. In D. D. Cummins & C. Allen (Eds.), *The evolution of mind* (pp. 28–50). New York: Oxford University Press.

Curtiss, S. (1977). *Genie: A psycholinguistic study of a modern day "wild child."* New York: Academic Press.

Custance, D. M., Whiten, A., & Bard, K. A. (1995). Can young chimpanzees imitate arbitrary actions? Hayes and Hayes (1952) revisited. *Behavior, 132,* 839–858.

Cutler, W. B., Krieger, A., Huggins, G. R., Garcia, C. R., & Lawley, H. J. (1986). Human axillary secretions influence women's menstrual cycles: The role of donor extracts from men. *Hormones and Behavior, 20,* 463–473.

Cutting, A. L., & Dunn, J. (1999). Theory of mind, emotion understanding, language, and family background: Individual differences and interrelations. *Child Development, 70,* 853–865.

Cziko, G. (1995). *Without miracles: Universal selection theory and the second Darwinian revolution.* Cambridge, MA: MIT Press.

Dabbs, J. M., Jr., Chang, E-L., Strong, R. A., & Milun, R. (1998). Spatial ability, navigation strategy, and geographic knowledge among men and women. *Evolution and Human Behavior, 19,* 89–98.

Daly, M., Salmon, C., & Wilson, M. (1997). Kinship: The conceptual hole in psychological studies of social cognition and close relationships. In J. A. Simpson & D. T. Kenrick (Eds.), *Evolutionary social psychology* (pp. 265–296). Mahwah, NJ: Erlbaum.

Daly, M., & Wilson, M. (1981). Abuse and neglect of children in evolutionary perspective. In R. D. Alexander & D. W. Tinkle (Eds.), *Natural selection and social behavior* (pp. 405–416). New York: Chiron.

Daly, M., & Wilson, M. I. (1982). Whom are newborn babies said to resemble? *Ethology and Sociobiology, 3,* 69–78.

Daly, M., & Wilson, M. (1983). *Sex, evolution and behavior* (2nd ed.). Boston: William Grant.

Daly, M., & Wilson, M. (1984). A sociobiological analysis of human infanticide. In G. Hausfater & S. B. Hrdy (Eds.), *Infanticide: Comparative and evolutionary perspectives* (pp. 487–502). New York: Aldine de Gruyter.

Daly, M., & Wilson, M. (1985). Child abuse and other risks of not living with both parents. *Ethology and Sociobiology, 6,* 197–210.

Daly, M., & Wilson, M. (1988a). Evolutionary social psychology and family homicide. *Science, 242,* 519–524.

Daly, M., & Wilson, M. (1988b). *Homicide.* New York: Aldine de Gruyter.

Daly, M., & Wilson, M. (1996). Violence against children. *Current Directions in Psychological Science, 5,* 77–81.

Damast, A. M., Tamis-LeMonda, C. S., & Bornstein, M. H. (1996). Mother–child play: Sequential interactions and the relation between maternal beliefs and behaviors. *Child Development, 67,* 1752–1766.

Daniel, M. H. (1997). Intelligence testing: Status and trends. *American Psychologist, 52,* 1038–1045.

Dansky, J., & Silverman, I. (1973). Effects of play on associative fluency of preschool-age children. *Developmental Psychology, 9,* 38–43.

Dansky, J., & Silverman, I.W. (1975). Play: A general facilitator of associative fluency. *Developmental Psychology, 11,* 104.

Darwin, C. (1859). *The origin of species.* New York: Modern Library.

Darwin, C. (1871). *The descent of man, and selection in relation to sex.* London: John Murray.

Davidson, P. E. (1914). *The recapitulation theory and human infancy.* New York: Columbia University.

Davies, P. S., Fetzer, J. H., & Foster, T. S. (1995). Logical reasoning and domain specificity: Critique of the social exchange theory of reasoning. *Biology and Philosophy, 10,* 1–37.

Davis, H., & Pérusse, R. (1988). Numerical competence in animals: Definitional issues, current evidence, and a new research agenda. *Behavioral and Brain Sciences, 11,* 561–615.

Dawkins, R. (1976). *The selfish gene.* New York: Oxford University Press.

Dawkins, R. (1982). *The blind watchmaker.* Essex, England: Longman.

Deacon, T. W. (1997). *The symbolic species: The co-evolution of language and the brain.* New York: Norton.

Dean, M. C., Stringer, C. B., & Bromage, T. G. (1986). Age at death of the Neanderthal child from Devil's Tower, Gibraltar, and the implications for studies of general growth and development in Neanderthals. *American Journal of Anthropology, 70,* 301–309.

de Beer, G. (1958). *Embryos and ancestors* (3rd ed.). Oxford, England: Clarendon Press.

DeCasper, A. J., & Fifer, W. P. (1980). Of human bonding: Newborns prefer their mother's voice. *Science, 208,* 1174–1176.

DeCasper, A. J., & Spence, M. J. (1986). Prenatal maternal speech influences newborns' perception of speech sounds. *Infant Behavior and Development, 9,* 133–150.

de Haan, M., Oliver, A., & Johnson, M. H. (1998). Electro physiological correlates of face processing by adults and 6-month-old infants. *Journal of Cognitive Neural Science* [Annual Meeting Suppl.], 36.

Dempster, F. N. (1985). Short-term memory development in childhood and adolescence. In C. J. Brainerd & M. Pressley (Eds.), *Basic processes in memory development: Progress in cognitive development research* (pp. 209–248). New York: Springer.

Dempster, F. N. (1992). The rise and fall of the inhibitory mechanism: Toward a unified theory of cognitive development and aging. *Developmental Review, 12*, 45–75.

Dempster, F. N. & Brainerd, C.J. (Eds.) (1995). *New perspectives on interference and inhibition in cognition.* New York: Academic Press.

Devlin, B., Daniels, M., & Roeder, K. (1997). The heritability of IQ. *Nature, 388*, 468–471.

DeVries, M. W. (1984). Temperament and infant mortality among the Masai of East Africa. *American Journal of Psychiatry, 141*, 1189–1193.

de Waal, F. B. M. (1982). *Chimpanzee politics: Power and sex among apes.* London: Jonathan Cape.

de Waal, F. B. M. (1989). *Peace making among primates.* Cambridge, MA: Harvard University Press.

de Winter, W., & Oxnard, C. E. (2001). Evolutionary radiations and convergences in the structural organization of mammalian brains. *Nature, 409*, 710–714.

Dickens, G., & Trethowan, W. H. (1971). Cravings and aversions during pregnancy. *Journal of Psychosomatic Research, 15*, 259–268.

DiPietro, J. A. (1981). Rough and tumble play: A function of gender. *Developmental Psychology, 17*, 50–58.

Dishion, T. J., Andrews, D. W., & Crosby, L. (1995). Anti-social boys and their friends in early adolescence: Relationships characteristics, quality, and interactional process. *Child Development, 66*, 139–151.

Dobzhansky, T. (1937). *Genetics and the origins of species.* New York: Columbia University Press.

Dodge, K. A. (1986). A social information processing model of social competence in children. In M. Perlmutter (Ed.), *Minnesota symposium on child psychology* (Vol. 18, pp. 77–125). Hillsdale, NJ: Erlbaum.

Dolhinow, P. J., & Bishop, N. H. (1970). The development of motor skills and social relationships among primates through play. In J. P. Hill (Ed.), *Minnesota symposia on child psychology* (pp. 180–198). Minneapolis: University of Minnesota Press.

Doman, G. (1984). *How to multiply your baby's intelligence.* Garden City, NY: Doubleday.

Donald, M. (1991). *Origins of the modern mind: Three stages in the evolution of culture and cognition.* Cambridge, MA: Harvard University Press.

Dore, F. Y., & Dumas, C. (1987). Psychology of animal cognition: Piagetian studies. *Psychological Bulletin, 102*, 219–233.

Doupe, A. J., & Kuhl, P. K. (1999). Birdsong and human speech: Common themes and mechanisms. *Annual Review of Neuroscience, 22*, 567–631.

Draper, P. (1976). Social and economic constraints on child life among the !Kung. In R. B. Lee & I. De Vore (Eds.), *Kalahari hunter–gatherers* (pp. 199–217). Cambridge, MA: Harvard University Press.

Draper, P., & Harpending, H. (1987). A sociobiological perspective on human reproductive strategies. In K. B. MacDonald (Ed.), *Sociobiological perspectives on human development* (pp. 340–372). New York: Springer-Verlag.

Drickamer, L. C. (1988). Preweaning stimulation with urinary chemosignals and age of puberty in female mice. *Developmental Psychobiology, 21,* 77–87.

Duberman, L. (1975). *The reconstituted family: A study of remarried couples and their children.* Chicago: Nelson-Hall.

Dunbar, R. I. M. (1988). *Primate social systems.* Ithaca, NY: Cornell University Press.

Dunbar, R. I. M. (1992). Neocortex size as a constraint on group size in primates. *Journal of Human Evolution, 20,* 469–493.

Dunbar, R. I. M. (1995). Neocortex size and group size in primates: A test of the hypothesis. *Journal of Human Evolution, 28,* 287–296.

Dunbar, R. I. M., & Spoors, M. (1995). Social networks, support cliques, and kinship. *Human Nature, 6,* 273–290.

Duncan, J., Seitz, R. J., Kolodny, J., Bor, D., Herzog, H., Ahmed, A., Newell, F. N., & Emslie, H. (2000). A neural basis for general intelligence. *Science, 289,* 457–460.

Dunn, J. (1988). *The beginnings of social understanding.* Cambridge, MA: Harvard University Press.

Dunn, J., & Kendrick, C. (1981). Interaction between young siblings: Association with the interaction between mother and firstborn child. *Developmental Psychology, 17,* 336–343.

Eagly, A. H. (1987). *Sex differences in social behavior: A social-role interpretation.* Hillsdale, NJ: Erlbaum.

Eals, M., & Silverman, I. (1994). The hunter–gatherer theory of spatial sex differences: Proximate factors mediating the female advantage in recall of object arrays. *Ethology and Sociobiology, 15,* 95–105.

Eaton, W. O., & Enns, L. R. (1986). Sex differences in human motor activity level. *Psychological Bulletin, 100,* 19–28.

Eaton, W. O., & Yu, A. P. (1989). Are sex differences in child motor activity level a function of sex differences in maturational status? *Child Development, 60,* 1005–1011.

Eccles, J. C. (1989). *Evolution of the brain: Creation of the self.* New York: Routledge.

Eccles, J. S., Midgley, C., Wigfield, A., Buchanan, C. M., Reuman, D., Flanagan, & MacIver, D. (1993). Development during adolescence: The impact of stage–environment fit on young adolescents' experiences in school and families. *American Psychologist, 48,* 90–101.

Eckerman, C. O., & Whitehead, H. (1999). How toddler peers generate coordinated action. *Early Education and Development, 10,* 241–266.

Edelman, G. M. (1987). *Neural Darwinism: The theory of neuronal group selection.* New York: Basic Books.

Edwards, C. P., & Whiting, B. B. (1988). *Children of different worlds.* Cambridge, MA: Harvard University Press.

Egan, S., & Perry, D. G. (2001). Gender identity: A multidimensional analysis with implications for psychological adjustment. *Developmental Psychology, 37*, 451–463.

Egeland, B., & Farber, E. A. (1984). Infant–mother attachment: Factors related to its development and changes over time. *Child Development, 55*, 753–771.

Egeland, B., & Sroufe, L. A. (1981). Attachment and early maltreatment. *Child Development, 52*, 44–52.

Eibl-Eibesfeldt, I. (1967). Concepts of ethology and their significance in the study of human behavior. In H. W. Stevenson, E. Hess, & H. L. Rheingold (Eds.), *Early behavior* (pp. 127–146). New York: Wiley.

Eibl-Eibesfeldt, I. (1989). *Human ethology.* New York: Aldine de Gruyter.

Eimas, P. D., Siqueland, E. R., Jusczyk, P., & Vigorito, J. (1971). Speech perception in infants. *Science, 71*, 303–306.

Eisenberg, A. R. (1988). Grandchildren's perspectives on relationships with grandparents: The influence of gender across generations. *Sex Roles, 19*, 295–217.

Eizenman, D. R., & Bertenthal, B. I. (1998). Infants' perception of object unity in translating and rotating displays. *Developmental Psychology, 34*, 426–434.

Eldredge, N., & Gould, S. J. (1972). Punctuated equilibria: An alternative to phyletic gradualism. In T. J. M. Schopf (Ed.), *Models in paleobiology* (pp. 83–115). San Francisco: Freeman, Cooper.

Ellis, B. J., & Graber, J. (2000). Psychosocial antecedents of variation in girls' pubertal timing: Maternal depression, stepfather presence, and marital and family stress. *Child Development, 71*, 485–501.

Ellis, B. J., McFadyen-Ketchum, S., Dodge, K. A., Pettit, G. S., & Bates, J. E. (1999). Quality of early family relationships and individual differences in the timing of pubertal maturation in girls: A longitudinal test of an evolutionary model. *Journal of Personality and Social Psychology, 77*, 387–401.

Elman, J. (1994). Implicit learning in neural networks: The importance of starting small. In C. Umilta & M. Moscovitch (Eds.), *Attention and performance XV: Conscious and nonconscious information processing* (pp. 861–888). Cambridge, MA: MIT Press.

Elman, J. L., Bates, E. A., Johnson, M. H., Karmiloff-Smith, A., Parisi, D., & Plunket, K. (1996). *Rethinking innateness: A connectionist perspective on development.* Cambridge, MA: MIT Press.

Ernst, C., & Angst, J. (1983). *Birth order: Its influence on personality.* Berlin: Springer-Verlag.

Euler, H. A., & Weitzel, B. (1996). Discriminative grandparental solicitude as reproductive strategy. *Human Nature, 7*, 39–59.

Fagen, R. (1981). *Animal play behavior.* New York: Oxford University Press.

Fagot, B. I. (1974). Sex differences in toddlers' behaviour and parental reaction. *Developmental Psychology, 10*, 554–558.

Fagot, B. I. (1985). Beyond the reinforcement principle: Another step toward understanding sex role development. *Developmental Psychology, 21*, 1097–1104.

Fairbanks, L. A. (1988). Vervet monkey grandmothers: Interactions with infant grandoffspring. *International Journal of Primatology, 9,* 425–441.

Fantz, R. L. (1961). The origin of form perception. *Scientific American, 204,* 66–72.

Fein, G. (1981). Pretend play: An integrative review. *Child Development, 52,* 1095–1118.

Feingold, A. (1992). Gender differences in mate selection preferences: A test of the parental investment model. *Psychological Bulletin, 112,* 125–139.

Feldman, R. S., & White, J. B. (1980). Detecting deception in children. *Journal of Communication, 30,* 121–128.

Fernald, A. (1992). Human maternal vocalizations to infants as biologically relevant signals: An evolutionary perspective. In J. H. Barkow, L. Cosmides, & J. Tooby (Eds.), *The adaptive mind: Evolutionary psychology and the generation of culture* (pp. 391–428). New York: Oxford University Press.

Fernald, A., & Mazzie, C. (1991). Prosody and focus in speech to infants and adults. *Developmental Psychology, 27,* 209–221.

Field, T. (1994). Infant day care facilitates later social behavior and school performance. In E. Jacobs & H. Goelman (Eds.), *Children's play in day care centers* (pp. 69–84). Albany: State University of New York Press.

Finlay, B. L., & Darlington, R. D. (1995). Linked regularities in the development and evolution of mammalian brains. *Science, 268,* 1579–1584.

Finlay, B. L., Darlington, R. B., & Nicastro, N. (2001). Developmental structure in brain evolution. *Behavioral and Brain Sciences, 24,* 263–308.

Fishbein, H. D. (1976). *Evolution, development, and children's learning.* Santa Monica, CA: Goodyear.

Fisher, C., & Tokuro, H. (1996). Acoustic cues to grammatical structure in infant-directed speech: Cross-linguistic evidence. *Child Development, 67,* 3192–3218.

Fisher, H. E. (1992). *Anatomy of love: The natural history of monogamy, adultery, and divorce.* New York: Norton.

Flavell, J. H., Green, F. L., & Flavell, E. R. (1986). Development of knowledge about the appearance–reality distinction. *Monographs of the Society for Research in Child Development, 51*(1, Serial No. 212).

Flaxman, S. M., & Sherman, P. W. (2000). Morning sickness: A mechanism for protecting mother and embryo. *Quarterly Review of Biology, 75,* 113–148.

Flinn, M. V. (1988). Step and genetic parent/offspring relationships in a Caribbean village. *Ethology and Sociobiology, 9,* 335–369.

Flinn, M. V., Leone, D. V., & Quinlan, R. J. (1999). Growth and fluctuating asymmetry of stepchildren. *Evolution and Human Behavior, 20,* 465–479.

Fodor, J. A. (1983). *The modularity of mind.* Cambridge, MA: MIT Press.

Fodor, J. (2000). *The mind doesn't work that way: The scope and limits of computational psychology.* Cambridge, MA: MIT Press.

Foley, M. A., & Ratner, H. H. (1998). Children's recoding in memory for collaboration: A way of learning from others. *Cognitive Development, 13,* 91–108.

Foley, M. A., Ratner, H. H., & Passalacqua, C. (1993). Appropriating the actions of another: Implications for children's memory and learning. *Cognitive Development, 8*, 373–401.

Folstad, I., & Karter, A. J. (1992). Parasites, bright males, and the immunocompetence handicap. *American Naturalist, 139*, 603–622.

Fontaine, R. (1984). Imitative skill between birth and six months. *Infant Behavior and Development, 7*, 323–333.

Freedman, D. G. (1974). *Human infancy: An evolutionary perspective.* New York: Wiley.

Freese, J., Powell, B., & Steelman, L. C. (1999). Rebel without a cause or effect: Birth order and social attitudes. *American Sociological Review, 64*, 207–231.

Freud, S. (1952). *Totem and taboo* (J. Strachey, Trans.). New York: Norton.

Fruman, W., & Buhrmester, D. (1985). Children's perceptions of the qualities of sibling relationships. Child Development, 56, 448–461.

Furlow, F. B. (1997). Human neonatal cry quality as an honest signal of fitness. *Evolution and Human Behavior, 18*, 175–193.

Furlow, F. B., Armijo-Prewitt, T., Gangstead, S. W., & Thornhill, R. (1997). Fluctuating asymmetry and psychometric intelligence. *Proceedings of the Royal Society of London, Series B, 264*, 823–829.

Fuster, J. M. (1984). The prefrontal cortex and temporal integration. In A. Peters & E. G. Jones (Eds.), *Cerebral cortex: Vol. 4. Association and auditory cortices* (pp. 151–177). New York: Plenum.

Gabunia, L., Vekua, A., Lordkipanidze, D., Swisher, C. C. III, Ferring, R., Justus, A., Nioradze, M., Tvalchrelidze, M., Antón, S. C., Bosinski, G., Jöris, O., de Lumley, M.-A., Majsuradze, G., & Mouskhelishvili, A. (2000). Earliest Pleistocene hominid cranial remains from Dmanisi, Republic of Georgia: Taxonomy, setting, and age. *Science, 288*, 1019–1025.

Galef, B. G., Jr. (1988). Imitation in animals: History, definition, and interpretation of data. In T. Zentall & B. C. Galef, Jr. (Eds.), *Social learning* (pp. 3–28). Hillsdale, NJ: Erlbaum.

Galef, B. G. (1996). Tradition in animals. In M. Bekoff & D. Jamieson (Eds.)., *Readings in animal cognition* (pp. 91–105). Cambridge, MA: MIT Press.

Gallistel, C. R. (1990). *The organization of learning.* Cambridge, MA: MIT Press.

Galton, F. (1869). Hereditary genius: An inquiry into its laws and consequences. London: Macmillian.

Gangestad, S. W., & Thornhill, R. (1997). Human sexual selection and developmental stability. In J. A. Simpson & D. T. Kenrick (Eds.), *Evolutionary social psychology* (pp. 169–195). Mahwah, NJ: Erlbaum.

Gangestad, S. W., Thornhill, R., & Yeo, R. A. (1994). Facial attractiveness, developmental stability, and fluctuating symmetry. *Ethology and Sociobiology, 15*, 73–85.

Garcia, J., Ervin, F. R., & Koelling, R. A. (1966). Learning with prolonged delay of reinforcement. *Psychonomic Science, 5*, 121–122.

Garcia, J., & Koelling, R. A. (1966). Relation of cue to consequence in avoidance learning. *Psychonomic Science, 4,* 123–124.

Garcia, M. M., Shaw, D. S., Winslow, E. B., & Yaggi, K. E. (2000). Destructive sibling conflict and the development of conduct problems in young boys. *Developmental Psychology, 36,* 44–53.

Gardner, H. (1983). *Frames of mind: The theory of multiple intelligences.* New York: Basic.

Gardner, H. (1999). Are there additional intelligences? The case for naturalist, spiritual, and existential intelligences. In J. Kane (Ed.), *Education, information and transformation* (pp. 111–131). Englewood Cliffs, NJ: Prentice-Hall.

Gardner, R. A., & Gardner, B. T. (1969). Teaching sign language to a chimpanzee. *Science, 165,* 664–672.

Gardner, B. T., & Gardner, R. A. (1971). Two-way communication with an infant chimpanzee. In A. M. Schrier & F. Stollnitz (Eds.), *Behavior of nonhuman primates,* (Vol 4, pp. 117–185). New York: Academic Press.

Garnham, W. A., & Ruffman, T. (2001). Doesn't see, doesn't know: Is anticipatory looking really related to understanding belief? *Developmental Science, 4,* 94–100.

Garstang, W. (1922). The theory of recapitulation: A critical restatement of the biogenetic law. *Journal of Linnaean Society (Zoology), 35,* 81–101.

Gaulin, S. J. (1980). Sexual dimorphism in the human post-reproductive lifespan: Possible causes. *Human Evolution, 9,* 227–232.

Gaulin, S. J. C., McBurney, D. H., & Brakeman-Wartell, S. L. (1997). Matrilateral biases in the investment of aunts and uncles: A consequence and measure of paternity certainty. *Human Nature, 8,* 139–151.

Geary, D. C. (1994). *Children's mathematical development: Research and practical applications.* Washington, DC: American Psychological Association.

Geary, D. C. (1995). Reflections of evolution and culture in children's cognition: Implications for mathematical development and instruction. *American Psychologist, 50,* 24–37.

Geary, D. C. (1996). Sexual selection and sex differences in mathematical abilities. *Behavioral and Brain Sciences, 19,* 229–284.

Geary, D. C. (1998). *Male, female: The evolution of human sex differences.* Washington, DC: American Psychological Association.

Geary, D. C. (1999). Evolution and developmental sex differences. *Current Directions in Psychological Science, 8,* 115–120.

Geary, D. C. (2000). Evolution and proximate expression of human paternal investment. *Psychological Bulletin, 126,* 55–77.

Geary, D. C. (2001). Sexual selection and sex differences in social cognition. In A. V. McGillicuddy-DeLisi & R. DeLisi (Eds.), *Biology, society, and behavior: The development of sex differences in cognition* (pp. 23–53). Greenwich, CT: Ablex.

Geary, D. C., & Bjorklund, D. F. (2000). Evolutionary developmental psychology. *Child Development, 71,* 57–65.

Geary, D. C., & Flinn, M. V. (2001). Evolution of human parental behavior and the human family. *Parenting: Science and Practice, 1,* 5–61

Geary D. C., & Huffman, K. (2001). Brain and cognitive evolution: Forms of modularity and functions of mind. Manuscript under review.

Geman, R., & Gallistel, R. (1978). *The child's understanding of number.* Cambridge, MA: Harvard University Press.

Gelman, R., & Williams, E. M. (1998). Enabling constraints for cognitive development and learning: Domain-specificity and epigenesis. In W. Damon (Gen. Ed.) & D. Kuhn & R. S. Siegler (Eds.), *Handbook of child psychology: Vol. 2. Cognition, perception, and language* (5th ed., pp. 575–630) New York: Wiley.

Gibbs, A. C., & Wilson, J. F. (1999). Sex differences in route learning by children. *Perceptual & Motor Skills, 88,* 590–594.

Gibson, K. R. (1990). New perspectives on instincts and intelligence: Brain size and the emergence of hierarchical mental construction skills. In S. T. Parker & K. R. Gibson (Eds.), *"Language" and intelligence in monkeys and apes* (pp. 97–128). New York: Cambridge University Press.

Gibson, K. R. (1991). Myelination and behavioral development: A comparative perspective on questions of neoteny, altriciality and intelligence. In K. R. Gibson & A. C. Petersen (Eds.), *Brain maturation and cognitive development: Comparative and cross-cultural perspectives* (pp. 29–63). New York: Aldine de Gruyter.

Goldberg, S., Blumberg, S. L., & Kriger, A. (1982). Menarche and interest in infants: Biological and social influences. *Child Development, 53,* 1544–1550.

Goldstein, S., & Barkley, R. A. (1998). ADHD, hunting, and evolution: "Just so stories." *ADHD Report,* Vol. 6, No. 5.

Gómez, J. C. (1990). The emergence of intentional communication as a problem-solving strategy in the gorilla. In S. T. Parker & K. R. Gibson (Eds.), *"Language" and intelligence in monkeys and apes* (pp. 333–355). Cambridge, England: Cambridge University Press.

Goodall, J. (1971). *In the shadow of man.* London: Collins.

Goodall, J. (1986). *The chimpanzees of Gombe.* Cambridge, MA: Belknap.

Goodman, J. F. (1992). *When slow is fast enough: Educating the delayed preschool child.* New York: Guilford.

Gopnik, A., & Astington, J. W. (1988). Children's understanding of representational change and its relation to the understanding of false belief and the appearance-reality distinction. *Child Development, 59,* 26–37.

Gopnik, A., & Meltzoff, A. N. (1997). *Words, thoughts, and theories.* Cambridge, MA: MIT Press.

Gottlieb, G. (1976). The roles of experience in the development of behavior and the nervous system. In G. Gottlieb (Ed.), *Neural and behavioral plasticity* (pp. 25–54). New York: Academic Press.

Gottlieb, G. (1983). The psychobiological approach to developmental issues. In J. J. Campos & M. Haith (Eds.), *Handbook of child psychology: Vol. II. Infancy and developmental psychobiology* (pp. 1–26). New York: Wiley.

Gottlieb, G. (1987). The developmental basis of evolutionary change. *Journal of Comparative Psychology, 101*, 262–271.

Gottlieb, G. (1991a). Experiential canalization of behavioral development: Theory. *Developmental Psychology, 27*, 4–13.

Gottlieb, G. (1991b). Experiential canalization of behavioral development: Results. *Developmental Psychology, 27*, 35–39.

Gottlieb, G. (1992). *Individual development and evolution: The genesis of novel behavior.* New York: Oxford University Press.

Gottlieb, G. (1995a). Some conceptual deficiencies in "developmental" behavior genetics. *Human Development, 38*, 131–141.

Gottlieb, G. (1995b). Reply. *Human Development, 38*, 165–169.

Gottlieb, G. (1996). Developmental psychobiological theory. In R. B. Cairns, G. H. Elder, Jr., & E. J. Costello (Eds.), *Developmental science* (pp. 63–77). Cambridge, MA: Cambridge University Press.

Gottlieb, G. (1997). *Synthesizing nature–nurture: Prenatal roots of instinctive behavior.* Mahwah, NJ: Erlbaum.

Gottlieb, G. (1998). Normally occurring environmental and behavioral influences on gene activity: From central dogma to probabilistic epigenesis. *Psychological Review, 105*, 792–802.

Gottlieb, G. (2000). Environmental and behavioral influences on gene activity. *Current Directions in Psychological Science, 9*, 93–102.

Gottlieb, G., Tomlinson, W. T., & Radell, P. L. (1989). Developmental intersensory interference: Premature visual experience suppresses auditory learning in ducklings. *Infant Behavior and Development, 12*, 1–12.

Gottlieb, G., Wahlsten, D., & Lickliter, R. (1998). The significance of biology for human development: A developmental psychobiological systems view. In W. Damon (Gen. Ed.) & R. M. Lerner (Ed.), *Handbook of child psychology: Vol. 1. Theoretical models of human development* (5th ed., pp. 233–273). New York: Wiley.

Gould, E., Reeves, A. J., Graziano, M. S. A., & Gross, C. G. (1999). Neurogenesis in the neocortex of adult primates. *Science, 286*, (15 October), 548–552.

Gould, S. J. (1977). *Ontogeny and phylogeny.* Cambridge, MA: Harvard University Press.

Gould, S. J. (1981). *The mismeasure of man.* New York: Norton.

Gould, S. J. (1991). Exaptation: A crucial tool for evolutionary psychology. *Journal of Social Issues, 47*, 43–65.

Gould, S. J. (1996). *Full house: The spread of excellence from Plato to Darwin.* New York: Harmony Books.

Gould, S. J. (1997). The exaptive excellence of spandrels as a term and prototype. *Proceedings of the National Academy of Sciences, 94*, 10750–10755.

Gould, S. J., & Lewontin, R. C. (1979). The spandrels of San Marco and the Panglossian paradigm: A critique of the adaptationist programme. *Proceedings of the Royal Society of London B, 205,* 581–598.

Gould, S. J., & Vrba, E. E. (1982). Exaptation: A missing term in the science of form. *Paleobiology, 8,* 4–15.

Gowaty, P. A. (1992). Evolutionary biology and feminism. *Human Nature, 3,* 217–249.

Graber, J. A., Brooks-Gunn, J., & Warren, M. P. (1995). The antecedents of menarcheal age: Heredity, family environment and stressful life events. *Child Development, 66,* 346–359.

Grant, B. R., & Grant, P. R. (1989). Natural selection in a population of Darwin's finches. *American Naturalist, 133,* 377–393.

Gredlein, J. M. (April, 2001). *Constructive play, problem solving, and tool use: Gender differences in preschooler's use of tools to solve simple problems.* Paper presented at meeting of the Society for Research in Child Development, Minneapolis, MN.

Greenfield, P., Maynard, A., Boehm, C., & Schmidtling, E. Y. (2000). Cultural apprenticeship and cultural change: Tool learning and imitation in chimpanzees and humans. In S. T. Parker, J. Langer, & M. L. McKinney (Eds.), *Biology, brains, and behavior: The evolution of human development* (pp. 237–277). Santa Fe, NM: School of American Research Press.

Greenough, W. T., Black, J. E., & Wallace, C. S. (1987). Experience and brain development. *Child Development, 58,* 539–559.

Greenough, W. T., McDonald, J., Parnisari, R., & Camel, J. E. (1986). Environmental conditions modulate degeneration and new dendrite growth in cerebellum of senescent rats. *Brain Research, 380,* 136–143.

Groos, K. (1898). *The play of animals.* New York: Appleton.

Groos, K. (1901). *The play of man.* New York: Appleton.

Grossmann, K., Grossmann, K. E., Spangler, G., Suess, G., & Unzer, L. (1985). Maternal sensitivity and newborns' orientation responses as related to quality of attachment in northern Germany. In I. Bretherton & E. Waters (Eds.), *Growing points of attachment theory and research* (pp. 233–256). *Monographs of the Society for Research in Child Development, 50*(1–2, Serial No. 209).

Groves, C. P. (1989). *A theory of human and primate evolution.* Oxford, England: Clarenden.

Haig, D. (1993). Genetic conflicts in human pregnancy. *Quarterly Review of Biology, 68,* 495–532.

Haight, W. L., & Miller, P. J. (1993). *Pretending at home: Early development in a sociocultural context.* Albany: State University of New York Press.

Haith, M. M. (1966). The response of the human newborn to visual movement. *Journal of Experimental Child Psychology, 3,* 235–243.

Haith, M. M., & Benson, J. B. (1998). Infant cognition. In N. Damon (Gen. Ed.) & D. Kuhn & R. S. Siegler (Eds.), *Handbook of child psychology: Vol. 2.*

Cognitive, language, and perceptual development (5th ed., pp. 199–254). New York: Wiley.

Halberstadt, J., & Rhodes, G. (2000). The attractiveness of nonface averages: Implications for an evolutionary explanation of the attractiveness of average faces. *Psychological Science, 11*, 285–289.

Hale, S., Fry, A. F., & Jessie, K. A. (1993). Effects of practice on speed of information processing in children and adults: Age sensitivity and age invariants. *Developmental Psychology, 29*, 880–892.

Hall, G. S. (1904). *Adolescence: Its psychology and its relation to physiology, anthropology, sociology, sex, crime, religion, and education* (Vols. 1–2). New York: Appleton.

Halpern, D. F. (1996). Sex, brains, and spatial cognition. *Developmental Review, 16*, 261–270.

Hamilton, C. E. (2000). Continuity and discontinuity of attachment from infancy through adolescence. *Child Development, 71*, 690–694.

Hamilton, W. D. (1964). The genetical theory of social behavior. *Journal of Theoretical Biology, 7*, 1–52.

Hamilton, W. D. (1966). The moulding of senescence by natural selection. *Journal of Theoretical Biology, 12*, 12–45.

Hare, B., Call, J., Agentta, B., & Tomasello, M. (2000). Chimpanzees know what conspecifics do and do not see. *Animal Behaviour, 59*, 771–785.

Hare, B., Call, J., & Tomasello, M. (2001). Do chimpanzees know what conspecifics know? *Animal Behaviour, 61*, pp. 139–151.

Harlow, H. (1959, December). The development of learning in the Rhesus monkey. *American Scientist*, pp. 459–479.

Harlow, H., & Harlow, M. (1962). Social deprivation in monkeys. *Scientific American, 207*, 136–146.

Harlow, H., Harlow, M., Dodsworth, R. O., & Arling, G. L. (1966). Maternal behavior in rhesus monkeys deprived of mothering and peer associations in infancy. *Proceedings of the American Philosophical Society, 110*, 58–66.

Harlow, H., & Zimmerman, R. R. (1959). Affectional responses in the infant monkey. *Science, 130*, 421–432.

Harnishfeger, K. K. (1995). The development of cognitive inhibition: Theories, definitions, and research evidence. In F. Dempster & C. Brainerd (Eds.), Interference and Inhibition in Cognition (pp. 175–204). New York: Academic Press.

Harris, J. R. (1995). Where is the child's environment? A group socialization theory of development. *Psychological Review, 102*, 458–489.

Harris, J. R. (1998). *The nurture assumption: Why children turn out the way they do.* New York: Free Press.

Harris, J. R. (2000). Context-specific learning, personality, and birth order. *Current Directions in Psychological Science, 9*, 174–177.

Harris, P. L. (1990). The work of the imagination. In A. Whiten (Ed.), *The emergence of mindreading* (pp. 283–304). Oxford, UK: Blackwell.

Harris, P. L. (1991, March). Natural simulation of mental states. In *Developmental processes underlying the acquisition of concepts of mind*, symposium conducted at the biennial meeting of the Society for Research in Child Development, Seattle, WA.

Harris, P. L., & Núñez, M. (1996). Understanding of permission rules by preschool children. *Child Development, 67*, 1572–1591.

Hartup, W. W. (1996). The company they keep: Friendships and their developmental significance. *Child Development, 67*, 1–13.

Hasher, L., & Zacks, R. T. (1979). Automatic and effortful processes in memory. *Journal of Experimental Psychology: General, 108*, 356–388.

Haskett, G. J. (1971). Modification of peer preferences of first-grade children. *Developmental Psychology, 4*, 429–433.

Hattori, K. (1998). Drivers of intelligence evolution in Homo: Sexual behavior, food acquisition and infant neoteny. *Mankind Quarterly, 39*, 127–146.

Hauser, M. D. (2000). *Wild minds: What animals really think.* New York: Holt.

Hauser, M. D., Carey, S., & Hauser, L. B. (2000). Spontaneous number representation in semi-free-ranging rhesus monkeys. *Proceeding of the Royal Society of London B, 267*, 829–833.

Haviland, J. J., & Malatesta, C. Z. (1981). The development of sex differences in nonverbal signals: Fallacies, facts, and fantasies. In C. Mayo & N. M. Henley (Eds.), *Gender and nonverbal behavior* (pp. 183–208). New York: Springer-Verlag.

Hawkes, K., O'Connell, J. F., & Blurton Jones, N. G. (1997). Hadza women's time allocation, offspring provisioning, and the evolution of post-menopausal lifespans. *Current Anthropology, 38*, 551–578.

Hawkes, K., Rogers, A. R., & Charnov, E. L. (1995). The male's dilemma: Increased offspring production is more paternity to steal. *Evolutionary Ecology, 9*, 662–677.

Hawley, P. A. (1999). The ontogenesis of social dominance: A strategy-based evolutionary perspective. *Developmental Review, 19*, 97–132.

Hayes, B. K., & Hennessy, R. (1996). The nature and development of nonverbal implicit memory. *Journal of Experimental Child Psychology, 63*, 22–43.

Hayes, C. (1951). *The ape in our house.* New York: Harper.

Haynes, H., White, B. L., & Held, R. (1965). Visual accommodation in human infants. *Science, 148*, 528–530.

Hazen, N. L. (1982). Spatial exploration and spatial knowledge: Individual and developmental differences in very young children. *Child Development, 53*, 826–833.

Hebb, D. O. (1949). *The organization of behavior.* New York: Wiley.

Heimann, M. (1989). Neonatal imitation gaze aversion and mother–infant interaction. *Infant Behavior and Development, 12*, 495–505.

Held, R., & Hein, A. (1963). Movement-produced stimulation in the development of visually guided behavior. *Journal of Comparative and Physiological Psychology, 56*, 872–876.

Herman, J. F., & Siegel, A. W. (1978). The development of cognitive mapping of the large-scale environment. *Journal of Experimental Child Psychology, 26*, 389–406.

Hess, E. H. (1973). Ethology and developmental psychology. In W. P. H. Mussen (Gen. Ed.), *Handbook of child psychology* (Vol. I, 3rd ed., pp. 1–38). New York: Wiley.

Hetherington, E. M., Henderson, S. H., & Reiss, D. (1999). Adolescent siblings in stepfamilies: Family functioning and adolescent adjustment. *Monographs of the Society for Research in Child Development, 64*(4, Serial No. 259).

Heyes, C. M. (1998). Theory of mind in nonhuman primates. *Behavioral and Brain Sciences, 21*, 101–148.

Hill, K., & Hurtado, A. M. (1991). The evolution of premature reproductive senescence and menopause in human females: An evaluation of the "grandmother hypothesis." *Human Nature, 2*, 313–350.

Hill, K., & Hurtado, A. M. (1996). *Ache life history: The ecology and demography of a foraging people.* New York: Aldine de Gruyter.

Hill, K., & Hurtado, A. M. (1999). Packer and colleagues model of menopause in humans. *Human Nature, 10*, 199–204.

Hinde, R. A. (1974). *Biological bases of human social behavior.* New York: McGraw-Hill.

Hinde, R. A. (1976). Interactions, relationships, and social structure. *Man, 11*, 1–17.

Hinde, R. A. (1980). *Ethology.* London: Fontana.

Hinde, R. A. (1982). Attachment: Some conceptual and biological issues. In C. M. Parkes & J. Stevenson-Hinde (Eds.), *The place of attachment in human behavior* (pp. 60–76). New York: Basic Books.

Hinde, R. (1983). Ethology and child development. In J. J. Campos & M. H. Haith (Eds.), *Handbook of child psychology: Vol. 2. Infancy and developmental psychobiology* (pp. 27–94). New York: Wiley.

Hines, M., & Kaufman, F. R. (1994). Androgens and the development of human sex-typical behavior: Rough-and-tumble play and sex of preferred playmates in children with congenital hyperplasia (CAH). *Child Development, 65*, 1042–1053.

Hirata, S., & Morimura, N. (2000). Naïve chimpanzees' (*Pan troglodytes*) observation of experienced conspecifics in a tool-using task. *Journal of Comparative Psychology, 114*, 291–296.

Ho, M. -W. (1998). Evolution. In G. Greenberg & M. M. Haraway (Eds.), *Comparative psychology: A handbook* (pp. 107–119). New York: Garland.

Ho, M. -W., Tucker, C., Keeley, D., & Saunder, P. T. (1983). Effects of successive generations of ether treatment on penetrance and expression of the bithorax phenocopy in Drosophila melanogaster. *Journal of Experimental Zoology, 225*, 357–368.

Hodges, E. V., Boivin, M., Vitaro, F., & Bukowski, W. M. (1999). The power of friendship: Protection against an escalating cycle of peer victimization. *Developmental Psychology, 35,* 94–101.

Hodges, E. V., & Perry, D. G. (1999). Personal and interpersonal antecedents of victimization by peers. *Journal of Personality and Social Psychology, 76,* 677–685.

Hoff, E. (2001). *Language development (2nd ed.).* Belmont, CA: Wadsworth.

Hoff-Ginsburg, E. (1985). Some contributions of mothers' speech to their children's syntactic growth. *Journal of Child Language, 12,* 367–385.

Hoff-Ginsburg, E. (2000). Soziale umwelt und sprachlernen. (The social environment and language learning.) In H. Grimm (Hrsg) *Enzyklopadie der Psychologie,* Vol. C3/3, *(Encyclopedia, Vol 3: Language Development).* Sprachentwicklung (pp. 463–494). Gottingen: Hogrefe.

Hoffman, E. (1978–1979). Young adults' relations with their grandparents: An exploratory study. *International Journal of Aging and Human Development, 10,* 299–310.

Hood, B., Carey, S., & Prasada, S. (2000). Predicting the outcomes of physical events: Two-year-olds fail to reveal knowledge of solidity and support. *Child Development, 71,* 1540–1554.

Hook, E. B. (1978). Dietary cravings and aversions during pregnancy. *American Journal of Clinical Nutrition, 31,* 1355–1362.

Howes, C. (1994). *The collaborative construction of pretend.* Albany: State University of New York Press.

Hrdy, S. B. (1986). Sources of variation in the reproductive success of female primates. *Proceedings of the International Meeting on Variability and Behavioral Evolution, 259,* 191–203.

Hrdy, S. B. (1999). *Mother nature: A history of mothers, infants, and natural selection.* New York: Pantheon Books.

Hughes, C., & Cutting, A. L. (1999). Nature, nurture, and individual differences in early understanding of mind. *Psychological Science, 10,* 429–432.

Huizinga, J. (1950). *Homo ludens: A study of the play-element in culture.* Boston: Beacon Press.

Hulanicka, B. (1999). Acceleration of menarcheal age of girls from dysfunctional families. *Journal of Reproductive and Infant Psychology, 17,* 119–132.

Hull, D. L., Langman, R. E., & Glenn, S. S. (2001). A general account of selection: Biology, immunology and behavior. *Behavioral and Brain Sciences, 24,* 511–528.

Humphrey, N. K. (1976). The social function of intellect. In P. P. G. Bateson & R. A. Hinde (Eds.), *Growing points in ethology* (pp. 303–317). Cambridge, England: Cambridge University Press.

Humphreys, A. P., & Smith, P. K. (1987). Rough-and-tumble play, friendship, and dominance in school children: Evidence for continuity and change with age. *Child Development, 58,* 201–212.

Hunt, J. M. (1961). *Intelligence and experience.* New York: Ronald Press.

Hutt, C. (1966). Exploration and play in children. *Symposia of the Zoological Society of London, 18,* 61–81.

Hutt, C. (1972). Sex differences in human development. *Human Development, 15,* 153–170.

Hymovitch, B. (1952). The effects of experimental variations on problem solving in the rat. *Journal of Comparative and Physiological Psychology, 45,* 313–321.

Hyson, M. C., Hirsh-Pasek, K., & Rescorla, L. (1990). Academic environments in preschool: Challenge or pressure? *Early Education and Development, 1,* 401–423.

Ingman, M., Kaessmann, H., Pääbo, S., & Gyllensten, U. (2000). Mitochondrial genome variation and the origin of modern humans. *Nature, 408,* 708–713.

Isabella, R. A., & Belsky, J. (1991). Interactional synchrony and the origins of infant–mother attachment. *Child Development, 62,* 373–384.

Itard, J.M.G. (1962). *The wild boy of Aveyron* (G. Humphrey & M. Humphrey, Trans.). New York: Appleton-Century-Crofts.

Jablonka, E., & Lamb, M. J. (1995). *Epigenetic inheritance and evolution: The Lamarckian dimension.* Oxford, England: Oxford University Press.

Jackson, J. F. (1993). Human behavioral genetics: Scarr's theory, and her views on intervention: A critical review and commentary on their implications for African American children. *Child Development, 64,* 1318–1332.

Jacobsen, S. W. (1979). Matching behavior in the young infant. *Child Development, 50,* 425–430.

Jacobsen, T., Edelstein, W., & Hofmann, V. (1994). A longitudinal study of the relation between representations of attachment in childhood and cognitive functioning in childhood and adolescence. *Developmental Psychology, 30,* 112–124.

Jacobson, M. (1969). Development of specific neuronal connections. *Science, 163,* 543-547.

Jankowiak, W., & Diderich, M. (2000). Sibling solidarity in a polygamous community in the USA: Unpacking inclusive fitness. *Evolution and Human Behavior, 21,* 125–139.

Jarrold, C., Carruthers, P., Smith, P. K., & Boucher, J. (1994). Pretend play: Is it metarepresentational? *Mind and Language, 9,* 445–468.

Jenkins, J. M., & Astington, J. W. (1996). Cognitive factors and family structure associated with theory of mind development in young children. *Developmental Psychology, 32,* 70–78.

Jensen, A. R. (1980). *Bias in mental testing.* New York: Free Press.

Jensen, A. R. (1998). *The g factor: The science of mental ability.* Westport, CT: Praeger.

Jensen, P. S., Mrazek, D., Knapp, P. K., Steinberg, L., Pfeffer, C., Schwalter, J., & Shapiro, T. (1997). Evolution and revolution in child psychiatry: ADHD as a disorder of adaptation. *Journal of the American Academy of Child & Adolescent Psychiatry, 36,* 1672–1681.

Jensvold, M. L. A., & Fouts, R. S. (1993). Imaginary play in chimpanzees (*Pan troglodytes*). *Human Evolution, 8,* 217–227.

Jerison, H. J. (1973). *Evolution of the brain and intelligence*. New York: Academic Press.

Joffe, T. H. (1997). Social pressures have selected for an extended juvenile period in primates. *Journal of Human Evolution, 32*, 593–605.

Johanson, D., & Edgar, B. (1996). *From Lucy to language*. New York: Simon & Schuster.

Johnson, D. W., & Johnson, R. T. (1989). *Cooperation and competition: Theory and research*. Edina, MN: Interaction.

Johnson, J. S., & Newport, E. L. (1989). Critical period effects in second language learning: The influence of instructional state on the acquisition of English as a second language. *Cognitive Psychology, 21*, 60–99.

Johnson, M. H. (1998). The neural basis of cognitive development. In W. Damon (Gen. Ed.) & D. Kuhn & R. S. Siegler (Eds.), *Handbook of child psychology: Vol. 2. Cognition, perception, and language* (5th ed., pp. 1–49). New York: Wiley.

Johnson, M. H. (2000). Functional brain development in infants: Elements of an interactive specialization framework. *Child Development, 71*, 75–81.

Johnson, M. H., Dziurawiec, S., Ellis, H. D., & Morton, J. (1991). Newborns' preferential tracking of faces and its subsequent decline. *Cognition, 40*, 1–19.

Johnson, M. H., & Morton, J. (1991). *Biology and cognitive development: The case of face recognition*. Oxford, England: Blackwell.

Johnson, S. P., & Aslin, R. N. (1995). Perception of object unity in 2-month-old infants. *Developmental Psychology, 31*, 739–745.

Johnson, S. P., & Aslin, R. N. (1996). Perception of object unity in young infants: The roles of motion, depth, and orientation. *Cognitive Development, 11*, 161–180.

Jolly, A. (1966). Lemur social behavior and primate intelligence. *Science, 153*, 501–506.

Kagan, J. (1996). Three pleasing ideas. *American Psychologist, 51*, 901–908.

Kail, R. V. (1991). Developmental changes in speed of processing during childhood and adolescence. *Psychological Bulletin, 109*, 490–501.

Kail, R. V. (1997). Processing time, imagery, and spatial memory. *Journal of Experimental Child Psychology, 64*, 67–78.

Kail, R. V., & Salthouse, T. A. (1994). Processing speed as a mental capacity. *Acta Psychologica, 86*, 199–225.

Kail, R. V., Jr., & Siegel, A. W. (1977). Sex differences in retention of verbal and spatial characteristics of stimuli. *Journal of Experimental Child Psychology, 23*, 341–347.

Kaler, S. R., & Freedman, B. J. (1994). Analysis of environmental deprivation: Cognitive and social development in Romanian orphans. *Journal of Child Psychology and Psychiatry, 35*, 769–781.

Kamin, L. J. (1974). *The science and politics of IQ*. Potomac, MD: Erlbaum.

Kaplan, H., Hill, K., Lancaster, J., & Hurtado, A. M. (2000). A theory of human life history evolution: Diet, intelligence, and longevity. *Evolutionary Anthropology, 9*, 156–185.

Karen, R. (1990, February). Becoming attached. *Atlantic Monthly*, pp. 35–70.

Karmiloff-Smith, A. (1979). *A functional approach to language*. Cambridge, England: Cambridge University Press.

Karmiloff-Smith, A. (1991). Beyond modularity: Innate constraints and developmental change. In S. Carey & R. Gelman (Eds.), *The epigenesis of mind: Essays on biology and cognition* (pp. 171–197). Hillsdale, NJ: Erlbaum.

Karmiloff-Smith, A. (1992). *Beyond modularity: A developmental perspective on cognitive science*. Cambridge, MA: MIT Press.

Karp, S. A., & Konstadt, N. (1971). *Children's Embedded Figures Test*. Palo Alto, CA: Consulting Psychologists Press.

Karzon, R. G. (1985). Discrimination of polysyllabic sequences by one- to four-month-old infants. *Journal of Experimental Child Psychology, 39*, 326–342.

Kawai, M. (1965). Newly acquired pre-cultural behavior of natural troop of Japanese monkeys. *Primates, 6*, 1–30.

Keeley, L. H. (1996). *War before civilization: The myth of the peaceful savage*. New York: Oxford University Press.

Keller, H. (2000). Human parent–child relationships from an evolutionary perspective. *American Behavioral Scientist, 43*, 957–969.

Kellman, P. J., & Banks, M. S. (1998). Infant visual perception. In W. Damon (Gen. Ed.) & D. Kuhn & R. S. Siegler (Eds.), *Handbook of child psychology: Vol. 2. Cognitive, language, and perceptual development* (5th ed., pp. 103–146). New York: Wiley.

Kellman, P. J., & Spelke, E. S. (1983). Perception of partly occluded objects in infancy. *Cognitive Development, 15*, 483–524.

Kenny, P., & Turkewitz, G. (1986). Effects of unusually early visual stimulation on the development of homing behavior in the rat pup. *Developmental Psychobiology, 19*, 57–66.

Kermoian, R., & Campos, J. J. (1988). Locomotor experience: A facilitator of spatial cognitive development. *Child Development, 59*, 908–917.

Kersten, A. W., & Earles, J. L. (2001). Less really is more for adults learning a miniature artificial language. *Journal of Memory and Language, 44*, 250–273.

Kettlewell, H. B. D. (1959). Darwin's missing evidence. *Scientific American, 200*, 48–53.

Kim, K. H. S., Relkin, N. R., Lee, K. -M., & Hirsch, J. (1997). Distinct cortical areas associated with native and second languages. *Nature, 388*, 171–174.

Kim, K., Smith, P. K., & Palermiti, A. (1997). Conflict in childhood and reproductive development. *Evolution and Human Development, 18*, 109–142.

Kline, J., Stein, Z., & Susser, M. (1989). *Conception to birth: Epidemiology of prenatal development*. New York: Oxford University Press.

Kluender, K. R., Diehl, R. L., & Killeen, P. R. (1987). Japanese quail can learn phonetic categories. *Science, 237*, 1195–1197.

Kochanska, G., Murray, K., Jacques, T. Y., Koenig, A. L., & Vandegeest, K. A. (1996). Inhibitory control in young children and its role in emerging internalization. *Child Development, 67*, 490–507.

Köhler, W. (1925). *The mentality of apes.* London: Kegan Paul.

Koluchova, J. (1976). The further development of twins after severe and prolonged deprivation: A second report. *Journal of Child Psychology & Psychiatry & Allied Disciplines, 17*, 181–188.

Konner, M. (1972). Aspects of the developmental ethology of a foraging people. In N. Blurton Jones (Ed), *Ethological studies of child behaviour* (pp. 285–304). Cambridge, England: Cambridge University Press.

Kotovsky, L., & Baillargeon, R. (1998). Calibration-based reasoning about collision events in 11 month-old infants. *Cognition, 67*, 311–351.

Kovacs, D. M., Parker, J. G., & Hoffman, L. W. (1996). Behavioral, affective, and social correlates of involvement in cross-sex friendship in elementary school. *Child Development, 67*, 2269–2286.

Krumhansl, C. L., & Jusczyk, P. W. (1990). Infants' perception of phrase structure in music. *Psychological Science, 1*, 70–73.

Kuhl, P. K., Andruski, J. E., Christovich, I. A., Christovich, L. A., Kozhevnikova, E. V., Ryskina, V. L., Stolyarova, E. I., Sundberg, U., & Lacerda, F. (1997). Cross-language analysis of phonetic units in language addressed to infants. *Science, 277*, 684–686.

Kuhl, P. K., & Miller, J. L. (1975). Speech perception by the chinchilla: Voiced–voiceless distinction in alveolar plosive consonants. *Science, 190*, 69–72.

Kuo, Z. Y. (1967). *The dynamics of behavior development: An epigenetic view.* New York, NY: Random House.

Lamb, M. E. (Ed.). (1997). *The role of the father in child development* (3rd ed). New York: Wiley.

Lamb, M. E., Frodi, A. M., Hwang, C.-P., & Frodi, M. (1982). Characteristics of maternal and paternal behavior in traditional and nontraditional Swedish families. *International Journal of Behavioral Development, 5*, 131–141.

Lancaster, J. B. (1989). Evolutionary and cross-cultural perspectives on single-parenthood. In R. W. Bell & N. J. Bell (Eds.), *Sociobiology and the social sciences* (pp. 63–72). Austin: University of Texas Press.

Lancaster, J. B., & Lancaster, C. S. (1983). Parental investment: The hominid adaptation. In D. J. Ortner (Ed.), *How humans adapt: A biocultural odyssey* (pp. 33–56). Washington, DC: Smithsonian Institution Press.

Lancaster, J. B., & Lancaster, C. S. (1987). The watershed: Change in parental-investment and family-formation strategies in the course of human evolution. In J. B. Lancaster, J. Altmann, A. S. Rossi, & L. R. Sherrod (Eds.), *Parenting across the life span: Biosocial dimensions* (pp. 187–205). New York: Aldine de Gruyter.

Langer, J. (1998). Phylogenetic and ontogenetic origins of cognition: Classification. In J. Langer & M. Killen (Eds.), *Piaget, evolution, and development* (pp. 33–54). Mahwah, NJ: Erlbaum.

Langer, J. (2000). The heterochronic evolution of primate cognitive development. In S. T. Parker, J. Langer, & M. L. McKinney (Eds.), *Biology, brains, and behavior: The evolution of human development* (pp. 215–235). Santa Fe, NM: School of American Research Press.

Langlois, J. H., Ritter, J. M., Roggman, L. A., & Vaughn, L. S. (1991). Facial diversity and infant preferences for attractive faces. *Developmental Psychology, 27,* 79–84.

Langlois, J. H., & Roggman, L. A. (1990). Attractive faces are only average. *Psychological Science, 1,* 115–121.

Langlois, J. H., Roggman, L. A., Casey, R. J., Ritter, J. M., Rieser-Danner, L. A., & Jenkins, V. Y. (1987). Infant preferences for attractive faces: Rudiments of a stereotype? *Developmental Psychology, 23,* 363–369.

Laslett, P. (1965). *The world we have lost.* New York: Scribner.

Laursen, B., & Hartup, W. W. (1989). The dynamics of preschool children's conflicts. *Merrill–Palmer Quarterly, 35,* 281–297.

Leakey, M. G., Spoor, F., Brown, F. H., Gathogo, P. N., Kiarie, C., Leakey, L. N., & McDougall, I. (2001). New hominid genus from eastern Africa shows diverse middle Pliocene lineages. *Nature, 410,* 433–440.

Lee, B. J., & George, R. M. (1999). Poverty, early childbearing and child maltreatment: A multinomial analysis. *Children & Youth Services Review, 21,* 755–780.

Legerstee, M. (1991). The role of person and object in eliciting early imitation. *Journal of Experimental Child Psychology, 51,* 423–433.

Lehrman, D. S. (1970). Semantic and conceptual issues in the nature-nurture problem. In L. R. Aronson, D. S. Lehrman, E. Tobach, & J. S. Rosenblatt (Eds.), *Development and evolution of behavior* (pp. 17–52). San Francisco: Freeman.

Leinbach, M. D., & Fagot, B. I. (1993). Categorical habituation to male and female faces: Gender schematic processing in infancy. *Infant Behavior and Development, 16,* 317–332.

Lerner, R. M., & von Eye, A. (1992). Sociobiology and human development: Arguments and evidence. *Human Development, 35,* 12–33.

Leslie, A. M. (1987). Pretense and representation: The origins of "theory of mind." *Psychological Review, 94,* 412–426.

Leslie, A. (1994). ToMM, ToBY, and agency: Core architecture and domain specificity. In L. Hirschfeld & S. Gelman (Eds.), *Mapping the mind: Domain specificity in cognition and culture* (pp. 119–148). Cambridge, England: Cambridge University Press.

Lever, J. (1976). Sex differences in the games children play. *Social Problems, 23,* 470–487.

Lever, J. (1978). Sex differences in the complexity of children's play and games. *American Sociological Review, 43,* 471–483.

Levine, S. C., Huttenlocher, J., Taylor, A., & Langrock, A. (1999). Early sex differences in spatial skills. *Developmental Psychology, 35,* 940–949.

Lewis, C., Freeman, N. H., Kyriakidou, C., Maridaki-Kassotaki, K., & Berridge, D. M. (1996). Social influence on false belief access: Specific sibling influences or general apprenticeship? *Child Development, 67,* 2930–2947.

Lewis, M., Feiring, C., McGuffog, C., & Jaskir, J. (1984). Predicting psychopathology in six-year-olds from early social relations. *Child Development, 55,* 123–136.

Lewis, M., Feiring, C., & Rosenthal, S. (2000). Attachment over time. *Child Development, 71,* 707–720.

Lewis, M., Young, G., Brooks, J., & Michalson, L. (1975). The beginning of friendship. In M. Lewis & L. A. Rosenblum (Eds.), *Friendship and peer relations* (pp. 27–66). New York: Wiley.

Lewontin, R. C. (1982). Gene, organism, and environment. In D. S. Bendell (Ed.), *Evolution: From molecules to men* (pp. 273–285). London: Cambridge University Press.

Lickliter, R. (1990). Premature visual stimulation accelerates intersensory functioning in bobwhite quail neonates. *Developmental Psychobiology, 23,* 15–27.

Lickliter, R. (1996). Structured organisms and structured environments: Development systems and the construction of learning capacities. In J. Valsiner & H. Voss (Eds.), *The structure of learning processes* (pp. 86–107). Norwood, NJ: Ablex.

Lickliter, R., & Berry, T. D. (1990). The phylogeny fallacy: Developmental psychology's misapplication of evolutionary theory. *Developmental Review, 10,* 348–364.

Lickliter, R., & Hellewell, T. B. (1992). Contextual determinants of auditory learning in bobwhite quail embryos and hatchlings. *Developmental Psychobiology, 25,* 17–24.

Lickliter, R., & Lewkowitz, D. J. (1995). Intersensory experience and early perceptual development: Attenuated prenatal sensory stimulation affects postnatal auditory and visual responsiveness in bobwhite quail chicks (*Colinus virginianus*). *Developmental Psychology, 31,* 609–618.

Lie, E., & Newcombe, N. S. (1999). Elementary school children's explicit and implicit memory for faces of preschool classmates. *Developmental Psychology, 35,* 102–112.

Lillard, A. S. (1993a). Pretend play skills and the child's theory of mind. *Child Development, 64,* 348–371.

Lillard, A. S. (1993b). Young children's conceptualization of pretense: Action or mental representational state. *Child Development, 64,* 372–386.

Lillard, A. S. (in press). Pretend play as Twin Earth: A social–cognitive analysis. *Developmental Review.*

Locke, J. L. (1993). *The child's path to spoken language.* Cambridge, MA: Harvard University Press.

Locke, J. L. (1994). Phases in the child's development of language. *American Scientist, 82*, 436–445.

Locke, J. L. (1996). Why do infants begin to talk? Language as an unintended consequence. *Journal of Child Language, 23*, 251–268.

Logue, A. W., & Chavarro, A. (1992). Self-control and impulsiveness in preschool children. *Psychological Record, 42*, 189–204.

Loizos, C. (1967). Play behavior in higher primates: A review. In D. Morris (Ed.), *Primate ethology* (pp. 176–218). Chicago: Aldine.

Lorenz, K. (1937). The companion in the bird's world. *Auk, 54*, 245–273.

Lorenz, K. (1943). Die angeborenen Formen möglicher Erfahrung. (The innate forms of experience). *Zeitschrift fur Tierpsychologie, 5*, 235–409.

Lorenz, K. (1965). *Evolution and modification of behavior.* Chicago: University of Chicago Press.

Lukas, W. D., & Campbell, B. C. (2000). Evolutionary and ecological aspects of early brain malnutrition in humans. *Human Nature, 11*, 1–26.

Lummaa, V., Vuorisalo, T., Barr, R. G., & Lehtonen, L. (1998). Why cry? Adaptive significance of intensive crying in human infants. *Evolution and Human Behavior, 19*, 193–202.

Luria, A. R. (1973). *The working brain: An introduction to neuropsychology.* New York: Basic Books.

Lussier, L., & Buskirk, E. R. (1977). Effects of an endurance training regimen on assessment of work capacity in pre-pubertal children. *Annals of the New York Academy of Sciences, 301*, 734–747.

Lynch, M. P., Eilers, R. E., Oller, K., & Urbano, R. C. (1990). Innateness, experience, and music perception. *Psychological Science, 1*, 272–276.

Maccoby, E. E. (1988). Gender as a social category. *Developmental Psychology, 24*, 755–756.

Maccoby, E. E. (1998). *The two sexes: Growing up apart, coming together.* Cambridge, MA: Harvard University Press.

Maccoby, E., & Jacklin, C. (1974). *The psychology of sex differences.* Stanford, CA: Stanford University Press.

Maccoby E. E., & Jacklin, C. N. (1987). Gender segregation in childhood. In H. W. Rose (Ed.), *Advances in child development and behavior* (Vol. 20, pp. 239–287). New York: Academic Press.

MacDonald, K. B. (Ed.). (1988). *Sociobiological perspectives on human development.* New York: Springer-Verlag.

MacDonald, K. (1997). Life history theory and human reproductive behavior: Environmental/contextual influences and heritable variation. *Human Nature, 8*, 327–359.

MacDonald, K., & Geary, D. C. (2000, June). *The evolution of general intelligence: Domain-general cognitive mechanisms and human adaptation.* Paper presented at meeting of Human Evolution and Behavior Society, Amherst, MA.

MacDonald, K., & Parke, R. (1986). Parent–child physical play. *Sex Roles, 15,* 367–378.

Macfarlane, A. (1975). Olfaction in the development of social preferences in the humane neonate. *CIBA Foundation Symposium 33: Parent–Infant Interaction,* Amsterdam, The Netherlands: Elsevier.

Mackintosh, H. J. (1995). *Cyril Burt: Fraud or framed?* Oxford, England: Oxford University Press.

Makin, J. W., & Porter, R. H. (1989). Attractiveness of lactating females' breast odors to neonates. *Child Development, 60,* 803–810.

Mandel, D. R., Jusczyk, P., & Pisoni, D. B. (1995). Infants' recognition of the sound patterns of their own names. *Psychological Science, 6,* 314–317.

Mandler, J. M. (1992). How to build a baby: II. Conceptual primitives. *Psychological Review, 99,* 587–604.

Mandler, J. M. (1998). Representation. In W. Damon (Gen. Ed.) & D. Kuhn & R. S. Siegler (Eds.), *Handbook of child psychology: Vol. 2. Cognition, perception, and language* (5th ed., pp. 255–308). New York: Wiley.

Mandler, J. M. (2000). Perceptual and conceptual processes in infancy. *Journal of Cognition and Development, 1,* 3–36.

Mandler, J. M., & McDonough, L. (1995). Long-term recall of event sequences in infancy. *Journal of Experimental Child Psychology, 59,* 457–474.

Manion, V., & Alexander, J. M. (1997). The benefits of peer collaboration on strategy use, metacognitive causal attribution, and recall. *Journal of Experimental Child Psychology, 67,* 268–289.

Mann, J. (1992). Nurture or negligence: Maternal psychology and behavioral preference among preterm twins. In J. Barkow, L. Cosmides, & J. Tooby (Eds.), *The adapted mind: Evolutionary psychology and the generation of culture* (pp. 367–390). New York: Oxford University Press.

Marks, I. M. (1969). *Fears and phobias.* New York: Academic Press.

Marlier, L., Schaal, B., & Soussignan, R. (1998). Neonatal responsiveness to the odor of amniotic and lacteal fluids: A test of perinatal chemosensory continuity. *Child Development, 69,* 611–623.

Marlowe, F. (1999). Showoffs or providers? The parenting effort of Hazda men. *Evolution and Human Behavior, 20,* 391–404.

Martin, H., Beezley, P. C., Conway, E., & Kempe, H. (1974). The development of abused children: A review of the literature. *Advances in Pediatrics, 21,* 119–134.

Martin, P. (1982). The energy costs of play. Definition and estimation. *Animal Behaviour, 30,* 294–295.

Martin, P., & Bateson, P. P. G. (1993). *Measuring behavior.* Cambridge, England: Cambridge University Press.

Martin, P., & Caro, T. M. (1985). On the function of play and its role in behavioral development. In J. Rosenblatt, C. Beer, M. Bushnel, & P. Slater (Eds.), *Advances in the study of behavior* (Vol. 15, pp. 59–103). New York: Academic Press.

Masataka, N. (1996). Perception of motherese in a signed language by 6-month-old deaf infants. *Developmental Psychology, 32,* 874–879.

Masataka, N. (1998). Perception of motherese in Japanese sign language by 6-month-old hearing infants. *Developmental Psychology, 34,* 241–246.

Mason, W. A. (1968a). Early social deprivation in the nonhuman primates: Implications for human behavior. In D. C. Glass (Ed.), *Environmental influence* (pp. 70–101). New York: Rockefeller University Press.

Mason, W. A. (1968b). Scope and potential of primate research. In J. H. Masserman (Ed.), *Science and psychoanalysis, Vol. 12: Animal and human* (pp. 101–118). New York: Grune & Stratton.

Masten, A. S., & Coatsworth, J. D. (1998). The development of competence in favorable and unfavorable environments. *American Psychologist, 53,* 205–220.

Mathews, W. S. (1977). Modes of transformation in the initiation of fantasy play. *Developmental Psychology, 13,* 212–216.

Matsuzawa, T. (1999). Communication and tool use in chimpanzees: Cultural and social contexts. In M. Hauser & M. Knoish (Eds.), *The design of animal communication* (pp. 645–671). Cambridge, MA: MIT Press.

Matthews, M. H. (1987). Sex differences in spatial competence: The ability of young children to map "primed" unfamiliar environments. *Educational Psychology, 7,* 77–90.

Matthews, M. H. (1992). *Making sense of place: Children's understanding of large-scale environments.* Savage, MD: Barnes & Noble Books.

Maynard Smith, J. (1972). *On evolution.* Edinburgh, Scotland: Edinburgh University Press.

Mayr, E. (1942). *Systematics and the origins of species from the viewpoint of a zoologist.* New York: Columbia University Press.

Mayr, E. (1982) *The growth of biological thought: Diversity, evolution, and inheritance.* Cambridge, MA: Belknap Press.

McBride, T., & Lickliter, R. (1994). Specific postnatal auditory stimulation interferes with species-typical responsiveness to maternal visual cues in bobwhite quail chicks. *Journal of Comparative Psychology, 107,* 320–327.

McCall, R. (1977). Challenges to a science of developmental psychology. *Child Development, 48,* 333–394.

McCartney, K., Harris, M. J., & Bernieri, F. (1990). Growing up and growing apart: A development meta-analysis of twin studies. *Psychological Bulletin, 97,* 226–237.

McCune-Nicholich, L., & Fenson, L. (1984). Methodological issues in the study of early pretend play. In T. D. Yawkey & A. D. Pellegrini (Eds.), *Child's play* (pp. 81–104). Hillsdale, NJ: Erlbaum.

McDonough, L., Mandler, J. M., McKee, R. D., & Squire, L. R. (1995). The deferred imitation task as a nonverbal measure of declarative memory. *Proceedings of the National Academy of Sciences, 92,* 7580–7584.

McGrew, W. C. (1972). *An ethological study of children's behaviour*. London: Academic Press.

McGrew, W. C. (1992). *Chimpanzee material culture: Implications for human evolution*. Cambridge. England: Cambridge University Press.

McGrew, W. C., & Tutin, C. E. G. (1978). Evidence for a social custom in wild chimpanzees? *Man, 13*, 234–251.

McGuffin, P., Riley, B., & Plomin, R. (2001). Toward behavioral genomics. *Science, 291*, 1232–1233.

McHale, S. M., & Gramble, W. C. (1989). Sibling relationships of children with disabled and nondisabled brothers and sisters. *Developmental Psychology, 25*, 421–429.

McKinney, M. L. (1998). Cognitive evolution by extending brain development: On recapitulation, progress, and other heresies. In J. Langer & M. Killen (Eds.), *Piaget, evolution, and development* (pp. 9–31). Mahwah, NJ: Erlbaum.

McKinney, M. L. (2000). Evolving behavioral complexity by extending development. In S. T. Parker, J. Langer, & M. L. McKinney (Eds.), *Biology, brains, and behavior: The evolution of human development* (pp. 25–40). Santa Fe, NM: School of American Research Press.

McKinney, M. L., & McNamara, K. (1991). *Heterochrony: The evolution of ontogeny*. New York: Plenum.

McLain, D. K., Setters, D., Moulton, M. P., & Pratt, A. E. (2000). Ascription of resemblance of newborns by parents and nonrelatives. *Evolution and Human Behavior, 21*, 11–23.

McLoyd, V. (1980). Verbally expressed modes of transformation in the fantasy and play of black preschool children. *Child Development, 51*, 1133–1139.

McNally, R. J. (1987). Preparedness and phobias: A review. *Psychological Bulletin, 101*, 283–303.

Meaney, M. J., Stewart, J., & Beatty, W.W. (1985). Sex differences in social play. In J. Rosenblatt, C. Beer, M. C. Bushnell, & P. Slater (Eds.), *Advances in the study of behavior: Vol. 15* (pp. 2–58). New York: Academic Press.

Mehler, J., Jusczyk, P., Lambertz, G., Halsted, N., Bertoncini, J., & Amiel-Tison, C. (1988). A precursor of language acquisition in young infants. *Cognition, 29*, 143–178.

Meltzoff, A. N. (1990). Towards a developmental cognitive science: The implications of cross-modal matching and imitation for the development of memory in infancy. In A. Diamond (Ed.), *The development and neural bases of higher cognitive functions* (pp. 1–37), Vol 608, Annals of the New York Academy of Sciences.

Meltzoff, A. N. (1995). Understanding the intentions of others: Re-enactment of intended acts by 18-month-old children. *Developmental Psychology, 31*, 838–850.

Meltzoff, A. N., & Moore, M. K. (1977). Imitation of facial and manual gestures by human neonates. *Science, 198*, 75–78.

Meltzoff, A. N., & Moore, M. K. (1985). Cognitive foundations and social functions of imitation and intermodal representation in infancy. In J. Mehler & R. Fox (Eds.), *Neonate cognition: Beyond the booming buzzing confusion* (pp. 139–156). Hillsdale, NJ: Erlbaum.

Meltzoff, A. N., & Moore, M. K. (1992). Early imitation within a functional framework: The importance of person identity, movement, and development. *Infant Behavior and Development, 15,* 479–505.

Meltzoff, A. N., & Moore, M. K. (1997). Explaining facial imitation: A theoretical model. *Early Development and Parenting, 6,* 179–192.

Menzel, E. W., Jr. (1974). A group of young chimpanzees in a 1-acre field: Leadership and communication. In A. M. Schrier & F. Stollnitz (Eds.), *Behavior of nonhuman primates* (Vol. 5, pp. 83–153). New York: Academic Press.

Meulemans, T., Van der Linden, M., & Perruchet, P. (1998). Implicit sequence learning in children. *Journal of Experimental Child Psychology, 69,* 199–221.

Michalski, R. L., & Shackelford, T. K. (June, 1999). *Does birth order predict mating strategy?* Paper presented at Conference of the Human Behavior and Evolution Society, Salt Lake City, UT.

Miles, H. L. (1990). The cognitive foundations for reference in a signing orangutan. In S. T. Parker & K. K. Gibson (Eds.), *"Language" and intelligence in monkeys and apes* (pp. 511–539). Cambridge, England: Cambridge University Press.

Miles, H. L. (1994). ME CHANTEK: The development of self-awareness in a signing orangutan. In S. T. Parker, R. W. Mitchell, & M. L. Broccia (Eds.), *Self-awareness in animals and humans: Developmental perspectives* (pp. 254–272). Cambridge, England: Cambridge University Press.

Miles, H. L., Mitchell, R. W., & Harper, S. E. (1996). Simon says: The development of imitation in an enculturated orangutan. In A. E. Russon, K. A. Bard, & S. T. Parker (Eds.), *Reaching into thought: The minds of the great apes* (pp. 278–299). New York: Cambridge University Press.

Miller, L. T., & Vernon, P. A. (1996). Intelligence, reaction time, and working memory in 4- to 6-year-old children. *Intelligence, 22,* 155–190.

Miller, L. T., & Vernon, P. A. (1997). Developmental changes in speed of information processing in young children. *Developmental Psychology, 33,* 549–554.

Miller, N. E., & Dollard, J. (1941). *Social learning and imitation.* New Haven, CT: Yale University Press.

Milner, B. (1964). Some effects of frontal lobectomy in man. In J. M. Warren & K. Akert (Eds.), *The frontal granular cortex and behavior* (pp. 313–334). New York: McGraw-Hill.

Mineka, S. (1992). Evolutionary memories, emotional processing, and the emotional disorders. In D. L. Medin (Ed.), *The psychology of learning and motivation: Advances in research and theory* (Vol. 28, pp. 161–206). San Diego, CA: Academic Press.

Mineka, S., Davidson, M., Cook, M., & Keir, R. (1984). Observational conditioning of snake fear in rhesus monkeys. *Journal of Abnormal Psychology, 93,* 355–372.

Mithen, S. (1996). *The prehistory of the mind: The cognitive origins of art, religion and science*. London: Thames & Hudson.

Mix, K., Huttenlocher, J., & Levine, S. C. (1996). Do preschool children recognize auditory-visual numerical correspondence? *Child Development, 67,* 1592–1608.

Moerk, E. L. (1986). Environmental factors in early language acquisition. In G. J. Whitehurst (Ed.), *Annals of child development* (Vol. 3, pp. 191–236). Greenwich, CT: JAI.

Moffat, S. D., Hampson, E., & Hatzipantelis, M. (1998). Navigation in a "virtual" maze: Sex differences and correlation with psychometric measures of spatial ability in humans. *Evolution & Human Behavior, 19,* 73–87.

Moffitt, T. E., Caspi, A., Belsky, J., & Silva, P. A. (1992). Childhood experience and the onset of menarche: A test of a sociobiological hypothesis. *Child Development, 63,* 47–58.

Molenaar, P. C. M., Boomsma, D. I., & Dolan, C. V. (1993). A third source of developmental differences. *Behavior Genetics, 23,* 519–524.

Mondloch, C. J., Lewis, T. L., Budreau, D. R., Maurer, D., Dannemiller, J. L., Stephens, B. R., & Kleiner-Gathercoal, K. A. (1999). Face perception during early infancy. *Psychological Science, 10,* 419–422.

Money, J., Alexander, D., & Walker, H. T., Jr. (1965). *A standardized road-map test of direction sense*. Baltimore: Johns Hopkins University Press.

Money, J., & Ehrhardt, A. A. (1972). *Man and woman, boy and girl*. Baltimore: Johns Hopkins University Press.

Montagu, A. (1989). *Growing young* (2nd ed.). Grandy, MA: Bergin & Garvey.

Montagu, M. F. A. (1962). Time, morphology, and neoteny in the evolution of man. In M. F. A. Montagu (Ed.), *Culture and the evolution of man* (pp. 324–342). New York: Oxford University Press.

Moore, D. S., Spence, M. J., & Katz, G. S. (1997). Six-month-olds' categorization of natural infant-directed utterances. *Developmental Psychology, 33,* 980–989.

Morss, J. R. (1990). *The biologising of childhood: Developmental psychology and the Darwinian myth*. Hillsdale, NJ: Erlbaum.

Morton, J., & Johnson, M. H. (1991). CONSPEC and CONLERN: A two-process theory of infant face recognition. *Psychological Review, 98,* 164–181.

Moskowitz, B. A. (1978, November). The acquisition of language. *Scientific American,* 92–108.

Müller-Schwarze, D. (1984). Analysis of play behavior. In P. K Smith (Ed.), *Play in animals and humans* (pp. 147–158). Oxford, England: Blackwell.

Murray, A. D. (1985). Aversiveness is in the mind of the beholder: Perception of infant crying by adults. In B. M. Lester & C. F. Boukydis (Eds.), *Infant crying: Theoretical and research perspectives* (pp. 217–240). New York: Plenum.

Nagell, K., Olguin, K., & Tomasello, M., (1993). Processes of social learning in the tool use of chimpanzees (*Pan troglodytes*) and human children (*Homo sapiens*). *Journal of Comparative Psychology, 107,* 174–186.

National Center for Health Statistics. (1999). *Health, United States*. Hyattsville, MD: U.S. Department of Health and Human Services.

Nelson, C. A., & Bloom, F. E. (1997). Child development and neuroscience. *Child Development, 68*, 970–987.

Nelson J., & Aboud F. E. (1985). The resolution of social conflict between friends. *Child Development, 56*, 1009–1017.

Nelson, K. (1993). The psychological and social origins of autobiographical memory. *Psychological Science, 4*, 7–14.

Nelson, K. (1996). *Language in cognitive development: The emergence of the mediated mind*. New York: Cambridge University Press.

Nesse, R. M. (1990). Evolutionary explanations of emotions. *Human Nature, 1*, 261–289.

Nesse, R. M., & Williams, G. C. (1994). *Why we get sick: The new science of Darwinian medicine*. New York: Times Books.

Newcombe, A. F., & Bagwell, C. L. (1995). Children's friendship relationships: A meta-analytic review. *Psychological Bulletin, 117*, 306–347.

Newcombe, N., Bandura, M. M., & Taylor, D. C. (1983). Sex differences in spatial ability and spatial activities. *Sex Roles, 9*, 377–386.

Newcombe, N., & Fox, N. A. (1994). Infantile amnesia: Through a glass darkly. *Child Development, 65*, 31–40.

Newcombe, N., Huttenlocher, J., Drummey, A. B., & Wiley, J. (1998). The development of spatial location coding: Use of external frames of reference and dead reckoning. *Cognitive Development, 13*, 185–200.

Newcombe, N., Huttenlocher, J., & Learmonth, A. (1999). Infants' coding of location in continuous space. *Infant Behavior and Development, 22*, 483–510.

Newport, E. L. (1990). Maturational constraints on language learning. *Cognitive Science, 14*, 11–28.

Newport, E. L. (1991). Constraining concepts of the critical period for language. In S. Carey & R. Gelman (Eds.), *The epigenesis of mind: Essays on biology and cognition* (pp. 111–130). Hillsdale, NJ: Erlbaum.

Nunn, C. L. (1999). The evolution of exaggerated sexual swellings in primates and the graded signal hypothesis. *Animal Behaviour, 58*, 229–246.

Nunn, C. L., Gittleman, J. L., & Antonovics, J. (2000). Promiscuity and the primate immune system. *Science, 290*, 1168–1170.

Ober, C., Weitkamp, L. R., Cox, N., Dytch, H., Kostyu, D., & Elias, S. (1997). HLA and mate choice in humans. *American Journal of Human Genetics, 61*, 497–504.

O'Brien, M., & Huston, A. C. (1985). Development of sex-typed play behavior in toddlers. *Developmental Psychology, 21*, 866–871.

O'Connell, J. F., Hawkes, K., & Blurton Jones, N. G. (1999). Grandmothering and the evolution of *Homo erectus*. *Journal of Human Evolution, 36*, 461–485.

O'Connor, T. G., Rutter, M., Beckett, C., Keaveney, L., Kreppner, J. M., & the English and Romanian Adoptees Study Team. (2000). The effects of global severe privation on cognitive competence: Extension and longitudinal follow-up. *Child Development, 71*, 376–390.

Öhman, A. (1986). Face the beast and fear the face: Animal and social fears as prototypes for evolutionary analyses of emotion. *Psychophysiology, 23*, 123–145.

Olds, D. L., Henderson, C. R., & Tatelbaum, R. (1994). Intellectual impairment in children of women who smoke cigarettes during pregnancy. *Pediatrics, 93*, 221–227.

Oliver, M. B., & Hyde, J. S. (1993). Gender differences in sexuality: A meta-analysis. *Psychological Bulletin, 114*, 29–36.

Olweus, D. (1993). *Bullying at school*. Cambridge, MA: Blackwell.

Oppenheim, R. W. (1981). Ontogenetic adaptations and retrogressive processes in the development of the nervous system and behavior. In K. J. Connolly & H. F. R. Prechtl (Eds.), *Maturation and development: Biological and psychological perspectives* (pp. 73–108). Philadelphia: International Medical.

Overpeck, M. D., Brenner, R. A., Trumble, A. C., Trifiletti, L. B., & Berendes, H. W. (1998). Risk factors for infant homicide in the United States. *New England Journal of Medicine, 339*, 1211–1216.

Owens, K., & King, M-C. (1999). Genomic views of human history. *Science, 286*, 451–453.

Oyama, S. (2000a). *The ontogeny of information: Developmental systems and evolution* (2nd ed.). Durham, NC: Duke University Press.

Oyama, S. (2000b). *Evolution's eye: A systems view of biology-culture divide*. Durham, NC: Duke University Press.

Packer, C., Lewis, S., & Pusey, A. (1992). A comparative analysis of non-offspring nursing. *Animal Behaviour, 43*, 265–281.

Packer, C., Tatar, M., & Collins, A. (1998). Reproductive cessation in female mammals. *Nature, 392*, 807–811.

Panksepp, J. (1998). Attention deficit hyperactivity disorders, psychostimulants, and intolerance of childhood playfulness: A tragedy in the making? *Current Directions in Psychological Science, 7*, 91–98.

Parke, R. D., & Suomi, S. J. (1981). Adult male–infant relationships: Human and nonhuman primate evidence. In K. Immelman, G. W. Barlow, L. Petronovitch, & M. Main (Eds.), *Behavioural development* (pp. 700–725). New York: Cambridge University Press.

Parker, S. T. (1996). Using cladistic analysis of comparative data to reconstruct the evolution of cognitive development in humans. In E. P. Martins (Ed.), *Phylogenies and the comparative method in animal behavior* (pp. 361–398). New York: Oxford University Press.

Parker, S. T, & Gibson, K. R. (1979). A developmental model for the evolution of language and intelligence in early hominids. *Behavioral and Brain Sciences, 2*, 367–408.

Parker, S. T., & McKinney, M. L. (1999). *Origins of intelligence: The evolution of cognitive development in monkeys, apes, and humans*. Baltimore: Johns Hopkins University Press.

Parker, W. D. (1998). Birth order effects in the academically talented. *Gifted Child Quarterly, 42*, 29–36.

Parkin, A. J. (1997). The development of procedural and declarative memory. In N. Cowan (Ed.), *The development of memory in childhood* (pp. 113–137). Hove East Essex, England: Psychology Press.

Parten, M. (1932). Social participation among preschool children. *Journal of Abnormal and Social Psychology, 27*, 243–269.

Pascual-Leone, J., & Johnson, J. (1999). A dialectical constructivist view of representation: Role of mental attention, executives, and symbols. In I. Sigel (Ed.), *Development of mental representation: Theories and applications* (pp. 169–200). Mahwah, NJ: Erlbaum.

Pashos, A. (2000). Does paternal uncertainty explain discriminative grandparental solicitude? A cross-cultural study in Greece and Germany. *Evolution and Human Behavior, 21*, 97–109.

Patterson, F. (1978). Linguistic capabilities of a lowland gorilla. In F. Peng (Ed.), *Sign language and language acquisition in man and ape* (pp. 161–201). Boulder, CO: Westview Press.

Paulhus, D. L., Trapnell, P. D., & Chen, D. (1999). Birth order effects on personalities and achievement within families. *Psychological Science, 10*, 482–488.

Pellegrini, A. D. (1982). Explorations in preschoolers' construction of cohesive text in two play contexts. *Discourse Processes, 5*, 101–108.

Pellegrini, A. D. (1984). Identifying causal elements in the thematic fantasy play paradigm. *American Education Research Journal, 19*, 443–452.

Pellegrini, A. D. (1985).The narrative organization of children's play. *Educational Psychology, 5*, 17–25.

Pellegrini, A. D. (1988). Elementary school children's rough-and-tumble play and social competence. *Developmental Psychology, 24*, 802–806.

Pellegrini, A. D. (1989). Elementary school children's rough-and-tumble play. *Early Childhood Research Quarterly, 4*, 245–260.

Pellegrini, A. D. (1993). Boys' rough-and-tumble play, social competence and group composition. *British Journal of Developmental Psychology, 11*, 237–248.

Pellegrini, A. D. (1995a). A longitudinal study of boys' rough-and-tumble play and dominance during early adolescence. *Journal of Applied Developmental Psychology, 16*, 77–93.

Pellegrini, A. D. (1995b). *School recess and playground behavior*. Albany: State University of New York Press.

Pellegrini, A. D., & Bartini, M. (2000). A longitudinal study of bullying, victimization, and peer affiliation during the transition from primary to middle school. *American Educational Research Journal, 37*, 699-726.

Pellegrini, A. D., Bartini, M., & Brooks, F. (1999). School bullies, victims, and aggressive victims: Factors relating to group affiliation and victimization in early adolescence. *Journal of Educational Psychology, 91,* 216–224.

Pellegrini, A. D., & Bjorklund, D. F. (1997). The role of recess in children's cognitive performance. *Educational Psychologist, 32,* 35–40.

Pellegrini, A. D., & Bjorklund, D. F. (in press). The ontongeny and phylogeny of children's object and fantasy play. *Human Nature.*

Pellegrini, A. D., & Davis, P. (1993). Relations between children's playground and classroom behavior. *British Journal of Educational Psychology, 63,* 86–95.

Pellegrini, A. D., & Galda, L. (1982). The effects of thematic fantasy play training on the development of children's story comprehension. *American Educational Research Journal, 19,* 443–452.

Pellegrini, A. D., Galda, L., Bartini, M., & Charak, D. (1998). Oral language and literacy learning in context: The role of social relationships. *Merrill–Palmer Quarterly, 44,* 38–54.

Pellegrini, A .D., Galda, L., Shockley, B., & Stahl, S. (1995). The nexus of social and literacy events at home and school: Implications for primary school oral language and literacy. *British Journal of Educational Psychology, 65,* 273–285.

Pellegrini, A. D., & Horvat, M. (1995). A developmental contextual critique of attention deficit hyperactivity disorder. *Educational Researcher, 24,* 13–20.

Pellegrini, A. D., Horvat, M., & Huberty, P. D. (1998). The relative costs of children's physical activity play. *Animal Behaviour, 55,* 1053–1061.

Pellegrini, A. D., Huberty, P. D., & Jones, I. (1995). The effects of recess timing on children's playground and classroom behaviors. *American Educational Research Journal, 32,* 845–864.

Pellegrini, A. D., & Perlmutter, J. C. (1987). A re-examination of the Smilansky–Parten matrix of play behavior. *Journal of Research in Childhood Education, 2,* 89–96.

Pellegrini, A. D., & Perlmutter, J. C. (1989). Classroom contextual effects on children's play. *Developmental Psychology, 25,* 289–296.

Pellegrini, A. D., & Smith, P. K. (1998). Physical activity play: The nature and function of neglected aspect of play. *Child Development, 69,* 577–598.

Pellis, S. M., Field, E. F., Smith, L, K., & Pellis, V. C. (1996). Multiple differences in the play fighting of male and female rats. *Neuroscience and Biobehavioral Reviews, 21,* 105–120.

Pellis, S. M., & Pellis, V. C. (1998). Structure–function interface in the analysis of play fighting. In M. Bekoff & J. A. Byers (Eds.), *Animal play* (pp. 115–140). New York: Cambridge University Press.

Pennisi, E. (2001). Tracking the sexes by their genes. *Science, 291,* 1733–1734.

Pepperberg, I. M. (1994). Numerical competence in an African gray parrot (*Psittacus erithacus*). *Journal of Comparative Psychology, 198,* 36–44.

Pereira, M. E., & Macedonia, J. M. (1991). Ringtailed lemur anti-predator calls denote predator class, not response urgency. *Animal Behaviour, 41,* 543–544.

Perner, J., Frith, U., Leslie, A., & Leekam, S. (1989). Exploration of the autistic child's theory of mind: Knowledge, belief, and communication. *Child Development, 60,* 689–700.

Perner, J., & Lang, B. (2000). Theory of mind and executive function: Is there a developmental relationship? In S. Baron-Cohen, H. Tager-Flusberg, & D. Cohen (Eds.), *Understanding other minds: Perspectives from autism and developmental cognitive neuroscience* (2nd ed.). Oxford, England: Oxford University Press.

Perner, J., Leekam, S. R., & Wimmer, H. (1987). Three-year-olds' difficulty with false belief: The case for a conceptual deficit. *British Journal of Developmental Psychology, 5,* 125–137.

Perner, J., Ruffman, T., & Leekam, S. R. (1994). Theory of mind is contagious: You catch it from your sibs. *Child Development, 67,* 1228–1238.

Perner, J., Stummer, S., & Lang, B. (1999). Executive functions and theory of mind: Cognitive complexity or functional dependence? In P. D. Zelazo, J. W. Astington, & D. R. Olson (Eds.), *Developing theories of intention: Social understanding and self control* (pp. 133–152). Mahwah, NJ: Erlbaum.

Perry, D. G., Hodges, E. V. E., & Egan, S. K. (in press). Risk and protective factors in peer victimization: A review and a new theory. In J. Juvonen & S. H. Graham (Eds.) *Peer harassment: The plight of the vulnerable and victimized.* New York: Guilford.

Peskin, J. (1992). Ruse and representations: On children's ability to conceal information. *Developmental Psychology, 28,* 84–89.

Peterson, L., Brazeal, T., Oliver, K., & Bull, C. (1997). Gender and developmental patterns of affect, belief, and behavior in simulated injury events. *Journal of Applied Developmental Psychology, 18,* 531–546.

Pettit, G. S., Dodge, K. A., & Brown, M. M. (1988). Early family experience, social problem solving patterns, and children's social competence. *Child Development, 59,* 107–120.

Piaget, J. (1952). *The origins of intelligence in children.* New York: Norton.

Piaget, J. (1962). *Play, dreams, and imitation.* New York: Norton.

Piaget, J. (1965). *The moral judgment of the child.* New York: Free Press.

Piaget, J. (1976). *Behaviour and evolution.* London: Routledge & Kegan Paul.

Piaget, J., & Inhelder, B. (1973). *Memory and intelligence.* New York: Basic Books.

Pinker, S. (1994). *The language instinct: How the mind creates language.* New York: Morrow.

Pinker, S. (1997). *How the mind works.* New York: Norton.

Pipp, S., Easterbrooks, M. A., & Harmon, R. J. (1992). The relation between attachment and knowledge of self and mother in one- to three-year-old infants. *Child Development, 63,* 738–750.

Pitcher, E. G., & Schultz, L. H. (1983). *Boys and girls at play: The development of sex roles.* South Hadley, MA: Bergin & Garvey.

Pleck, J. H. (1997). Paternal involvement: Levels, sources, and consequences. In M. E. Lamb (Ed.), *The role of the father in child development* (3rd ed., pp. 66–103). New York: Wiley.

Plomin R., & Daniels, D. (1987). Why are children in the same family so different from one another? *Behavioral and Brain Sciences, 10*, 1–60.

Plomin, R., DeFries, J. C., & Loehlin, J. C. (1977). Genotype-environment interaction and correlation in the analysis of human behavior. *Psychological Bulletin, 84*, 309–32.

Plomin, R., DeFries, J. C., McClearn, G. E., & Rutter, M. (1997). *Behavioral genetics* (3rd ed.). New York: Freeman.

Plomin, R., & Petrill, S. A. (1997). Genetics and intelligence: What's new? *Intelligence, 24*, 53–77.

Plomin, R., & Rutter, M. (1998). Child development, molecular genetics, and what to do with genes once they are found. *Child Development, 69*, 1221–1240.

Plumert, J. M. (1995). Relation between children's overestimation of their physical abilities and accident proneness. *Developmental Psychology, 31*, 866–876.

Potegal, M., & Einon, D. (1989). Aggressive behavior in adult rats deprived of playfighting experience as juveniles. *Developmental Psychobiology, 22*, 159–172.

Potts, R. (1998). Variability selection in Hominid evolution. *Evolutionary Anthropology, 7*, 81–96.

Potts, R. (2000, June). *The adaptive crunch: Habitat instability as the context of early human behavioral evolution.* Paper presented at meeting of Human Evolution and Behavior Society, Amherst, MA.

Povinelli, D. J., Bering, J., & Giambrone, S. (2000). Toward a science of other minds: Escaping the argument by analogy. *Cognitive Science, 24*, 509–541.

Povinelli, D. J., Bering, J., & Giambrone, S. (2001). Chimpanzee "pointing": Another error of the argument by analogy? In S. Kita (Ed.), *Pointing: Where language, culture, and cognition meet.* Mahwah, NJ: Erlbaum.

Povinelli, D. J., & Eddy, T. J. (1996). What young chimpanzees know about seeing. *Monograph of the Society for Research in Child Development, 61* (3, Serial No. 247).

Povinelli, D. J., Nelson, K. E., & Boysen, S. T. (1992). Comprehension of role reversal in chimpanzees: Evidence of empathy. *Animal Behaviour, 43*, 633–640.

Povinelli, D. J., Reaux, D. J., Bierschwale, D. T., Allain, A. D., & Simon, B. B. (1997). Exploitation of pointing as a referential gesture in young children, but not adolescent chimpanzees. *Cognitive Development, 12*, 423–461.

Power, T. G. (2000). *Play and exploration in children and animals.* Mahwah, NJ: Erlbaum.

Powlishta, K. K. (1995). Intergroup processes in childhood: Social categorization and sex role development. *Developmental Psychology, 31*, 781–788.

Premack, D., & Woodruff, G. (1978). Does the chimpanzee have a theory of mind? *Behavioral and Brain Sciences, 4*, 515–526.

Preuss, T. M. (2001). The discovery of cerebral diversity: An unwelcome scientific revolution. In D. Falk & K. Gibson (Eds.), *Evolutionary anatomy of the primate cerebral cortex* (pp. 138–164). Cambridge, England: Cambridge University Press.

Profet, M. (1992). Pregnancy sickness as adaptation: A deterrent to maternal ingestion of teratogens. In J. H. Barkow, L. Cosmides, & J. Tooby (Eds.), *The adaptive mind: Evolutionary psychology and the generation of culture* (pp. 327–365). New York: Oxford University Press.

Pylyshyn, Z. (1980). Computation and cognition: Issues in the foundations of cognitive science. *Behavioral and Brain Sciences, 3,* 111–132.

Quadagno, D. M., Briscoe, R., & Quadagno, J. S. (1977). Effects of perinatal gonadal hormones on selected nonsexual behavior patterns: A critical assessment of the nonhuman and human literature. *Psychological Bulletin, 84,* 62–82.

Quartz, S. R., & Sejnowski, T. J. (1997). The neural basis of cognitive development: A constructivist manifesto. *Behavioral and Brain Sciences, 20,* 537–596.

Quinn, P. C. (1994). The categorization of above and below spatial relations by young infants. *Child Development, 65,* 58–69.

Quinn, P. C., & Eimas, P. D. (1996). Perceptual cues that permit categorical differentiation of animal species by infants. *Journal of Experimental Child Psychology, 63,* 189–211.

Radke-Yarrow, M., & Kochanska, G. (1990). Anger in young children. In N. L. Stein, B. Leventhal, & T. Trabasso (Eds.), *Psychological and biological approaches to emotion* (pp. 297–310). Hillsdale, NJ: Erlbaum.

Ramus, F., Hauser, M. D., Miller, C., Morris, D., & Mehler, J. (2000). Language discrimination by human newborns and by cotton-top tamarin monkeys. *Science, 288,* 349–351.

Ratner, H. H., Foley, M. A., & Gimpert, N. (in press). The role of collaborative planning in children's source-monitoring errors and learning. *Journal of Experimental Child Psychology.*

Reaux, J. E., Theall, L. A., & Povinelli, D. J. (1999). A longitudinal investigation of chimpanzee's understanding of visual perception. *Child Development, 70,* 275–290.

Rebers, A. S. (1992). The cognitive unconscious: An evolutionary perspective. *Consciousness and Cognition, 1,* 93–133.

Reid, A. (1997). Locality of class? Spatial and social differentials in infant and child mortality in England and Wales, 1895–1911. In C. A. Corsini & P. P. Viazzo (Eds.), *The decline of infant and child mortality: The European experience: 1750–1990* (pp. 129–154). The Hague: Martinus Nijhoff.

Reinisch, J. M., Sanders, S. A., Mortensen, E. L., & Rubin, D. B. (1995). In utero exposure to phenobarbital and intelligence deficits in adult men. *Journal of the American Medical Association, 274,* 1518–1525.

Rescorla, L., Hyson, M. C., & Hirsh-Pasek, K. (Eds.). (1991). *Academic instruction in early childhood: Challenge or pressure?* San Francisco: Jossey-Bass.

Resnick, S. M., Berenbaum, S. A., Gottesman, I. I., & Bouchard, T. J. (1986). Early hormonal influences on cognitive functioning in congenital adrenal hyperplastia. *Developmental Psychology, 22,* 191–198.

Rhodes, G., & Tremewan, T. (1996). Averageness, exaggeration, and facial attractiveness. *Psychological Science, 7,* 105–110.

Rice, S. H. (1997). The analysis of ontogenetic trajectories: When a change in size or shape is not heterochrony. *Proceedings of the National Academy of Sciences, 94,* 907–912.

Ridley, M. (1993). *The red queen.* London: Viking.

Rilling, J. K., & Insel, T. R. (1999). The primate neocortex in comparative perspective using magnetic resonance imaging. *Journal of Human Evolution, 37,* 191–223.

Rivera, S. M., Wakeley, A., & Langer, J. (1999). The drawbridge phenomenon: Representational reasoning or perceptual preference? *Developmental Psychology, 35,* 427–435.

Robinson, B. F., & Mervis, C. B. (1998). Disentangling early language development: Modeling lexical and grammatical acquisition using an extension of case-study methodology. *Developmental Psychology, 34,* 363–375.

Rodkin, P. C., Farmer, T. W., Pearl, R., & Van Acker, R. (2000). Heterogeneity of popular boys: Antisocial and prosocial configurations. *Developmental Psychology, 36,* 14–24.

Roggman, L., & Langlois, J. (1987). Mothers, infants, and toys: Social play correlates of attachment. *Infant Behavior and Development, 10,* 233–237.

Rogoff, B. (1990). *Apprenticeship in thinking: Cognitive development in social context.* New York: Oxford University Press.

Rogoff, B. (1998). Cognition as a collaborative process. In W. Damon (Gen. Ed.) & D. Kuhn & R. S. Siegler (Eds.), *Handbook of child psychology: Vol. 2. Cognition, language, and perceptual development* (5th ed., pp. 679–744). New York: Wiley.

Roopnarine, J. L., Hooper, F. H., Ahmeduzzaman, M. & Pollack, B. (1993). Gentle play partners: Mother-child and father-child play in New Delhi, India. In K. MacDonald (Ed), *Parent-child play* (pp. 287–304). New York: SUNY Press.

Rosen, R. C. (1973). Suppression of penile tumescence by instrumental conditioning. *Psychometric Medicine, 35,* 509–514.

Rosenberg, A. (1990). Is there an evolutionary biology of play? In M. Bekoff & D. Jamieson (Eds.), *Interpretation and explanation in the study of animal behavior, Vol. 1: Interpretation, intentionality, and communication* (pp. 180–196). Boulder, CO: Westview Press.

Rosenthal, M. K. (1994). Social and non-social play of infants and toddlers in family day care. In E. V. Jacobs & H. Goelman (Eds.), *Children's play in child care settings* (pp. 163–192). Albany: State University of New York Press.

Rossi, A. S., & Rossi, P. H. (1990). *Of human bonding: Parent–child relations across the life course.* Hawthorne, NY: Aldine de Gruyter.

Routh, D., Schroeder, C., & O'Tuama, L. (1974). Development of activity levels in children. *Developmental Psychology, 10*, 163–168.

Rowe, D. C., Jacobson, K. C., & van der Oord, E. J. C. G. (1999). Genetic and environmental influences on vocabulary IQ: Parental education level as a moderator. *Child Development, 70*, 1151–1162.

Rowher, S., Herron, J. C., & Daly, M. (1999). Stepparental behavior as mating effort in birds and other animals. *Evolution and Human Behavior, 20*, 367–390.

Rozin, P. (1976). The evolution of intelligence and access to the cognitive unconscious. *Progress in Psychobiology and Physiological Psychology, 6*, 245–280.

Rubin, K. H. (1982). Nonsocial play in preschoolers: Necessarily evil? *Child Development, 53*, 651–657.

Rubin, K. H., Bukowki, W., & Parker, J. G. (1998). Peer interactions, relationships, and groups. In W. Damon (Gen. Ed.) & N. Eisenberg (Ed.), *Handbook of child psychology: Vol. 3. Social, emotional, and personality development* (5th ed., pp. 619–700). New York: Wiley.

Rubin, K. H., Fein, G., & Vandenberg B. (1983). Play. In E. M. Hetherington (Ed.), *Handbook of child psychology* (Vol. 4, pp. 693–774). New York: Wiley.

Rubin, K. H., & Krasnor, L. R. (1986). Social cognitive and social behavior perspectives on problem solving. In M. Perlmutter (Ed.), *Minnesota Symposium on Child Psychology* (Vol. 18, pp. 1–86). Hillsdale, NJ: Erlbaum.

Rubin, K. H., & Maioni, T. (1975). Play reference and its relationship to egocentrism, popularity and classification skills in preschoolers. *Merrill–Palmer Quarterly, 21*, 171–179.

Rubin, K. H., Watson, R., & Jambor, T. (1978). Free-play behaviors in preschool and kindergarten children. *Child Development, 49*, 534–546.

Ruda, M. A., Ling, Q. -D., Hohmann, A. G., Bo Peng, Y., & Tachibana, T. (2000). Altered nociceptive neuronal circuits after neonatal peripheral inflammation. *Science, 289*, (July 28), 628–630.

Rudy, J. W., Vogt, M. B., & Hyson, R. L. (1984). A developmental analysis of the rat's learned reactions to gustatory and auditory stimulation. In R. Kail & N. E. Spear (Eds.), *Memory development: Comparative perspectives* (pp. 181–208). Hillsdale, NJ: Erlbaum.

Ruff, C. B., Trinkaus, E., & Holliday, T. W. (1997). Body mass and encephalization in Pleistocene *Homo. Nature, 387*, 173–176.

Ruff, H. A., & Birch, H. G. (1974). Infant visual fixation: The effect of concentricity, curvilinearity, and number of directions. *Journal of Experimental Child Psychology, 17*, 460–473.

Ruffman, T., Perner, J., Naito, M., Parkin, L., & Clements, W. A. (1998). Older (but not younger) siblings facilitate false belief understanding. *Developmental Psychology, 34*, 161–174.

Russell, J., Mauthner, N., Sharpe, S., & Tidswell, T. (1991). The "windows tasks" as a measure of strategic deception in preschoolers and autistic subjects. *British Journal of Developmental Psychology, 9*, 331–349.

Russell, R. J. H., & Wells, P. A. (1987). Estimating paternity confidence. *Ethology and Sociobiology, 8,* 215–220.

Russo, R. Nichelli, P., Gibertoni, M., & Cornia, C. (1995). Developmental trends in implicit and explicit memory: A picture completion study. *Journal of Experimental Child Psychology, 59,* 566–578.

Russon, A. E. (1996). Imitation in everyday use: Matching and rehearsal in the spontaneous imitation of rehabilitant orangutans (*Pongo pygmaeus*). In A. E. Russon, K. A. Bard, & S. T. Parker (Eds.), *Reaching into thought: The minds of the great apes* (pp. 152–176). New York: Cambridge University Press.

Russon, A. E., Bard, K. A., & Parker, S. T. (Eds.). (1996). *Reaching into thought: The minds of the great apes.* New York: Cambridge University Press.

Rutter, M., & the English and Romanian Adoption Adoptees Study Team. (1998). Developmental catch-up, and delay, following adoption after severe global early privation. *Journal of Child Psychology and Psychiatry, 39,* 465–476.

Saarni, C. (1984). An observational study of children's attempts to monitor their expressive behavior. *Child Development, 55,* 1504–1513.

Sabbagh, M. A., & Taylor, M. (2000). Neural correlates of theory-of-mind reasoning: An event-related potential study. *Psychological Science, 11,* 46–50.

Sagi, A., Lanb, M. E., Lewkowitz, K. S., Shoham, R., Dvir, R., & Estes, D. (1985). Security of infant–mother,–father, and –metapelet attachments among kibbutz-reared Israeli children. In I. Bretherton & E. Waters (Eds.), Growing points of attachment theory and research (pp. 257–275). *Monographs of the Society for Research in Child Development, 50*(1–2, Serial No. 209).

Salmon, C. A. (1999). On the impact of sex and birth order on contact with kin. *Human Nature, 10,* 183–197.

Salmon, C. A., & Daly, M. (1998). The impact of sex and birth order on familial sentiment: Middleborns are different. *Evolution and Human Behavior, 19,* 299–312.

Sameroff, A. J., & Suomi, S. J., (1996). Primates and persons: A comparative developmental understanding of social organization. In R. B. Cairns, G. H. Elder, Jr., & E. J. Costello (Eds.), *Developmental Science* (pp. 97–120). New York: Cambridge University Press.

Sandberg, D. E., & Meyer-Bahlburg, H. F. (1994). Variability in middle childhood play behavior: Effects of gender, age, and family background. *Archives of Sexual Behavior, 23,* 645–663.

Sanders, S. A., & Reinisch, J. M. (1990). Biological and social influences on the endocrinology of puberty: Some additional considerations. In J. Bancroft & J. M. Reinisch (Eds.), *Adolescence and puberty* (pp. 50–62). New York: Oxford University Press.

Savage-Rumbaugh, E. S. (1986). *Ape language: From condition response to symbol.* New York: Columbia University Press.

Savage-Rumbaugh, E. S., McDonald, K., Sevcik, R. A., Hopkins, W. D., & Rubert, E. (1986). Spontaneous symbol acquisition and communicative use by pygmy chimpanzees (*Pan paniscus*). *Journal of Experimental Psychology: General, 115*, 211–235.

Savage-Rumbaugh, E. S., Murphy, J., Sevcik, R. A., Brakke, K. E., Williams, S. L., & Rumbaugh, D. M. (1993). Language comprehension in ape and child. *Monographs of the Society for Research in Child Development, 58*(Serial No. 233).

Savin-Williams, R. C. (1979). Dominance hierarchies in groups of early adolescents. *Child Development, 50*, 923–935.

Scarr, S. (1992). Developmental theories for the 1990s: Development and individual differences. *Child Development, 63*, 1–19.

Scarr, S. (1993). Biological and cultural diversity: The legacy of Darwin for development. *Child Development, 64*, 1333–1353.

Scarr, S. (1995a). Commentary to Gottlieb's "Some conceptual deficiencies in "developmental" behavior genetics." *Human Development, 38*, 154–158.

Scarr, S. (1995b). Psychology will be truly evolutionary when behavior genetics is included. *Psychological Inquiry, 6*, 68–71.

Scarr, S., & McCartney, K. (1983). How people make their own environments: A theory of genotype à environment effects. *Child Development, 54*, 424–435.

Scarr, S., & Weinberg, R. A. (1976). IQ test performance of Black children adopted by White families. *American Psychologist, 31*, 726–739.

Schacter, D. L. (1992). Understanding implicit memory. *American Psychologist, 47*, 559–569.

Schacter, D. L., Norman, K. A., & Koustaal, W. (2000). The cognitive neuroscience of constructive memory. In D. F. Bjorklund (Ed.)., *False-memory creation in children and adults: Theory, research, and implications* (pp. 129–168). Mahwah, NJ: Erlbaum.

Schneider, M. A., & Hendrix, L. (2000). Olfactory sexual inhibition and the Westermark effect. *Human Nature, 11*, 65–91.

Schneider, W. (1998). Performance prediction in young children: Effects of skill, metacognition and wishful thinking. *Developmental Science, 1*, 291–197.

Schneider, W., & Bjorklund, D. F. (1998). Memory. In W. Damon (Gen. Ed.) & D. Kuhn & R. S. Siegler (Eds.), *Handbook of child psychology: Vol. 2. Cognitive, language, and perceptual development* (5th ed., pp. 467–521). New York: Wiley.

Schneider, W., & Bjorklund, D. F. (in press). Memory and knowledge development. In J. Valsiner & K. Connolly (Eds.), *Handbook of developmental psychology*. London: Sage.

Schneider, W., Korkel, J., & Weinert, F. E. (1987). The effects of intelligence, self-concept, and attributional style on metamemory and memory behaviour. *International Journal of Behavioral Development, 10*, 281–299.

Schneider, W., Perner, J., Bullock, M., Stefanek, J., & Ziegler, A. (1999). The development of intelligence and thinking. In F. E. Weinert & W. Schneider (Eds.), *The Munich Longitudinal Study on the Genesis of Individual Competencies (LOGIC)* (pp. 9–28). Cambridge, England: Cambridge University Press.

Schneider, W., & Pressley, M. (1997). *Memory development between 2 and 20* (2nd ed.). Mahwah, NJ: Erlbaum.

Schneirla, T. C. (1961). Instinctive behavior, maturation-experience and development. In B. Kaplan & S. Wapner (Eds.), *Perspectives in psychological theory: Essays in honor in Heinz Werner* (pp. 303–334). New York: International Universities Press.

Schwartz, J. H. (1999). *Sudden origins: Fossils, genes, and the emergence of species.* New York: Wiley.

Segal, N. (2000). Virtual twins: New finiding on within-family environmental influences on intelligence. *Journal of Educational Psychology, 92,* 442–448.

Seielstad, M. T., Minch, E., & Luca Cavalli-Sforza, L. (1998). Genetic evidence for a higher female migration rate in humans. *Nature Genetics, 20,* 278–280.

Seligman, M. E. (1971). Phobias and preparedness. *Behavior Therapy, 2,* 307–320.

Senghas, A., & Coppola, M. (2001). Children creating language: How Nicaraguan Sign Language acquired a spatial grammar. *Psychological Science, 12,* 323–326.

Shackelford, T. K., & Larsen, R. J. (1997). Facial asymmetry as an indicator of psychological, emotional, and physiological distress. *Journal of Personality and Social Psychology, 72,* 456–466.

Shea, B. T. (1989). Heterochrony in human evolution: The case for neoteny revisited. *Yearbook of Physical Anthropology, 32,* 69–101.

Shea, B. T. (2000). Current issues in the investigation of evolution by heterochrony, with emphasis on the debate over human neoteny. In S. T. Parker, J. Langer, & M. L. McKinney (Eds.), *Biology, brains, and behavior: The evolution of human development* (pp. 181–213). Santa Fe, NM: School of American Research Press.

Shepher, J. (1983). *Incest: A biosocial view.* New York: Academic Press.

Sherif, M. H., Harvey, O. J., White, B. J., Hood, W. R., & Sherif C. W. (1961). *Inter-group conflict and cooperation: The Robbers Cave experiment.* Norman: University of Oklahoma Press.

Sherman, P. (1998). The evolution of menopause. Nature, 392, 759–761.

Sherrod, K. B., O'Connor, S., Vietze, P. M., & Altemeier, W. A. (1984). Child health and maltreatment. *Child Development, 55,* 1174–1183.

Sherry, D. F. (2000). What sex differences in spatial ability tell us about the evolution of cognition. In M. S. Gazzaniga (Ed.), *The new cognitive neurosciences* (2nd ed., pp. 1209–1217). Cambridge, MA: MIT Press.

Sherry, D. F., & Schacter, D. L (1987). The evolution of multiple memory systems. *Psychological Review, 94,* 439–454.

Shipman, P. (1994). *The evolution of racism: Human differences and the use and abuse of science.* New York: Simon & Schuster.

Siegler, R. S., & Ellis, S. (1996). Piaget on childhood. *Psychological Science, 7,* 211–215.

Sigman, M., Newmann, C., Carter, E., Cattle, D., D'Souza, S., & Bwido, N. (1988). Home interactions and the development of Embu toddlers in Kenya. *Child Development, 59,* 1251–1261.

Silverman, I., Choi, J., Mackewn, A., Fisher, M., Moro, J., & Olshansky, E. (2000). Evolved mechanisms underlying wayfinding: Further studies on the hunter-gatherer theory of spatial sex differences. *Evolution and Human Behavior, 21,* 201–213.

Silverman, I., & Eals, M. (1992). Sex differences in spatial abilities: Evolutionary theory and data. In. J. H. Barkow, L. Cosmides, & J. Tooby (Eds.), *The adapted mind: Evolutionary psychology and the generation of culture* (pp. 533–549). New York: Oxford University Press.

Simon, T., & Smith, P. K. (1985). Play and problem solving: A paradigm questioned. *Merrill–Palmer Quarterly, 31,* 73–86.

Simon, T. J. (1997). Reconceptualizing the origins of number knowledge: A "non-numerical" account. *Cognitive Development, 12,* 349–372.

Simon, T. J., Hespos, S. J., & Rochat, P. (1995). Do infants understand simple arithmetic? A replication of Wynn (1992). *Cognitive Development, 10,* 253–269.

Simons-Morton, B. G., O'Hara, N. M., Parcel, G .S., Huang, I. W., Baranowski, T., & Wilson, B. (1990). Children's frequency of participation in moderate to vigorous physical activities. *Research Quarterly for Exercise and Sport, 61,* 307–314.

Simpson, G. G. (1944). *Tempo and mode in evolution.* New York: Columbia University Press.

Skeels, H. M. (1966). Adult status of children with contrasting early life experiences. *Monographs of the Society for Research in Child Development, 31*(3, Serial No. 105).

Skodak, M., & Skeels, H. M. (1945). A follow-up study of children in adoptive homes. *Journal of Genetic Psychology, 66,* 21–58.

Slaby, R. G., & Parke, R. D. (1971). Effects of resistance to deviation of observing a model's affective reaction to response consequence. *Developmental Psychology, 5,* 40–47.

Slater, A. M., Morison, V., Somers, M., Mattock, A., Brown, E., & Taylor, D. (1990). Newborn and older infants' perception of partly occluded objects. *Infant Behavior and Development, 13,* 33–49.

Slater, A. M., Quinn, P. C., Hayes, R., & Brown, E. (2000). The role of facial orientation in newborn infants' preference for attractive faces. *Developmental Science, 3,* 181–185.

Smilansky, S. (1968). *The effects of sociodramatic play on disadvantaged preschool children.* New York: Wiley.

Smith, M. S. (1988). Research in developmental sociobiology: Parenting and family behavior. In K. B. MacDonald (Ed.), *Sociobiological perspectives on human development* (pp. 271–292). New York: Springer.

Smith, P. K. (1973). Temporal clusters and individual differences in the behavior of preschool children. In R. Michaels & J. Crook (Eds.), *Comparative ecology and the behavior of primates* (pp. 752–798). London: Academic.

Smith, P. K. (1974). Social and fantasy play in young children. In B. Tizard & D. Harvey (Eds.), *Biology of play* (pp. 123–145). London: SIMP/Heinemann.

Smith, P. K. (1982). Does play matter? Functional and evolutionary aspects of animal and human play. *Behavioral and Brain Sciences, 5,* 139–184.

Smith, P. K. (1988). Children's play and its role in early development: A re-evaluation of the "play ethos." In A. D. Pellegrini (Eds.), *Psychological bases for early education* (pp. 207–226). Chichester, UK: Wiley.

Smith, P. K. (1998, June). *The theory of mind acquisition support system: Social origins of theory of mind.* Paper presented at Hang Seng Conference on Evolution of Mind, Sheffield, UK.

Smith, P. K. (in press). Pretend play, metarepresentation, and theory of mind. In R. Mitchell (Ed.), *Pretending in animals and humans.* Cambridge, England: Cambridge University Press.

Smith, P. K., & Connolly, K. (1980). *The ecology of preschool behaviour.* London: Cambridge University Press.

Smith, P. K., & Hagan, T. (1980). Effects of deprivation on exercise play in nursery school children. *Animal Behaviour, 28,* 922–928.

Smith, P. K., & Syddall, S. (1978). Play and group play tutoring in preschool children: Is it play or tutoring that matters? *British Journal of Educational Psychology, 48,* 315–325.

Smith, P. K., & Vollstedt, R. (1985). On defining play. *Child Development, 56,* 1042–1050.

Smith, P. K., & Whitney, S. (1987). Play and associative fluency: Experimenter effects may be responsible for previous findings. *Developmental Psychology, 23,* 49–53.

Smuts, B. B. (1985). *Sex and friendship in baboons.* Hawthorne, NY: Aldine de Gruyter.

Smuts, B. B. (1995). The evolutionary origins of patriarchy. *Human Nature, 6,* 1–32.

Snow, C. (1972). Mother's speech to children learning language. *Child Development, 43,* 549–565.

Snow, C. E., Arlman-Rupp, A., Hassing, Y., Jobse, J., Joosten, J., & Vorster, J. (1976). Mothers' speech in three social classes. *Journal of Psycholinguistic Research, 5,* 1–20.

Soltys, J., Brown, H., Thompson, K., Pietrzak, R. H., & Thompson, N. S. (2000, June). *Do infant cries simulate respiratory distress?* Paper presented at meeting of Evolution and Human Behavior Society, Amherst, MA.

Sophian, C. (1997). Beyond competence: The significance of performance for conceptual development. *Cognitive Development, 12,* 281–303.

Spear, N. E. (1984). Ecologically determined dispositions control the ontogeny of learning and memory. In R. V. Kail, Jr., & N. E. Spear (Eds.), *Comparative perspectives on the development of memory* (pp. 325–358). Hillsdale, NJ: Erlbaum.

Spear, N. E., & Hyatt, L. (1993). How the timing of experience can affect the ontogeny of learning. In G. Turkewitz & D. A. Devenny (Eds.), *Developmental time and timing* (pp. 167–209). Hillsdale, NJ: Erlbaum.

Spelke, E. S. (1985). Perception of unity, persistence and identity: thoughts on infants' conceptions of objects. In J. Mehler & R. Fox (Eds.), *Neonate cognition: Beyond the booming, buzzing confusion* (pp. 89–114). Hillsdale, NJ: Erlbaum.

Spelke, E. S. (1991). Physical knowledge in infancy: Reflections on Piaget's theory. In S. Carey & R. Gelman (Eds.), *Epigenesis of mind: Essays in biology and knowledge* (pp. 133–169). Hillsdale, NJ: Erlbaum.

Spelke, E. S. (1994). Initial knowledge: Six suggestions. *Cognition, 50,* 431–455.

Spelke, E. S., Breinlinger, K., Macomber, J., & Jacobson, K. (1992). Origins of knowledge. *Psychological Review, 99,* 605–632.

Spelke, E. S., & Newport, E. L. (1998). Nativism, empiricism, and the development of knowledge. In W. Damon (Gen. Ed.) & R. Learner (Ed.), *Handbook of child psychology: Vol. 1. Theories of theoretical models of human development* (5th ed., pp. 275–340). New York: Wiley.

Spelke, E. S., Vishon, P., & von Hofsten, C. (1995). Object perception, object-directed action, and physical knowledge in infancy. In M. Gazzaniga (Ed.) *The cognitive neurosciences* (pp. 165–179). Cambridge, MA: MIT Press.

Spence, M. J., & Freeman, M. S. (1996). Newborn infants prefer the maternal low-pass filtered voice, but not the maternal whispered voice. *Infant Behavior and Development, 19,* 199–212.

Spinka, M., Newbury, R. C., & Bekoff, M. (2001). Mammalian play: Can training for the unexpected be fun? *Quarterly Review of Biology, 76,* 141–168.

Starkey, P., & Gelman, R. (1982). The development of addition and subtraction abilities prior to formal schooling in arithmetic. In T. P. Carpenter, J. M. Moser, & T. A. Romberg (Eds.), *Addition and subtraction: A cognitive perspective* (pp. 99–116). Hillsdale, NJ: Erlabum.

Starkey, P., Spelke, E. S., & Gelman, R. (1983). Detection of intermodal numerical correspondences by human infants. *Science, 222,* 179–181.

Starkey, P., Spelke, E. S., & Gelman, R. (1990). Numerical abstraction by human infants. *Cognition, 36,* 97–127.

Steele, E. J., Lindley, R. A., & Blanden, R. V. (1998). *Lamarck's signature: How retrogenes are changing Darwin's natural selection paradigm.* Reading, MA: Perseus Books.

Stein, Z. A., Susser, M. W., Saenger, G., & Marolla, F. (1975). *Famine and human development: The Dutch hunger winter of 1944–1945.* New York: Oxford University Press.

Steinberg, L., Dornbusch, S. M., & Brown, B. B. (1992). Ethnic differences in adolescent achievement: An ecological perspective. *American Psychologist, 4,* 723–729.

Steinberg, L., & Silk, J. S. (2002). Parenting adolescents. In M. Bornstein (Ed.), *Handbook of parenting* (2nd ed.). Mahwah, NJ: Erlbaum.

Steiner, J. E. (1979). Human facial expressions in response to taste and smell stimulation. In H. W. Reese & L. P. Lipsitt (Eds.), *Advances in child development and behavior*, Vol. 13 (pp. 257–295). New York: Academic.

Stern, D. N., Spieker, S., & MacKain, K. (1982). Intonation contours as signals in maternal speech to prelinguistic infants. *Developmental Psychology, 18*, 727–735.

Stevenson, H. W., & Lee, S. Y. (1990). Context of achievement. *Monographs of the Society for Research in Child Development, 55*(1 & 2, Serial No. 221).

Stevenson, J. C., & Williams, D. C. (2000). Parental investment, self-control, and sex differences in the expression of ADHD. *Human Nature, 11*, 405–422.

Stewart, R. B., & Marvin, R. S. (1984). Sibling relations: The role of conceptual perspective-taking in the ontogeny of sibling caregiving. *Child Development, 55*, 1322–1332.

Stipek, D. (1984). Young children's performance expectations: Logical analysis or wishful thinking? In J. G. Nicholls (Ed.), *Advances in motivation and achievement, Vol. 3: The development of achievement motivation* (pp. 33–56). Greenwich, CT: JAI.

Stoolmiller, M. (1999). Implications of the restricted range of family environments for estimates of heritability and nonshared environment in behavior genetic adoption studies. *Psychological Bulletin, 125*, 393–407.

Strauss, M. S., & Curtis, L. E. (1984). Development of numerical concepts in infancy: In C. Sophian (Ed.), *Origins of cognitive skills: The Eighteenth Annual Carnegie Symposium on Cognition* (pp. 131–155). Hillsdale, NJ: Erlbaum.

Strayer, F. F., & Noel, J. M. (1986). The prosocial and antisocial functions of aggression. In C. Zahn-Waxler, E. M. Cummings, & R. Iannoti (Eds.), *Altruism and aggression* (pp. 107–131). New York: Cambridge University Press.

Streissguth, A. P., Barr, H. M., Sampson, P. D., Darby, B. L., & Martin, D. C. (1989). IQ at age 4 in relation to maternal alcohol use and smoking during pregnancy. *Developmental Psychology, 25*, 3–11.

Stringer, C. B., Dean, M. C., & Martin, R. D. (1990). A comparative study of cranial and dental development within recent British samples among Neanderthals. In C. J. DeRousseau (Ed.), *Primate live history and evolution* (pp. 115–152). New York: Wiley-Liss.

Sulkowski, G. M., & Hauser, M. D. (2001). Can rhesus monkeys spontaneously subtract? *Cognition, 79*, 239–262.

Sulloway, F. J. (1996). *Born to rebel: Birth order, family dynamics, and creative lives.* New York: Pantheon.

Suomi, S. J. (1982). Why play matters. *Behavioral and Brain Sciences, 5*, 169–170.

Suomi, S., & Harlow, H. (1972). Social rehabilitation of isolate-reared monkeys. *Developmental Psychology, 6*, 487–496.

Surbey, M. K. (1990). Family composition, stress, and the timing of human menarche. In T. E. Ziegler & F. B. Bercovitvch (Eds.), *Socioendocrinology of primate reproduction* (pp. 11–32). New York: Wiley-Liss.

Surbey, M. K. (1998a). Developmental psychology and modern Darwinism. On C. B. Crawford & D. Krebs (Eds.), *Handbook of evolutionary psychology: Ideas, issues, and applications* (pp. 369–404). Hillsdale, NJ: Erlbaum.

Surbey, M. K. (1998b). Parent and offspring strategies in the transition at adolescence. *Human Nature, 9,* 67–94.

Sutton, J., Smith, P. K., & Swettenham, J. (1999). Socially undesirable need not be incompetent: A response to Crick and Dodge. *Social Development, 8,* 132–134.

Sutton-Smith, B. (1997). *The ambiguity of play.* Cambridge, MA: Harvard University Press.

Sutton-Smith, B., Rosenberg, B. G., & Morgan, E. F., Jr. (1963). Development of sex differences in play choices during adolescence. *Child Development, 34,* 119–126.

Swanson, H. L. (1999). What develops in working memory? A life span perspective. *Developmental Psychology, 35,* 986–1000.

Swisher, C. C. III, Rink, W. J., Antón, S. C., Schwarcz, H. P., Curtis, G. H., Suprijo, A., & Widiasmoro (1996). Latest *Homo erectus* of Java: Potential contemporaneity with *Homo sapiens* in Southeast Asia. *Science, 274,* 1870–1874.

Sylva, K., Bruner, J. S., & Genova, P. (1976). The role of play in the problem-solving behavior of children 3–5 years old. In J. S. Bruner, A. Jolly, & K. Sylva (Eds.), *Play: Its role in development and evolution* (pp. 244–261). New York: Basic Books.

Symons, D. (1978). *Play and aggression: A study of rhesus monkeys.* New York: Columbia University Press.

Tanner, J. M. (1978). *From fetus into man: Physical growth from conception to maturity.* Cambridge, MA: Harvard University Press.

Tardif, T., & Wellman, H. M. (2000). Acquisition of mental state language in Mandarin- and Cantonese-speaking children. *Developmental Psychology, 36,* 25–43.

Tattersall, I. (1998). *Becoming human: Evolution and human intelligence.* San Diego, CA: Harcourt Brace.

Taylor, M., & Carlson, S. M. (1997). The relation between individual differences in fantasy and theory of mind. *Child Development, 68,* 436–455.

Teasley, S. D. (1995). The role of talk in children's peer collaborations. *Developmental Psychology, 31,* 207–220.

Temerlin, M. K. (1975). *Lucy: Growing up human.* London: Souvenir Press

Teti, D. M., Nakagawa, M., Das, R., & Wirth, O. (1991). Security of attachment between preschoolers and their mothers: Relations among social interaction, parenting stress, and mothers' sorts of the Attachment Q-Set. *Developmental Psychology, 27,* 440–447.

Teti, D. M., Sakin, J. W., Kucera, E., Corns, K. M., & Eiden, R. D. (1996). And baby makes four: Predictors of attachment security among preschool-age firstborns during the transition to siblinghood. *Child Development, 67,* 579–596.

Thelen, E. (1979). Rhythmical stereotypes in normal human infants. *Animal Behaviour, 27,* 699–715.

Thelen, E. (1980). Determinants of amounts of stereotyped behaviour in normal human infants. *Ethology and Sociobiology, 1,* 141–150.

Thompson, N. S., Dessureau, B. N., & Olson, C. (1998). Infant cries as evolutionary melodrama: Extortion or deception? *Evolution of Communication, 2,* 25–43.

Thompson, N. S., Olson, C., & Dessureau, B. N., (1996). Babies cries: Who's listening? Who's being fooled? *Social Research, 63,* 763–784.

Thomson, K. W. (1988). *Morphogenesis and evolution.* New York: Oxford University Press.

Thorndike, E. L. (1913). *Educational psychology: Vol. 1. The original nature of man.* New York: Teachers College.

Thornhill, R., & Palmer, C. T. (2000). *A natural history of rape: Biological bases of sexual coercion.* Cambridge, MA: MIT Press.

Thorpe, W. (1956). *Learning and instincts in animals.* London: Methuen.

Tierson, F. D., Olsen, C. L., & Hook, E. B. (1985). Influence of cravings and aversion on diet in pregnancy. *Ecology of Food and Nutrition, 17,* 117–129.

Tierson, F. D., Olsen, C. L., & Hook, E. B. (1986). Nausea and vomiting of pregnancy and association with pregnancy outcome. *American Journal of Obstetrics ad Gynecology, 155,* 1017–1022.

Tinbergen, N. (1951). *The study of instinct.* New York: Oxford University Press.

Tinbergen, N. (1963). On the aims and methods of ethology. *Zeitschrift fur Tierpsychologie, 20,* 410–433.

Tobias, P. V. (1987). The brain of *Homo habilis*: A new level of organization in cerebral evolution. *Journal of Human Evolution, 16,* 741–761.

Tomasello, M. (1992). The social bases of language acquisition. *Social Development, 1,* 67–87.

Tomasello, M. (1996). Do apes ape? In C. Heyes & B. Galef (Eds.), *Social learning in animals: The role of culture* (pp. 319–346). San Diego, CA: Academic Press.

Tomasello, M. (1999). *The cultural origins of human cognition.* Cambridge, MA: Harvard University Press.

Tomasello, M. (2000). Culture and cognitive development. *Current Directions in Psychological Science, 9,* 37–40.

Tomasello, M., & Call, J. (1997). *Primate cognition.* New York: Oxford University Press.

Tomasello, M., Call, J., & Hare, B. (1998). Five primate species follow the visual gaze of conspecifics. *Animal Behavior, 55,* 1063–1069.

Tomasello, M., Davis-Dasilva, M., Camak, L., & Bard, K. A. (1987). Observational learning of tool use by young chimpanzees. *Journal of Human Evolution, 2,* 175–183.

Tomasello, M., Kruger, A. C., & Ratner, H. H. (1993). Cultural learning. *Behavioral and Brain Sciences, 16,* 495–511.

Tomasello, M., Savage-Rumbaugh, S., & Kruger, A. C. (1993). Imitative learning of actions on objects by children, chimpanzees, and enculturated chimpanzees. *Child Development, 64,* 1688–1705.

Tooby, J., & Cosmides, L. (1992). The psychological foundations of culture. In J. H. Barkow, L. Cosmides, & J. Tooby (Eds.), *The adapted mind: Evolutionary psychology and the generation of culture* (pp. 19–139). New York: Oxford University Press.

Townsend, F. (in press). Birth order and rebelliousness: Reconstructing the research in Born to Rebel. *Politics and the Life Sciences.*

Trainor, L. J., Austin, C. M., & Desjardins, R. N. (2000). Is infant-directed speech prosody a result of the vocal expression of emotion? *Psychological Science, 11,* 188–195.

Trinkaus, E., & Tompkins, R. L. (1990). The Neanderthal life cycle: The possibliity, probablity and perceptibility of contrasts with recent humans. In C. J. DeRousseau (Ed.), *Primate life history and evolution* (pp. 153–180). New York: Wiley-Liss.

Trivers, R. L. (1971). The evolution of reciprocal altruism. *Quarterly Review of Biology, 46,* 35–57.

Trivers, R. (1972). Parental investment and sexual selection. In B. Campbell (Ed.), *Sexual selection and the descent of man* (pp. 136–179). New York: Aldine de Gruyter.

Trivers, R. L. (1974). Parent-offspring conflict. *American Zoologist, 14,* 249–264.

Trivers, R. L. (1985). *Social evolution.* Menlo Park, CA: Benjamin/Cummings.

Tudge, J. R. H. (1992). Processes and consequences of peer collaboration: A Vygotskian analysis. *Child Development, 63,* 1364–1379.

Tulving, E. (1985). Memory and consciousness. *Canadian Psychology, 26,* 1–12.

Turke, P. W. (1997). Hypothesis: Menopause discourages infanticide and encourages continued investment by agnates. *Evolution and Human Behavior, 18,* 3–13.

Turkewitz, G., & Kenny, P. (1982). Limitations on input as a basis for neural organization and perceptual development: A preliminary theoretical statement. *Developmental Psychobiology, 15,* 357–368.

Turkheimer, E. (2000). Three laws of behavior genetics and what they mean. *Current Directions in Psychological Science, 9,* 160–164.

Turkheimer, E., Goldsmith, H. H., & Gottesman, I. I. (1995). Commentary to Gottlieb's "Some conceptual deficiencies in "developmental" behavior genetics." *Human Development, 38,* 142–153.

Turkheimer, E., & Waldron, M. (2000). Nonshared environment: A theoretical, methodological, and quantitative review. *Psychological Bulletin, 126,* 78–108.

Turner, A. M., & Greenough, W. T. (1985). Differential rearing effects on rat visual cortex synapses. I. Synaptic and neuronal density and synapses per neuron. *Brain Research, 329,* 195–203.

Uller, C., Carey, S., Huntley-Fenner, G., & Klatt, L. (1999). What representations might underlie infant numerical knowledge? *Cognitive Development, 14,* 1–36.

Ungerer, J. A., Zelazo, P. R., Kearsley, R. B., & O'Leary, K. (1981). Developmental changes in the representation of objects in symbolic play from 18 to 34 months of age. *Child Development, 52,* 186–195.

United Nations. (1985). *Socio-economic differentials in child mortality in developing countries.* New York: Author.

Uzgiris, I. C., & Hunt, J. M. (1975). *Assessment in infancy: Ordinal scales of psychological development.* Urbana: University of Illinois Press.

Vandenberg, B. (1981). Play: Dormant issues and new perspectives. *Human Development, 24,* 357–365.

van Loosbroek, E., & Smitsman, A. W. (1990). Visual perception of numerosity in infancy. *Developmental Psychology, 26,* 916–922.

van Schaik, C. P., & Paul, A. (1996). Male care in primates: Does it ever reflect paternity? *Evolutionary Anthropology, 5,* 152–156.

Vaughn, B. E. (1999). Power is knowledge (and vice versa): A commentary on "Winning some and losing some: A social relations approach to social dominance in toddler." *Merrill–Palmer Quarterly, 45,* 215–225.

Vaughn, B. E., & Water, E. (1981). Attention structure, sociometric status, and dominance: Inter-relations, behavioral correlates, and relationships to social competence. *Developmental Psychology, 17,* 275–288.

Venter, J. C., et al. (2001). The sequence of the human genome. *Science, 291,* 1304–1351.

Vinter, A., & Perruchet, P. (2000). Implicit learning in children is not related to age: Evidence from drawing. *Child Development, 71,* 1223–1240.

Visalberghi, E., Fragaszy, D. M., & Savage-Rumbaugh, E. S. (1995). Performance in a tool-using task by chimpanzees (*Pan troglodytes*), bonobos (*Pan paniscus*), orangutans (*Pongo pygmaeus*), and capuchin monkeys (*Cebus apella*). *Journal of Comparative Psychology, 109,* 52–60.

Vollenweider, M., Vaughn, B. E., Azria, M. R., Bost, K. K., & Krzysik, L. (in press). Reconsidering the meaning of aggression for preschool children: Negative interactions and social competence for children attending Head Start. *Monographs for the Society for the Study of Child Development.*

von Bertalanffy, L. (1968). *General systems theory.* New York: Braziller.

Vygotsky, L. S. (1978). *Mind in society.* Cambridge, MA: Harvard University Press.

Wachs, T. D. (2000). *Necessary but not sufficient: The respective roles of single and multiple influences on individual development.* Washington, DC: American Psychological Association.

Waddington, C. H. (1975). *The evolution of an evolutionist.* Ithaca, NY: Cornell University Press.

Wakeley, A., Rivera, S., & Langer, J. (2000). Can young infants add and subtract? *Child Development, 71,* 1525–1534.

Walden, T. & Passaretti, C. (1998). "*I don't care what you say, I am not touching that snake*": Biologically relevant fear and infant social referencing. Presented at the Biennial meeting of the International Society on Infant Studies, Atlanta, GA.

Walk, R. D., & Gibson, E. J. (1961). A comparative and analytical study of visual depth perception. *Psychological Monographs, 75*(No. 519).

Walton, G. E., Bower, N. J. A., & Bower, T. G. R. (1992). Recognition of familiar faces by newborns. *Infant Behavior and Development, 15,* 265–269.

Walton, G. E., & Bower, T. G. R. (1993). Newborns form prototypes in less than 1 minute. *Psychological Science, 4,* 203–205.

Want, S. C., & Harris, P. L. (2001). Learning from other peoples' mistakes: Causal understanding in learning to use a tool. *Child Development, 72,* 431–443.

Wason, P. (1966). Reasoning. In B. M. Foss (Ed.), *New horizons in psychology.* (pp. 133–151). London: Penguin.

Waters, E., Merrick, S., Treboux, D., Crowell, J., & Albersheim, L. (2000). Attachment security in infancy and early adulthood: A twenty-year longitudinal study. *Child Development, 71,* 684–689.

Watson, A. C., Nixon, C. L., Wilson, A., & Capage, L. (1999). Social interaction skills and theory of mind in young children. *Developmental Psychology, 35,* 386–391.

Wechsler, D. (1974). *Manual for the Wechsler Intelligence Scale for Children-Revised.* New York: Psychological Corporation.

Wedekind, C., & Füri, S. (1997). Body odour preferences in men and women: Do they aim for specific MHC combinations of simply heterozygosity? *Proceedings of the Royal Society of London, Series B, 265,* 1471–1479.

Wedekind, C., Seebeck, T., Bettens, F., & Paepke, A. J. (1995). MCH-dependent mate preferences in humans. *Proceedings of the Royal Society of London, Series B, 260,* 345–349.

Weigel, R. M., & Weigel, M. M. (1989). Nausea and vomiting of early pregnancy and pregnancy outcome: A meta-analytic review. *British Journal of Obstetrics and Gynecology, 96,* 1304–1318.

Weikart, D. P., & Schweinhart, L. J. (1991). Disadvantaged children and curriculum effects. In L. Rescorla, M. C. Hyson, & K. Hirsh-Pasek (Eds.), *Academic instruction in early childhood: Challenge or pressure?* (pp. 57–64). San Francisco: Jossey-Bass.

Weinberg, R. A., Scarr, S., & Waldman, I. D. (1992). The Minnesota Transracial Adoption Study: A follow-up of IQ test performance at adolescence. *Intelligence, 16,* 117–135.

Weinfield, N. S., Sroufe, A., & Egeland, B. (2000). Attachment from infancy to adulthood in a high-risk sample: Continuity, discontinuity, and their correlates. *Child Development, 71,* 695–702.

Weisfeld, G. E., & Billings, R. (1988). Observations on adolescence. In K. B. MacDonald (Ed.), *Sociobiological perspectives on human development* (pp. 207–233). New York: Springer-Verlag.

Wellman, H. M. (1990). *The child's theory of mind.* Cambridge, MA: MIT Press.

Wellman, H. M. (in press) Enablement and constraint. In U. M. Staudinger & U. Lindenberger (Eds.), *Understanding human development*. Dordrecht, Netherlands: Kluwer.

Wellman, H. M., & Gelman, S. A. (1998). Knowledge acquisition in foundational domains. In W. Damon (Gen. Ed.) & D. Kuhn & R. S. Siegler (Eds.), *Handbook of child psychology: Vol. 2. Cognitive, language, and perceptual development* (5th ed., pp. 523–573). New York: Wiley.

Wesson, R. (1991). *Beyond natural selection*. Cambridge, MA: MIT Press.

Westermark, E. A. (1891). *The history of human marriage*. New York: Macmillan.

White, B. L. (1978). *Experience and environment: Major influences on the development of the young child* (Vol. 2). Englewood Cliffs, NJ: Prentice-Hall.

Whiten, A. (1996). Imitation, pretense, and mindreading: Secondary representation in primatology and developmental psychology? In A. E. Russon, K. A. Bard, & S. T. Parker (Eds.), *Reaching into thought: The minds of the great apes* (pp. 300–324). New York: Cambridge University Press.

Whiten, A. (1998a). Evolutionary and developmental origins of the mind reading system. In J. Langer & M. Killen (Eds.), *Piaget, evolution, and development* (pp. 73–99). Mahwah, NJ: Erlbaum.

Whiten, A. (1998b). Imitation of sequential structure of actions by chimpanzees (*Pan troglodytes*). *Journal of Comparative Psychology, 112*, 270–281.

Whiten, A., & Byrne, R. W. (1988). The manipulation of attention in primate tactical deception. In R. W. Byrne & A. Whiten (Eds.), *Machiavellian intelligence: Social expertise and the evolution of intellect in monkeys, apes, and humans* (pp. 211–223). Oxford, England: Clarendon Press.

Whiten, A., Custance, D. M., Gómez, J. C., Teixidor, P., & Bard, K. A. (1996). Imitative learning of artificial fruit processing in children (*Homo sapiens*) and chimpanzees (*Pan troglodytes*). *Journal of Comparative Psychology, 110*, 3–14.

Whiten, A., Goodall, J., McGrew, W. C., Nishida, T., Reynolds, V., Sugiyama, Y., Tutin, C. E. G., Wrangham, R. W., & Boesch, C. (1999). Cultures in chimpanzees. *Nature 399*, 682–685.

Whiting, B. B., & Whiting, J. W. (1975). *Children of six cultures: A psycho-cultural analysis*. Cambridge, MA: Harvard University Press.

Wierson, M., Long, P. J., & Forehand, R. L. (1993). Toward a new understanding of early menarche: The role of environmental stress in pubertal timing. *Adolescence, 23*, 913–924.

Wiley, A. S., & Carlin, L. C. (1999). Demographic contexts and the adaptive role of mother–infant attachment: A hypothesis. *Human Nature, 10*, 135–161.

Williams, C. J. (2000, March 31). Germany opens its first drop-off site to save unwanted babies. *Miami Herald*.

Williams, G. C. (1957). Pleiotropy, natural selection, and the evolution of senescence. *Evolution, 11*, 398–411.

Williams, G. C. (1966). *Adaptation and natural selection*. Princeton, NJ: Princeton University Press.

Wilson, D. S. (1997). Incorporating group selection into the adaptationist program: A case study involving human decision making. In J. A. Simpson & D. T. Kenrick (Eds.), *Evolutionary social psychology* (pp. 345–386). Mahwah, NJ: Erlbaum.

Wilson, E. O. (1975). *Sociobiology: The new synthesis.* Cambridge, MA: Harvard University Press.

Wilson, E. O. (1998). *Consilience: The unity of knowledge.* New York: Knopf.

Wilson, M., & Daly, M. (1985). Competitiveness, risk taking, and violence: The young male syndrome. *Ethology and Sociobiology, 6,* 59–73.

Wilson, M., Daly, M., & Weghorst, S. J. (1980). Household composition and the risk of child abuse and neglect. *Journal of Biosocial Science, 12,* 333–340.

Wimmer, H., & Perner, J. (1983). Beliefs about beliefs: Representation and constraining function of wrong beliefs in young children's understanding of deception. *Cognition, 13,* 103–128.

Witelson, S. F. (1987). Neurobiological aspects of language in children. *Child Development, 58,* 653–688.

Wolf, A. (1995). *Sexual attraction and childhood association: A Chinese brief for Edward Westermark.* Stanford, CA: Stanford University Press.

Wood, B. A. (1994). The oldest hominid yet. *Nature, 317,* 280–281.

Wood-Gush, D. G., & Vestergaard, K. (1991). The seeking of novelty and its relation to play. *Animal Behaviour, 42,* 599–606.

Wood-Gush, D. G., Vestergaard, K., & Petersen, H. V. (1990). The significance of motivation and environment in the development of exploration in pigs. *Biology and Behaviour, 15,* 39–52.

Worthman, C. M. (1993). Biocultural interactions in human development. In M. F. Pereira & L. A. Fairbanks (Eds.), *Juvenile primates: Life history, development, and behaviors* (pp. 339–358). New York: Oxford University Press.

Wrangham, R., & Peterson, D. (1996). *Demonic males.* Boston: Houghton-Mifflin.

Wrangham, R., & Pilbeam, D. (in press). African apes as time machines. In B. M. F. Galdikas, N. Briggs, L. K. Sheeran, & G. L. Shapiro (Eds.), *Great and small apes of the world: Gorillas, orangutans, chimpanzees, bonobos, and gibbons.* New York: Plenum/Kluwer.

Wyles, J. S., Kunkel, J. G., & Wilson, A. C. (1983). Birds, behavior, and anatomical evolution. *Proceedings of the National Academy of Sciences, USA, 80,* 4394–4397.

Wynn, K. (1992). Addition and subtraction by human infants. *Nature, 358,* 749–750.

Wynne-Edwards, V. C. (1962). *Animal dispersion in relation to social behaviour.* Edinburgh, Scotland: Oliver & Boyd.

Yakovlev, P. I., & Lecours, A. R. (1967). The myelenogenetic cycles of regional maturation of the brain. In A. Minkowski (Ed.), *Regional development of the brain in early life* (pp. 3–70). Oxford, England: Blackwell.

Yan, J. H., Thomas, J. R., & Downing, J. H. (1998). Locomotion improves children's spatial search: A meta-analytic review. *Perceptual and Motor Skills, 87,* 67–82.

Youngblood, L. M., & Dunn, J. (1995). Individual differences in young children's pretend play with mother and sibling: Links to relationships and understanding of other people's feelings and beliefs. *Child Development, 66*, 1472–1492.

Youniss, J. (1996). Development in reciprocity through friendship. In C. Zahn-Waxler, E. M. Cummings, & R. Iannotti (Eds.), *Altruism and aggression* (pp. 88–106). New York: Cambridge University Press.

Zahn-Waxler, C., Radke-Yarrow, M., Wagner, E., & Chapman, M. (1992). Development of concern for others. *Developmental Psychology, 28*, 126–136.

Zelazo, P. R., Kearsley, R. B., & Stack, D. M. (1995). Mental representation for visual sequences: Increased speed of central processing from 22 to 32 months. *Intelligence, 20*, 41–63.

Zollikofer, C. P. E., Ponce de León, M. S., Martin, R. D., & Stucki, P. (1995). Neanderthal computer skulls. *Nature, 375*, 283–285.

Züberbühler, K. (2000). Referential labeling in Diana monkeys. *Animal Behaviour, 59*, 917–927.

Zvoch, K. (1999). Family type and investment in education: A comparison of genetic and stepparent families. *Evolution and Human Behavior, 20*, 453–464.

AUTHOR INDEX

Das, R., 281
Davidson, M., 149
Davidson, P. E., 49
Davies, P. S., 18n
Davis, H., 163
Davis, P., 326–327
Dawkins, R., 16, 48, 61, 66, 252, 263
Deacon, T. W., 92, 95, 96, 97
Dean, M. C., 100, 100–101
de Beer, G., 49, 50, 51, 56, 88
DeBois, S., 158
DeCasper, A. J., 151, 162
DeFries, J. C., 72, 76
DeHann, M., 70
Dempster, F. N., 95, 107, 133
Desjardins, R. N., 182
Dessureau, B. N., 280
Devlin, B., 82
DeVries, M. W., 279
de Waal, F. B. M., 208, 269, 289, 291, 293, 300
de Winter, 137–138
Dickens, G., 29
Diderich, M., 248, 252, 253, 254
DiPietro, J. A., 307
Dishion, T. J., 288
Dobzhansky, T., 47
Dodge, K. A., 194, 271, 272, 273, 282, 284
Dodo, Y., 100
Dodsworth, R. O., 265
Dolan, C. V., 83
Dolhinow, P. J., 322
Dollard, J., 194
Doman, G., 188
Donald, M., 123
Dore, F. Y., 160
Dornbusch, S. M., 282
Doupe, A. J., 179
Downing, J. H., 174, 306
Draper, P., 3, 84, 282, 283, 284, 318
Drickamer, L. C., 285
Dropik, P. L., 116
Drummey, A. B., 170
Duberman, L., 244
Dumas, C., 160
Dunbar, R. I. M., 102, 102n, 137, 263, 267, 289
Duncan, J., 133
Dunn, J., 207, 251, 317
Dziurawiec, S., 153

Eals, M., 171, 173, 307
Earles, J. L., 184
Easterbrooks, M. A., 281
Eaton, W. O., 304, 306
Eccles, J. C., 24, 95
Eccles, J. S., 272
Eckerman, C. O., 127, 270
Eddy, T. J., 210, 211
Edelman, G. M., 34, 105
Edelstein, W., 281
Edgar, B., 22, 24
Edwards, C. P., 290, 308, 325
Egan, S. K., 288, 291
Egeland, B., 281
Eibl-Eibesfeldt, I., 65, 221, 252, 303, 310, 325
Eicher, S. A., 288
Eilers, R. E., 151n
Eimas, P. D., 154, 179
Einon, D., 326
Eisenberg, A. R., 242
Eizenman, D. R., 156
Eklund, A., 258
Elliott, E. M., 133
Ellis, B. J., 284, 285, 286
Ellis, H. D., 153
Ellis, S., 114
Elman, J. L., 18, 20, 68, 69, 70, 71, 114, 136, 136–137, 182, 183, 184
English and Romanian Adoptees Study Team, 108, 109n
Enns, L. R., 306
Erhardt, A. A., 308
Ernst, C., 254
Ervin, F. R., 148
Euler, H. A., 41, 242

Fagen, R., 39, 270, 297, 298, 301, 311, 314, 315, 322, 323, 331
Fagot, B. I., 272, 289, 307
Fairbanks, L. A., 240
Fantz, R. L., 152, 155
Farber, E. A., 281
Farmer, T. W., 269
Fein, G., 299, 301, 312, 314, 317
Feingold, A., 224
Feiring, C., 281
Fenson, L., 301
Ferdinandsen, K., 152
Fernald, A., 3, 126, 181, 182

Graziano, M. S. A., 105
Gredlein, J. M., 310–311, 329
Green, B. L., 39, 107, 185, 190, 203,
    204, 322, 327
Green, F. L., 136
Greenfield, P. M., 124, 201, 215
Greenough, W. T., 80, 105
Groos, K., 39, 298, 312
Gross, C. G., 105, 152
Grossman, K., 281
Grossman, K. E., 281
Grotpeter, J. K., 272, 294
Guiness, F. E., 308
Gyllensten, U., 24

Hagan, T., 30, 326
Haig, D., 233
Haight, W. L., 317
Haith, M. M., 152, 167
Halberstadt, J., 153, 161
Hale, S., 134
Hall, G. S., 59
Halpern, D. F., 171
Hamilton, C. E., 281
Hamilton, W. D., 15, 37, 48, 61, 62, 233,
    240, 263, 288, 297
Hampson, E., 172
Hanisee, J., 108
Hare, B., 210, 211
Harlow, H., 108, 188, 189, 265, 277
Harlow, M., 265
Harmon, R. J., 281
Harnishfeger, K. K., 95, 102, 105, 107,
    114, 127, 133, 137, 208, 209
Harpending, H., 284
Harper, S. E., 117, 198
Harris, J. R., 75, 77, 78, 79, 249, 254,
    255, 256, 266, 292, 331
Harris, M. J., 77, 266
Harris, P. L., 200, 206, 217, 313
Hartup, W. W., 267, 269, 287, 288, 318
Harvey, P. H., 95
Hasenstaub, A., 238, 239
Hasher, L., 107
Haskett, G. J., 290
Hattori, K., 51, 104
Hatzipantelis, M., 172
Hauser, L. B., 165
Hauser, M. D., 160, 165, 167, 169, 179
Haviland, J. J., 270

Hawkes, K., 237, 240, 241, 241n, 244
Hawley, P. A., 268, 289, 292
Hayes, B. K., 119
Hayes, C., 314
Hayes, R., 153
Haynes, H., 152
Hazen, N. L., 173
Hebb, D. O., 105
Heimann, M., 39
Hein, A., 170
Held, R., 152, 170
Hellewell, T. B., 186
Henderson, C. R., 83
Henderson, S. H., 221
Hendrix, L., 258, 259
Hennessy, R., 119
Herman, J. F., 172
Herron, J. C., 244
Hespos, J., 167
Hess, E. H.,, 63, 64
Hetherington, E. M., 219, 221, 227
Heyes, C. M., 142
Hill, E., 223n
Hill, K., 22, 98, 227, 237, 238, 241n
Hinde, R. A., 13, 49, 61, 64, 65, 261,
    263, 266, 281, 289, 290, 297, 300
Hines, M., 308, 309, 310
Hirata, S., 197
Hirsh-Pasek, K., 188
Ho, M.-W., 16, 52, 56, 68
Hodes, R., 149
Hodges, E. V. E., 273, 288
Hoff, E., 175, 176, 181
Hofferth, S., 222
Hoff-Ginsburg, E., 180
Hoffman, E., 242
Hoffman, L. W., 290
Hoffmann, V., 281
Holliday, T. W., 24
Hood, B., 158, 164
Hood, K. E., 106
Hood, L., 175
Hook, E. B., 28, 29
Horvat, M., 33, 306, 309, 311, 323
Howes, 287
Howes, C., 288, 318, 319
Hrdy, S. B., 104, 226, 227, 228, 234,
    235, 252
Huberty, P. D., 305, 306, 311, 323, 327
Huffman, K., 137
Huggins, G. R., 285

SUBJECT INDEX

425

interactions in, 271–272
and language acquisition, 19–20,
107, 175–185 (*see also* Language
acquisition)
and learning, 107–108, 129
and meta-imitation, 202–203, 204
from orphanages, 108–109
and parenting, 219–220, 227–231,
233, 247–250 (*see also* Parenting)
and play, 39, 298, 321, 331 (*see also*
Play)
as preparation for adulthood,
337–338
relationships in, 287–288
resilient, 249
siblings, 250–259 (*see also* Siblings)
and theory of mind, 204–205, 207
See also Immaturity
Chimpanzees, 22, 139
brain growth of, 94
brain microcircuitry of, 137
and cognition, 140, 142, 143
counting, 167–168
object permanence, 160
diet of, 103
enculturated (human-reared), 111
(*see also* Enculturated chimpanzees)
and episodic culture, 123
extended juvenile period of, 101
and friendship, 287
and humans, 87, 88–89
warfare, 25
and imitation, 197–199, 212
deferred, 117
incest avoidance of, 256
numerosity ability of, 163
and orthogenesis, 49
and paternal investment, 239
paternity discrepancy among, 237
and play, 314, 315, 328
symbolic, 314
and pointing, 213
post-menarche period of, 99
sex differences in tool use in, 310
and social cognition
emulation, 195–196
instructed learning, 201
theory of mind, 208–212, 215
as social species, 25
Chimpanzees, pigmy. *See* Bonobos

Chomsky, Noam, 61, 174–175, 180
Chronotopic constraints, 70–71
Cinderella myth, 246
Climate, and play, 305–306
Clinton, Hillary, 226
Cognition, 113–114, 145
in children
immediate vs. deferred benefits
of, 126–127
Piaget's perspective on, 138–143
and processing speed, 133–135
and continuity of mind, 62
domain-specific and domain-general
mechanisms in, 16–17, 127–144,
180
Piaget's theory on, 138–143
explicit, 114–115, 145
higher level, 6–7
in human-reared apes, 111–112
immature, 38, 185–190, 203
implicit, 114, 118–122, 145
of infants, 160–162
intuitive mathematics, 162–168
as outstanding human characteristic,
113
of primates, 60, 138–140, 142–143
spatial, 168–174, 306–307
See also Intelligence; Knowledge;
Language acquisition; Memory
Cognitive evolution, 123
Cognitive immaturity, 39
adaptive nature of, 38, 185–190, 203
Cognitive plasticity (flexibility), 111–
112, 215
Cognitive primitive, 134
Collaborative learning, 201–202
Communication, gestural, 213
Comparative data, and evolutionary developmental psychology, 6
Comparative psychology, 334
Conditioned taste aversion, 148
Congenital variation, 53
Constraints, 19, 68–69, 71
architectural, 69–70, 106, 162,
182–183
chronotopic, 70–71
representational, 69, 106, 161
Constructive play. *See* Object play
Cooperation
and adolescent friendship, 288–289
and dominance, 269

and evolutionary developmental psychology, 85, 334

evolutionary perspective for, 8

and evolutionary psychology, 10, 42

origins of, 58

Developmental systems approach, 5, 9, 21, 32–37, 58, 335

and behavioral genetics approach, 78–79

and domain-general vs. domain-specific abilities, 144

and environmental effects, 83

and individual differences, 76, 80

and species-typical environemnt, 336

Developmental timing, 87–91

and brain plasticity, 104–112

and brain size, 94–95

delayed maturation, 97–104, 112

and social behavior, 106–107

Diseases, and natural selection, 40

DNA, 88

junk or intergenic, 57

Dogs, and object permanence, 160

Domain-general mechanisms or abilities, 5, 16–17, 19, 127–128, 339

and domain-specific mechanisms, 135–136, 143–144

as interacting with, 136–138

in evolutionary developmental psychology, 133–135

and Geary on mind, 128–132

and infants' cognitions, 161, 162

and intelligence, 113–114, 127, 128, 143

in language acquisition, 182–185

in "Piagetian intelligence," 138–143

Domain-specific mechanisms or abilities, 5, 17–18, 19–20, 127–128, 145, 147, 339

and domain-general mechanisms, 135–136, 143–144

as interacting with, 136–138

and Geary on mind, 128–132

in infants' understanding, 160–161, 162

and intelligence, 113–114, 128, 143

for mathematics, 162

and neo-nativists on language acquisition, 179–180

and social intelligence, 193, 206

and starting-state nativism, 148

Dominance and dominance hierarchies, 268–269, 289–290, 293–294

in adolescence, 292

in childhood, 291

and deontic reasoning ability, 217

and social intelligence, 269

in social relations, 218

Dominance theory, and theory-of-mind skills, 207

Donald, Merlin, 123–124

Ducklings, prehatching experience of, 35–36

Early deprivation, reversibility of, 108–109

Early experience or environment

in alteration of species-typical behavior, 35–36, 111, 186

attachment style and reproductive strategies influenced by, 84, 282–286

and early learning vs. late learning, 188

emphasis on, 65–66

and spatial cognition, 170

Education, 188

break time in, 327–328

"cultural epoch" theory of, 60

and genetic makeup, 27

readiness for, 188–190

Emotional develoment, and I-D speech, 182

Emulation, 195–196

Encephalization quotient, 91–92, 314

Enculturated bonobos, 111, 117, 213

Enculturated chimpanzees, 111

arithmetic abilities of, 167

and deferred imitation, 117

and pointing, 213

and social cognition, 218

and symbolic play, 314

Enculturated gorillas, 213

Enculturated orangutans, 111, 117, 218

Enculturation hypothesis, 212–215, 218

Enlightenment, and "progress," 13

Enrichment studies, on function of play, 328–330

Environment

alternative strategies in, 84

Environment, *continued*

and brain development, 105–106

and children's "expectation" of play, 331

and developmental pattern, 338

in different periods of ontogeny, 36–37

and domain-general vs. domain-specific mechanisms, 144

of evolutionary adaptedness, 21–28

and evolutionary psychology, 16, 335

and evolutionary vs. evolutionary developmental psychology, 4–5

for exercise play, 305

family as, 259

and hominid populations, 26

and IQ, 81–83

maternal (prenatal), 82

and natural selection, 33

nonshared, 78, 79

shared, 78

and social strategies, 261

species-typical, 35–37, 79, 80, 128, 294

and theory of mind, 206–214

Epigenesis, 33–34

Epigenetic perspective, 80

Epigenetic processes, 4

Epigenetic programs, 335–336

Epigenetic rules, 84

Epigenetic theories of evolution, 52–58

Episodic culture, 123

Episodic memory, 115

Equifinality, 261

Equilibrium, punctuated, 50

ESS (evolutionarily stable strategy), 264

Estimation, of abilities, 202–203

Ethologists, 63–64

Eugenics, 72

Evocative effects, 77

Evolution

and complexity, 111

and developmental systems perspective, 35

epigenetic theories of, 52–58

and exaptations, 31–32

and family relationships, 259

fossil record of, 22–25, 87, 92, 139–140

and Haeckel on races, 74

and information-processing mechanisms, 18–21

phylogenetic tree of, 23

and plasticity, 110–112

behavioral, 56–58, 110

cognitive, 111–112, 215

and play, 297, 331 (*see also* Play)

role of development in, 6, 48–58

in modern synthesis, 48

role of in development, 58–61

and slow rate of growing up, 97–104, 112

and social behavior, 294–295

Evolution, Development, and Children's Learning (Fishbein), 3, 61

Evolutionarily stable strategy (ESS), 264

Evolutionary developmental psychology, 4

basic principles of, 335–340

and changes in environment-organism interaction, 36–37

and developmental psychology, 85, 334

domain-general mechanisms in, 133–135 (*see also* Domain-general mechanisms or abilities; Domain-specific mechanisms or abilities)

evolution of, 340–341

and evolutionary psychology, 4–7, 43, 85

as hybrid, 10

and individual differences, 83–85

and social learning research, 194

Evolutionary psychology, 3, 8, 11, 13–16, 41–42, 61, 334

and adaptive thought, 114

and development, 37–41, 42, 48, 61

and developmental systems approach, 32–37

and domain-general vs. domain-specific domains, 144

and environment of evolutionary adaptedness, 21–28

and evolutionary developmental psychology, 4–7, 43, 85

and evolved psychological mechanisms, 16–21

and functional analysis, 28–32

gradual change argued by, 50

and individual differences, 75–76

Genetic determinism, *continued*
   and deviations from species-typical
      pattern, 71
   and evolutionary psychologists, 335
   See also Biological determinism
Genetic diversity, 24
Genetic influence
   and correlations vs. mean values, 80
   on IQ, 81–82
Genetics
   and Darwinian evolution, 47, 62
   and innateness, 68
   and natural selection, 47
   See also Behavioral genetics
Genie (socially deprived child), 177
Genius, Galton on, 72
Gesell, Arnold, 59
Gestural communication, 213
"Good-enough" parenting, 77, 249
Gorillas, 139
   brain size of, 101
   and cognition, 140, 142, 143
      and object permanence, 160
   enculturated, 213
   and paternal investment, 239
   and pointing, 213
Gottlieb, Gilbert, 3, 68, 71, 76, 79
Gould, Stephen Jay, 31, 51
Grammar
   on domain-general view, 183
   universal, 175–177
Grandmother hypothesis, 240–241
Great apes
   and cognition, 140
   enculturated (human-reared), 111,
      117
   and episodic culture, 123
   gestural communication of, 213
   imitation in, 197, 199
      deferred, 117
   and play, 316
      symbolic, 314
   and pointing, 214
   and theory of mind, 208–212, 214
   See also Bonobos; Chimpanzees;
      Orangutans
Groups, 267–268
   in adolescence, 292–294
   in childhood, 290–292
   in infancy and toddler years,
      289–290

Group selection, 15, 15n, 47, 262, 292
Group socialization theory, 266

Haeckel, Ernst, 49, 59, 74, 90
Hall, G. Stanley, 59–60, 66
Hamilton, William, 15, 48
Hamilton's rule, 15
Healthy-baby hypothesis, 229
Hearing, and infants' preferences, 151
Hereditary Genius (Galton), 72
Heritability, 75, 75n
   and broadened definition of environ-
      ment, 83
   of IQ (intelligence), 78, 81–82
   of personality, 78
   See also Inheritance
Heterochrony, 50, 52, 88, 143
Hierarchical organization
   of domain-specific modules (Geary),
      128–132
   See also Dominance and dominance
      hierarchies; Social hierarchies
Higher level cognitions, and evolutionary
      vs. evolutionary developmental
      psychology, 6–7
History of Human Marriage, The (Wester-
      marck), 257
H.M., case of, 115–116
Hominids, 21
Homosexuality, and sociobiology, 298
Hormones
   and girls' interests in infants and
      play parenting, 325
   and locomotion, 306
   and sex differences in play, 308
   and sex differences in spatial cogni-
      tion, 174
Human cognition. *See* Cognition
Human development. *See* Development
*Human Infancy: An Evolutionary Perspec-
      tive* (Freedman), 61
Human nature, 334–335
   developmental origins of, 41–43
   and Standard Social Science Model,
      65
Human Relations Area Files (HRAF),
      229–230
Humans
   fossil record of evolution of, 22–25,
      139–140

intelligence as outstanding trait of, 91

as youngest species, 112

Hunter-gatherer societies, 25

child mortality rates in, 100

effect of father's absence in, 237

food-gathering procedures in, 103

and genetic makeup, 27

male-male competition in, 324

paternal provisioning in, 236

percentage of time for play in, 311

play parenting in, 325

pregnancy sickness in, 29

sexual division of labor in, 171–172

See also Foragers

Huntington's disease, 40

Hypermorphosis, 88, 88n. *See also* Acceleration

Imitation, 123, 196–202

benefits of for newborns and older infants, 126–127

deferred, 111, 116–118, 198, 199, 212

in mimetic culture, 123

neonatal, 38–39, 339

Imitative learning, 196–197

Immaturity

as adaptive, 37–39, 203

cognitive, 38, 185–190, 203

extended period of, 91, 97, 98, 336–337

See also Childhood or children; Delayed maturation

Immunological research, and mating style, 25–26

Implicit cognition, 114, 118–122, 145

Implicit knowledge, 119–120, 121–122, 132, 158

Implicit memory, 115–118, 122–123

Implicit representations, 119–121

Imprinting, 63, 64, 108, 277

Incest avoidance, 256–259

Inclusive fitness, 15, 219, 263

and family bonding, 104

and grandmother hypothesis, 240

and male competition, 277

and parental investment, 232

Individual Development and Evolution (Gottlieb), 3

Individual differences

and behavioral genetics, 76–78, 80

and Darwinian thinking, 71–72, 74–75

and developmental systems perspective, 76, 80

and evolutionary approach to development, 76

and evolutionary developmental psychology, 83–85

and evolutionary vs. evolutionary developmental psychology, 5–6

and evolutionary psychologists, 75–76

in intelligence testing, 72–75

and similar environments, 81

Industrialized world, development of fantasy play in, 317

Industrial Revolution, and "progress," 13

Infancy and infants, 37–39

adaptive characteristics of, 302

immediate vs. deferred benefits from, 37–39, 338–339

cognition in, 160–162

arithmetic, 166

counting, 165–166

knowledge of physical objects, 155–160

numerosity judgments, 163–164

and ordinality, 164–165

spatial abilities, 169–170, 171

exploration in, 302

and groups, 289–290

and imitation, 126–127, 197, 199–200

and innate characteristics, 190–191

interactions of, 270–271

and language, 126, 175, 179–180, 185

and parental investment theory, 224–225 (*see also* Parenting)

and paternal resemblance, 223n

and perception of painful stimuli, 36

perceptual preferences in, 150–155

for faces, 152–155, 161

positive feelings toward, 104

premature, 186–187

relationships of, 277–282

See also Childhood or children

Infant-directed (I-D) speech, 181, 182, 184

Language acquisition, 174–175
    by children, 19–20, 107, 175
    and constraints, 70–71
    domain-specific abilities in, 136–137
    and general animal characteristics, 179–180
    neo-nativist perspective on, 175–179, 185
    and sensitive period, 131
    social-pragmatist perspectives on, 180–182
Language acquisition device, 175
Language acquisition support system, 180–181
Language development
    biologically based accounts of, 67
    without domain-specific mechanisms, 182–185
Language Instinct: How the Mind Creates Language, The (Pinker), 175
Learning
    as adaptive specialization, 122
    and childhood, 107–108, 129
    collaborative, 201–202
    cultural, 196
    and domain-specific perspective, 19, 147
    and "flexible" brain, 100
    imitative, 196–197
    instructed, 200–201
    prenatal, 151
    prepared, 148–162
    and prolonged juvenile period, 129
    social, 194–203, 308
Learning theorists, and development, 65
Lemurs
    and play, 315
    and social example, 127
Lesser apes, 139
Life expectancy, 98
Lifestyles, in environment of evolutionary adaptedness, 25–28
Litter size, and brain size, 97
Local enhancement, 194–195, 196
Locke, John (17th-century philosopher), 190
Locomotor play, 301, 303–309
    rough-and-tumble, 307–309
Longevity
    and brain size, 96–97

and natural selection, 40–41
Lorenz, Konrad, 63, 89, 108, 277, 278

Make-believe, feature-dependent, 316
Marital infidelity, sex differences in response to, 224
Maternal effects, 82–83
Maternal psychopathology, and rate of maturation, 284–285
Mathematics
    as biologically secondary skill, 125–126
    intuitive, 162–168
Mating
    and incest avoidance, 256–259
    and parenting, 220
Mating style or strategies
    and history of parental interaction, 249
    and immune system, 25–26
    See also Reproductive strategies; Sexual strategies
Maturation, delayed, 97–104, 112. *See also* Immaturity
Meliorism, 66
Memory
    as adaptive specialization, 122
    explicit 115–118, 122, 123
    implicit, 115–118, 122–123
Mendel, Gregor, 46, 62
Metacognition, of young children, 39
Meta-imitation, 202–203, 204
Metarepresentation, 312
Microevolution, 14
Mimetic culture, 123
Mimetic representation, 124
Mimicry, 123, 195
Mind. *See* Cognition
Mind, theory of. *See* Theory of mind
Mindblindness, 206
"Minor marriages," 257
Modern synthesis, 16, 47–48, 59
    and epigenetic evolution, 58
Modular perspective, for theory of mind, 205–206
Modules of mind, 17–18. *See also* Domain-specific mechanisms or abilities
Monkeys, 139
    brain growth and size of, 94, 101

ABOUT THE AUTHORS

David F. Bjorklund received his PhD in developmental psychology from the University of North Carolina at Chapel Hill in 1976. He is a professor of psychology at Florida Atlantic University and has held visiting positions at the University of Georgia, Emory University, the University of Wuerzburg (Germany), the Max Planck Society (Germany), and James I University (Spain). He is the author of several books, including *Children's Thinking: Developmental Function and Individual Differences*, now in its third edition. He has served as associate editor of *Child Development*; has served on the editorial boards of *Cognitive Development*, *Developmental Psychology*, *Developmental Review*, *Educational Psychology Review*, *Journal of Cognition and Development*, *Cognitive Development*, and *School Psychology Quarterly*; and has served as a contributing editor to *Parents Magazine*. His current research interests include children's memory and strategy development, cognitive developmental primatology, and evolutionary developmental psychology. His has been the recipient of numerous teaching and research awards from Florida Atlantic University, of the "Progama Cátedra" from the Spanish Fundación BVV, and of the Alexander von Humboldt Research Award from the German government.

Anthony D. Pellegrini is a professor in the Department of Educational Psychology at the University of Minnesota, Twins Cities Campus. He received his PhD in 1978 from Ohio State University and has held faculty positions at the University of Rhode Island and the University of Georgia, as well as visiting positions at the Universities of British Columbia, Leiden, Sheffield, and Cardiff. As a student of educational and developmental psychology, he has spent much of his career studying issues around children's play, aggression, and language. He is Fellow of the American Psychological Association in Divisions 7 (Developmental Psychology) and 15 (Educational Psychology) and was a 2001 British Psychological Society Travelling Fellow. His work on play was awarded a Creative Research Medal by the vice president for research at the University of Georgia. At Minnesota he is a member of interdisciplinary programs on relationships in psychology and the Biological Bases of Behavior Group. He teaches courses on research methods, aggression, and observational methods.